# Ecological Studies

Analysis and Synthesis

---

Edited by

J. Jacobs, München · O. L. Lange, Würzburg

J. S. Olson, Oak Ridge · W. Wieser, Innsbruck

Volume 11

# Tropical Ecological Systems

## Trends in Terrestrial and Aquatic Research

Edited by

Frank B. Golley and Ernesto Medina

With 131 Figures

Springer-Verlag New York · Heidelberg · Berlin 1975

Frank B. Golley
Executive Director
Institute of Ecology
University of Georgia
Athens, Georgia 30601

Ernesto Medina
Instituto Venezolano de Investigaciones Cientificas
Departamento de Ecologia
Apartado 1827
Caracas 101, Venezuela

**Library of Congress Cataloging in Publication Data**

Golley, Frank B
Tropical ecological systems.
(Ecological studies, v. 11)
1. Ecology—Tropics. I. Medina, Ernesto, joint author. II. Title. III. Series.
QH541.5.T7G64 574.5′0913 74–8828



Printed in the United States of America

ISBN 0–387–06706–X Springer-Verlag New York • Heidelberg • Berlin
ISBN 3–540–06706–X Springer-Verlag Berlin • Heidelberg • New York
ISBN 0–412–13580–9 Chapman & Hall Limited London

# Preface

In 1971 the International Society of Tropical Ecology and the International Association for Ecology held a meeting on Tropical Ecology, with an emphasis on organic production in New Delhi, India. At this meeting a Working Group on Tropical Ecology was organized, consisting of K. C. Misra (India), F. Malaisse (Zaire), E. Medina (Venezuela) and F. Golley (U.S.A.). The object of this Working Group was to stimulate interaction between tropical ecologists through future scientific meetings and other exchanges and communications. A second meeting of ISTE and INTECOL was held in Caracas, Venezuela in 1973, under the direction of Medina and Golley and sponsored by the Department of Ecology, Institute Venezolano Investigaciones Cientificas (IVIC). The basic structure of the meeting was provided by series of invited papers which considered topics of special interest from both an applied and theoretical view. These included physiological ecology (Pannier), populations (Rabinovich), tropical savannas (Lamotte), rivers (Sioli), estuaries (Rodriguez), and island ecosystems (Mueller-Dombois). Contributed papers considered details of these and other ecological topics, including the application of ecology to human problems. The present volume includes the invited papers listed above and a sampling of contributed papers which together illustrate the trends of research in tropical ecology. The papers show that tropical ecology is a vigorous subject of research. While the papers in this volume do not provide reviews of all the topics of study in tropical ecology, they do present authoritative statements on progress in the major subject in the field.

We are grateful to the participants of the meeting on Tropical Ecology, to our Institutions (IVIC and the University of Georgia) for help in bringing this version of the proceedings to the scientific public. We especially thank N. Payne, M. Murphy, M. Shedd and L. Edwards for typing, proofreading and assisting in the editing of the manuscript.

E. Medina
F. Golley

# Contributors

BERNHARD-REVERSAT, France, ORSTOM, B.P. 20, Abidjan, Côte-d'Ivoire.

BRINSON, MARK M., Department of Botany, University of Florida, Gainesville, Florida, USA.

BROCE, ALTERTO, Rama de Entomologia, Escuela Nacional Agricultura, Chapingo, Mexico.

BRUNIG, E., Institute of World Forestry, Federal Research Center for Forest and Timber Research, Reinbeck (Hamburg), Federal Republic of Germany.

DE KRUIJF, H.A.M., Caribbean Marine Biological Institute, Curacao, Netherlands, Antilles.

EVINK, GARY, Center for Aquatic Studies, University of Florida, Gainesville, Florida, USA

FITTKAU, E. J., Department of Tropical Ecology, Max Planck Institute for Limnology, Plön, Federal Republic of Germany.

FRESON, R., Prof. Dr., Université Nationale du Zaïre, Lubumbashi, Zaïre.

GARG, R. K., Department of Botany, School of Basic Science and Humanities, University of Udaipur, Udaipur, India.

GOFFINET, G., Prof. Dr., Université Nationale du Zaïre, Lubumbashi, Zaïre.

GOLLEY, FRANK B., Institute of Ecology, University of Georgia, Athens, Georgia, USA.

HAINES, BRUCE, Department of Biological Sciences, Illinois State University, Normal, Illinois, USA.

HARTSHORN, GARY S., Organization for Tropical Studies, University of Costa Rica, Costa Rica.

HUTTEL, CHARLES, ORSTOM, B.P. 20, Abidjan, Côte-d'Ivoire.

IRMLER, U., Max Planck Institute for Limnology, Plön, Federal Republic of Germany.

JUNK, W. J., Max Planck Institute for Limnology, Plön, Federal Republic of Germany.

KARR, JAMES R., Department of Biological Sciences, Purdue University, West Lafayette, Indiana, USA.

KLINGE, H., Department of Tropical Ecology, Max Planck Institute for Limnology, Plön, Federal Republic of Germany.

KNIGHT, DENNIS H., Department of Botany, University of Wyoming, Laramie, Wyoming, USA.

KREBS, JULIA ELIZABETH, Institute of Ecology, University of Georgia, Athens, Georgia, USA.

LAMOTTE, MAXIME, Laboratoire de Zoologie, Ecole Normale Superieure, Université de Paris, Paris, France.

LUGO, ARIEL E., Department of Botany, University of Florida, Gainesville, Florida, USA.

MALAISSE, F., Prof. Dr., Miombo Project Leader, Université Nationale du Zaïre, Lubumbashi, Zaïre.

MALAISSE-MOUSSET, M., Chef de travaux, Université Nationale du Zaïre, Lubumbashi, Zaïre.

MEDINA, ERNESTO, Departamento de Ecologia, Instituto Venezolano de Investigaciones Cientificas, Caracas, Venezuela.

MONASTERIO, M., Facultad de Ciencias, Universidad de Los Andes, Merida, Venezuela.

MONTGOMERY, G. G., National Zoological Park, Smithsonian Institution, Washington, D. C., USA.

MUELLER-DOMBOIS, DIETER, Department of Botany, University of Hawaii, Honolulu, Hawaii, USA.

PANNIER, FEDERICO, Escuela de Biologia, Universidad Venezolano Central, Caracas, Venezuela.

RABINOVICH, JORGE E., Departamento de Ecologia, Instituto Venezolano de Investigaciones Cientificas, Caracas, Venezuela.

REISS, F., Max Planck Institute for Limnology, Plön, Federal Republic of Germany.

RODRIGUES, W. A., Istituto Nacional de Presquisas da Amazonia, Manaus, Amazonas, Brasil.

RODRÍGUEZ, G., Departamento de Ecologia, Instituto Venezolano de Investigaciones Cientificas, Caracas, Venezuela.

SAN JOSÉ, J. J., Departamento de Ecologia, Instituto Venezolano de Investigaciones Cientificas, Caracas, Venezuela.

SARMIENTO, G., Facultad de Ciencias, Universidad de Los Andes, Merida, Venezuela.

SCHMIDT, G. W., Max Planck Institute for Limnology, Plön, Federal Republic of Germany.

SHRIMAL, R. L., Department of Botany, V.B.R.I., Udaipur, India.

SIOLI, HARALD, Max Planck Institute for Limnologie, Plön, Federal Republic of Germany.

SNEDAKER, SAMUEL C., Center for Aquatic Sciences, University of Florida, Gainesville, Florida, USA.

SUNQUIST, M. E., National Zoological Park, Smithsonian Institution, Washington, D. C., USA.

TERBORGH, JOHN, Department of Biology, Princeton University, Princeton, New Jersey, USA.

VYAS, L. N., Department of Botany, School of Basic Sciences and Humanities, University of Udaipur, Udaipur, India.

# Table of Contents

# Tropical Ecological Systems

## Trends in Terrestrial and Aquatic Research

# CHAPTER 1

## Ecological Research in the Tropics

Frank B. Golley and Ernesto Medina

Scientific studies in the tropics have burgeoned over the past 20 years. In particular, ecology has been used in decision making, land use planning, evaluation of environmental impact, agriculture and forestry, and other areas of human concern. As a result, the number of ecological research workers has increased, ecology is more frequently taught in universities and colleges, and the number of reports of ecological research is greater. During 1972–1973 The Institute of Ecology (TIE), an international organization, sponsored a workshop to determine research priorities for tropical ecology in the Americas. A total of 472 ecologists were identified in tropical America, 145 in Africa, and 206 in Asia. While these are not complete counts of scientific workers, they do suggest the size of the research cadres available. And when we realize that many of these workers are relatively young, the future of the subject seems secure.

Just as research activity has increased within many tropical countries, there also has been an increased interest in the tropics by ecologists from other regions. The TIE study identified 642 ecologists in North America, Europe, and Australia active in tropical ecology. While this interest may be explained partly as a continuation of the drive for exploration that has motivated Europeans for so many centuries, it more significantly has another rationale. Certain tropical environments—for example, the very wet, the very warm, or the very constant—form ends of continua that must be considered in the development of rigorous ecological theory. Furthermore, ecologists are concerned with the biosphere, and recognize that no region is separate from another. Any action in one part of the biosphere may have repercussions elsewhere! Thus our research transcends political boundaries. The interests of UNESCO/MAB and IICA/Tropics in tropical ecology are expressions of these biospheric concerns.

The kaleidoscope of interest in tropical ecology, cutting across international-governmental-professional barriers, is truly impressive. The ecologists' challenge in this context is to develop mechanisms for working together, for exchanging information, and for developing concepts with high predictive value that will be useful to society. This kind of statement is usually introductory to a call for team research. However, we do not call for massive international teams of research workers focused on learning everything that can be known about one spot, although that method of research can be very useful. Rather, we need more effective communication, better publication opportunities, wider exchange of journals, and more opportunities to participate in international meetings. We require more penetrating questions, more critical thought directed at the theory that confines the way we think about the tropical world. We need more visits between institutions where the visitor can assist national research efforts, and less exploitation of research opportunities solely for private purposes.

So much for philosophy and politics, let us turn to science.

Tropical ecology is plagued with a number of generalizations: greater diversity, stablest climate, longest period without disturbance, closed nutrient cycles, and so on. These characteristics may be true for *specific* habitats and environments but they are hardly typical of the *tropics*. While these generalizations are faithfully repeated in many textbooks, especially in nontropical countries, modern studies have clearly shown the priority avenues for research. For example, interactions between plants and animals have been described as fundamental in regulating the distribution and abundance of tree species (Janzen, 1971). The subject of coevolution of plants and animals (Baker, 1970) has important implications to the theory of community integration and control, and is one of the most stimulating research areas under development. Thus studies of population dynamics and species interaction occupy an important place in this volume. Furthermore there are fundamental ecophysiological and evolutionary questions to be answered for which the tropical environment offers the proper framework. These include study of the behavior and physiology of plants and animals under alternation of wet and dry periods, sensitivity to climatic triggering factors as in photo- and thermoperiodism, variability of form and function within particular ecosystems, age and evolution of ecosystems, and ecophysiology of plant nutrition, especially that aspect dealing with the root-rhizosphere-soil system.

Another area of fruitful research concerns ecosystem function. Studies by ecologists in many different countries have begun to show the range of ecosystem function under tropical conditions. Much more data on tropical communities are required however; thus this volume includes chapters on the structure and function of a variety of tropical ecosystems.

We also are beginning to understand the processes of recovery of natural systems that have been radically perturbed or destroyed. Recovery or succession is a function of site, environment, presence of seed, abundance of seed predators, history of the area, and many other factors. Eventually we hope to be able to predict recovery for a specific location; but at this time we only know that there is a wide range of steady state values for natural communities, and that recovery may be very rapid or very slow depending upon the controlling factors in the environment. A major shortcoming is our lack of knowledge of long-term events and of regional patterns and processes. These are the most difficult kinds of data to gather, yet they also are most essential for the application of ecological science to human affairs.

The following papers will expand on these topics; therefore, let us move to the application of ecological theory to applied problems. The tropics face a multitude of serious environmental problems, but neither their nature nor scale is unique to this region. Nor, as some will assert, are tropical ecosystems more fragile than any other system. It is the task of applied ecologists to show clearly how tropical problems develop and how they can be avoided or corrected.

One of these problems concerns so-called "development." Some ecologists oppose development, but it is unreasonable to think that countries with agrarian traditions should reverse child-bearing strategies quickly and limit migration and colonization of new lands. This does not mean that populations should not be controlled and that resource allocation and environmental protection are not important. Rather, national leaders must understand the constraints on land use and the capacity of the local ecosystems to handle disturbance. They ask, for example, can certain landscapes tolerate agriculture at all? How much pollution can be handled by a river? These questions ecologists are prepared to help answer. Here is a point where we are all on equal footing, where international cooperative efforts can be especially fruitful, and where discoveries made in one place may have great utility in another.

In the tropics our goal is to manipulate ecosystems without increasing soil loss, exhausting fertility, increasing sediment load in streams, and accelerating other degrading processes. If we make errors in management we want recovery of the steady state to proceed as rapidly as possible. Our aim is to work within the natural processes and to maintain their conservative impact on the physical degrading processes of the planet, with adequate and sustained production for man's use.

This can be accomplished by judging each piece of the landscape for its inherent ability to provide organic production, to process waste, and to remain stable. Ecological land-use planning would include all aspects of the plants and animals and the environment in its purview. For example, in the state of Georgia, USA, Odum (1970) has argued that the distribution of land should be 1½ acre for production of food, 1 acre for fiber, ½ acre for urban and transportation development, and 2 acres for preservation of water, air, recreation, and maintenance of organisms to effect repair of damaged systems. This distribution obviously cannot be applied everywhere, but it does illustrate the kind of thinking that must be the basis for planning for any given area.

These objectives must be based on an adequate inventory of resources. Ecologists will play a central role in this work, but the orientation must be toward the needs of society and not mere cataloging of species populations. Inventory must be concerned with chemical abundance in the soil and vegetation, system complexity, and dominance, productive capacity, species interactions, and so on. Several chapters in this volume are examples of such inventories. This is a job for local ecologists and should require the services of numerous university graduates. Second, we must be able to predict the fragility, resilience, and dynamic properties of each landscape. We have rather poor data for these predictions, and this is a task for cooperative teams of research workers from a variety of countries.

Finally, all these efforts must be based on a continuing program of basic ecological research that pursues questions for the sake of the question.

As man brings yet larger and larger areas of the biosphere under direct management, the pressure for understanding our ecosystems will become more urgent. There is no question that ecological science will experience ever increasing demands. In a sense, our patrons (the governments and people who employ us), who have invested in a science that seemed often esoteric or obtuse, may now begin to earn on their investment as we tackle some of the severest and most disheartening problems human society faces today. Our object is not only to get greater insight into our subject of study but also to find solutions to those problems we face as members of the human race.

## References

Baker, H. G. 1970. Evolution in the tropics. *Biotropica* 2(2):101–111.

Janzen, D. H. 1971. Seed predation by animals. *Ann. Rev. Ecol. Syst.* 2:465–472.

Odum, E. P. 1970. Optimum population and environment: A Georgia microcosm. *Current Hist.* June:355–359.

# Physiological Ecology

Physiological ecology considers the interface between the disciplines of physiology and ecology. As Pannier will show, the field has a considerable history and an equally great potential to provide explanations of phenomena observed at the population level of ecology, as well as use in applied problems. Pannier focuses his attention on plants. The physiological ecology of animals is equally interesting but has been little studied in the tropics. Such workers as McNab (1970), Inger (1959), and Ruibal (1961) currently are interested in such topics as bioenergetics, reproductive biology, nutrition, and temperature regulation. This is a fruitful area of research and has application to livestock and animal production.

## References


Inger, R. R. 1959. Temperature responses and ecological relations of two Bornean lizards. Ecology 40:127–136.

McNab, B. K. 1970. Body weight and the energetics of temperature regulation. *J. Exptl. Biol.* 53:329–348.

Ruibal, R. 1961. Thermal relations of five species of tropical lizards. *Evolution* 15: 98–111.

# CHAPTER 2

# Physioecological Problems in the Tropics

FEDERICO PANNIER

## Origins of Plant Physioecology

Perhaps the original idea of physioecology can be found in the early phytogeographic works of Humboldt (1806) and Grisebach (1884), who recognized the existence of a correlation between the distribution of vegetation types and climatic regions representing the fact that plants respond directly to their climatic environment. Later, with the spread of Darwin's evolutionary theories, the idea of the existence of adaptations was introduced in experimental botany. By 1857, Lundegardh stated that "a plant species would be better adapted if it could utilize with more efficiency the energy and nutrients available. Thus structure and functions of plants are to be considered as continuously submitted to the changes of environment of a given habitat." This idea permits us to understand why Schimper (1898) titled his outstanding book, *Plant Geography on Physiological Basis*. Unfortunately, at the time this book was published, the physiological mechanisms of plants were comparatively unknown.

The field of experimental ecology known specifically as autecology was developed by ecologically minded botanists, such as Clements in the United States, Boysen-Jensen in Sweden, and Walter in Germany—to mention a few. Their studies were based on the determination of the responses of an individual to a variety of ecological factors. These studies eventually resulted in the isolation between field and laboratory research. It was largely plant physiologists who became more interested in applying their rigorously controlled laboratory methods to the complex natural environment. This was due to several important factors:

1. The design of durable, portable instruments to withstand field conditions
2. The development of simple, rapid methods of analyzing photosynthesis (Sesták et al., 1971) and water potential (Kozlowski, 1968)
3. The introduction of programmed simulation of climate in growth chambers to

test the effect of single environmental factors on plants (Chouard and Bilderling, 1972; Evans, 1963)

4. The development of methods for quantitative evaluation of natural microclimates (Gates, 1968; Salisbury et al., 1968)

The culmination of all these efforts became visible for the first time with the publication of the UNESCO symposium on methodology of plant ecophysiology (Eckardt, 1965). "Physiological ecology" was reviewed by Billings (1957). The wide acceptance of this term nowadays is reflected by several international meetings held under this name, the publication by Academic Press of a Physiological Ecology series of monographs, and by the proposal of several specific physioecological trends, such as physioecology of stress effects (Kreeb, 1971), biochemical physioecology (Kinzl, 1971), and physioecology of pollution effects (Härtel, 1971).

## Definition of Physioecology

Here physioecology is defined as the specific physiological process of the individual under the influence of quantified environmental factors, resulting in a description of these responses at the level of biochemical and biophysical mechanisms.

Following Eckardt (1965), there are two basic conceptions of physioecological research:

1. Teleological: the study of adaptative characters during ontogeny and phylogeny
2. Descriptive: the quantitative description of the energetic relationships between organisms and environment

It seems dangerous to concentrate solely on one of these concepts, since it could unrealistically limit the results of research.

The teleological orientation of physioecological research results in a theory in which environmental factors progressively modify the plant structures and functions, resulting in an adaptation of the individual to its particular habitat. Lang (1963) has classified the effect of environmental factors on plant responses:

1. Direct effects, corresponding to a direct physiological response of the plant to a single environmental factor
2. Inductive effects, through which the plant responds in a variable period of time, ranging from several hours (geotropical induction) to several weeks (floral induction)
3. Conditioning effects, where a response of the individual is conditioned by a previously acting environmental factor
4. Carryover effects, representing a specific plant response to environmental factors that remained fixed for several generations

On the other hand, the descriptive physioecological research on energy relationships between environment and individual is advantageous because it works with only a few fundamental energy equations, which entirely describe the plant microclimate.

## Physioecological Research in the Tropics

Without attempting to give a full account of research done in tropical countries, I want to select some important examples. In view of the fact that physioecological research is based on experimentation with individuals, I will arrange the studies according to the life form of the species and the vegetation type to which it belongs, with reference to the specific physiological process studied.

### Some Physioecological Problems in Species of Tropical Rain Forests

Trees are the relatively least studied from physioecological standpoint of the forest life forms. A few scattered studies exist on photosynthesis (Stocker, 1931; Sander-Viehbahn, 1962), transpiration (Vareschi, 1954), and photoperiodicity in flowering (Vaartaja, 1961), but they do not give a general picture of the way these processes are correlated with morphology and with the forest microclimatic gradients. The lack of information is undoubtedly due to the difficulties of measuring physiological parameters in nature. But even so, a number of interesting morphological observations on tropical trees exist that could be points for future research.

For example, since the observations of Haberlandt (1896), there has been controversy about periodicity of rain forest trees. Much discussion (Klebs, 1915) has concerned the dependence of periodicity of individual species on climatic factors, or the existence of an endogenous rhythm (Bunning, 1952). The extensive observations of Simon (1914) and Volkens (1912) on the periodicity of foliar bud development in tropical trees of Asia are pertinent. The fact that the concept of "evergreen tree" is not tied to continuous bud formation but rather to growth cycles characteristic of the species suggests two lines of investigation: (1) a search for rhythmic appearance of foliar buds, and (2) a long-term analysis of the biochemical activity of these buds to find correlations with climatic or edaphic factors. Also the correlation between foliage and flower development has been shown for a number of trees (*Tectona grandis, Tabebuia pentaphylla,* and *Mucuna monosperma*). In these species flowers develop only after complete leaf loss. Exceptions are observed only in individuals growing in the polluted environment of tropical cities.

Not only the periodical activity of the buds but also cambial activity in tropical trees is of physioecological interest. Annual rings have been found in species of extreme habitats such as temporary flooded "gallery forests" of tropical rivers (Gessner, 1968), the high "paramos" exposed to periodic temperature changes (Trautner-Jaeger, 1962), and the tidal zone of mangrove forest (Gill, 1971).

It would be useful to determine the response of cambial activity to other environmental fluctuations.

Another tropical rain forest life form of notable physioecological interest are the epiphytes. Probably the most studied are the orchids and the bromeliads. We will mention only a few of the physiological processes studied in both families.

The existence of special structures involved in the water absorption such as the velamen, a special absorbing structure in the outer region of aerial roots of orchids, and the suction scales of bromeliads are well known. The velamen was considered by Wallach (1938/39) to be a structure for accumulation and supply of water to the inner cortex of the aerial roots, while Dycus and Knudson (1958) concluded it was a protective structure against excessive water loss. The latter authors, utilizing radioactive tracers and flame spectrophotometric techniques, showed that solutes in the velamen of decapitated root segments cannot directly penetrate to the internal root cortex. Apparently, even in aerial roots not in contact with soil, the root apex has conserved its importance for the absorption of nutrients.

In orchid genera like *Taeniophyllum, Microcaelia*, and *Polirrhiza*, which completely lack leaves and stems, we assume the existence of a $CO_2$-fixation process by chloroplasts present in the roots. Dycus and Knudson also have shown local fixation of $^{14}CO_2$ in aerial roots of these orchids but gave no evidence of translocation of assimilates. On the other hand, Dueker and Arditti (1968) studied $CO_2$ fixation by the green flowers of *Cymbidium*, and showed differences in the photosynthetic fixation capacity of different floral parts, all of which showed a certain degree of dark-$CO_2$ fixation. Dark-$CO_2$ fixation by leaves of Brazilian orchids also was found by Coutinho (1965). The results of these preliminary researches in the field of orchid photosynthesis can be extended by considering: (1) the effect of environmental changes on photosynthesis; (2) the effect of structural changes of chloroplasts (senescence) on the same process; and (3) the chemical characterization of the assimilation products and their translocation to the different plant organs.

With respect to physioecological research on bromeliads, see the mineral nutrition studies by Benzing (1970), the photocontrol of seed germination established by Downs (1964), and the recent findings of Medina (1972) on dark-$CO_2$ fixation in great number of taxa.

Scholander (1958) has studied water translocation in the liana, *Tetracera.* By an ingenuous potometric device, he was able to measure changes of hydrostatic pressure in the wide xylem vessels of the liana, correlated with critical stages of experimentally simulated "gaseous emboly." These embolies represent ruptures of the water columns inside the conducting vessels that normally occur during the critical hours of the day. Scholander developed an hypothesis to explain how the water transport mechanism can overcome these difficulties.

Gessner (1965) found an extreme poverty of electrolytes and oxygen in the xylem water of the Brazilean liana of the genus *Bauhinia*, which may be an indication of intense oxygen-demanding processes within the living cells along the water translocation pathway. We know nothing, however, about the physiology of these cells.

Another group of life forms important in tropical rain forests are the holoparasitic and semiparasitic plants. Of the first, we are still ignorant of the most elemental physiological processes, even of their germination behavior, but we have some knowledge of transpiration of semiparasitic plants. Vareschi and Pannier (1951) found that several species of Loranthaceae, in contrast to its hosts, failed to limit their water loss during the critical hours of the day—even during the dry season—due to a very reduced sensitivity of stomatal regulation. Failure of transpiration limitation and a transpiration capacity double and triple that of their hosts could explain the catastrophic damages that tropical Loranthaceae can cause in forests disturbed by man.

## Some Physioecological Problems Related to Mangroves

Mangroves are attractive subjects due to their adaptations to an extreme poverty of oxygen in the soil and a high and fluctuating salinity. Early studies described special root structures known as "pneumatophores" in several species of mangrove. These pneumatophores provide the plant roots with oxygen from the open air. Scholander et al. (1955) proposed that oxygen diffusion inside the connecting parenchyma of the pneumatophores and neighboring roots was driven by tidal fluctuation.

More difficult to explain are the mechanisms for regulation of internal salt levels (Pannier, 1968). Leaves of *Avicennia* (Scholander et al., 1962), *Laguncularia* (Biebl and Kinzel, 1965), and *Aegialitis* (Atkinson et al., 1967) possess special salt-excreting glands, while *Rhizophora* have no salt-excreting mechanisms. This represents a remarkable research problem, as discussed by Scholander (1968).

Further physioecological problems concern the developmental stages of *Rhizophora*. Using $^{14}C$ tracers, Pannier and Pannier (in press), recently have proved that there is translocation of assimilates produced in the leaves of the parent to the attached developing seedling, but only in a late stage of development during which the initially abundant endosperm surrounding the embryo is completely consumed. Furthermore a correlation between light intensity and salt tolerance in young mangrove seedlings has also been found recently (Pannier, 1971; Anno, 1972). Finally, there is the necessity of studying germination of nonviviparous mangrove species for mechanisms to explain temporary salt tolerance in early germination stages, as shown by Brown et al. (1969).

## Some Physioecological Problems Related to Herbaceous Tropical Vegetation

No doubt the most spectacular advance in the physioecology of tropical savanna plants was the discovery of the $CO_2$-fixation mechanism known as the "$C_4$-pathway." Medina (1972) has recently reviewed these research findings.

Water balance of savanna plants has been intensively investigated (Ferri, 1960; Ferri and Labouriau, 1952; Rashid, 1947; Medina, 1967; Vareschi,

1960). The latter two workers studied representative shrubs of the Venezuelan llanos *Byrsonyma crassifolia* and *Curatella america*, which, due to the high water table, are constantly well supplied with water (Foldats and Rutkis, 1961), and thus are independent of rainfall.

Investigations on germination of *Hyptis suaveolens*, a remarkable invader of previously burned sites in the llanos, showed a phytochrome-mediated mechanism. Seeds of this species have a characteristic size distribution, and light requirements change with size and age, but seed populations maintain about the same proportion of light germinations during a year, indicating after-ripening processes (Wulff and Medina, 1971).

The physioecology of the tropical high mountain species is largely unknown. Except for some transpiration studies performed under natural conditions on the megaphyte *Espeletia schultzii* at altitudes above 3,700 m which showed stomatal regulation with temporary limitation of water loss during the midday hours during the dry season (Pannier, unpublished), all other information comes from research done in a phytotron. Using this facility the optimal germination conditions for *Espeletia* were determined, as was a possible explanation for germination induction based on mechanical rupture of the seed coat (Pannier, 1969). Morphological plasticity of the adult plant to changes in environmental conditions was found indicating a dependence of meristematic activity of the terminal bud on the plant microclimate. Still unpublished results showed a good correlation between the qualitative and quantitative composition of foliar amino acids and temperature changes of the environment. The results suggest a relationship between proline content and climatic dryness as postulated in other plants.

Certain species of *Espeletia* have an arboreal growth form, which is exceptional for plants of high altitudes. We suspect the existence of a system of hormonal growth regulation that may interact with specific microclimatic parameters. In this sense the unpublished observations of Allan Smith on the altitudinal distribution and morphological changes of several well-defined species of *Espeletia* in the paramos will be very important for a physioecological analysis of this growth problem.

## Some Problems Related to the Physioecology of Aquatic Species

Perhaps there is no tropical aquatic higher plant species better known than *Victoria regia*, which was discovered in 1801 by the Spanish missionary Fray de la Cueva in one of the Amazon tributaries near the border of Bolivia, and later found in the Esequibo River in Guayana. This plant was the object of an interesting study on physioecology of floral development (Gessner, 1960). The enormous floral buds showed a perfect synchronization of the process of anthesis with the light intensity of the environment. Experimental study of development showed that after 16 hr of illumination a darkening time of only 30 min is able to induce unfolding. Once induced, floral opening can be inhibited temporarily by high light intensities. These interesting findings could be analyzed more rigorously today using the knowledge we have of mechanisms of photoperiodic leaf movements.

Another group of tropical river plants of outstanding physioecological interest is the Podostemonacea. These plants generally develop in rapid, flowing waters and cataracts of black water rivers. Seemingly, they are not ruled by the laws of symmetry, and, contrary to all higher water plants, lack intercellular spaces. Although well known from systematic and morphological viewpoints, physioecological investigations are scarce because it is difficult to maintain the plants detached from their natural substrate. This is the reason why experimental work has to be performed near the natural habitat, which is difficult under tropical conditions. Nevertheless, work has been done by Gessner and Hammer (1962), Pannier (1960), and Grubert (1970) on photosynthesis, respiration, and germination of a diversity of Podostemonaceae of the Venezuelan Caroni Falls. These studies have demonstrated that certain species of Podostemonaceae are not able to utilize the bicarbonate ions as sources of photosynthetic carbon, but only dissolved carbon dioxide, which is plentiful in turbulent waterfalls. The fact that these plants, as well as certain Asclepiadaceae of the region of Casiquiare-Rio Negro described by Mägdefrau and Wutz (1961) and Medina (1969), are living in waters of extreme low electrolyte content pose a substantial problem regarding the source of nutrients.

Sioli (1954) and Gessner (1968) have discussed the survival of plants subjected to prolonged submergence along the shores of tropical rivers. Two approaches to the problem of flooding tolerance are application of polarograph techniques (Armstrong, 1967) for measuring the oxygen-diffusion capacity of plants as an index of tolerance to long submergence, and metabolic studies like those of Crawford (1969) to determine mechanisms that reduce the accumulation of toxic metabolites produced by prolonged flooding.

Hopefully, this short summary will attract attention to research opportunities on physioecological problems in tropical plant species. These problems are not only important in their own right but their solutions are also needed to explain the behavior of tropical ecosystems.

## References


Anno, A. 1972. Etude eco-physiologique des jeunes plantes de *Rhizophora racemosa* en rapport avec la salinité, la temperature et la duree de l'eclairement. Thése, Faculty of Science, Paris.

Armstrong, W. 1967. The use of polarography in the assay of oxygen diffusing from roots in anaerobic media. *Physiol. Plant.* 20:540–553.

Atkinson, M. R., G. P. Findlay, A. B. Hope, M. G. Pitman, H. W. Saddler, and K. R. West. 1967. Salt regulation in the mangroves *Rhizophora mucronata* Lam. and *Aegialitia annulata* R. B. *Australian J. Biol. Sci.* 20:589–599.

Benzing, P. H. 1970. Availability of exogenously supplied nitrogen to seedlings of ten Bromeliaceae. *Bull. Torrey Bot. Club* 97:154–159.

Biebl, R., and H. Kinzel. 1965. Blattbau und Salzhaushalt von *Laguncularia racemosa* (L.) Gaertn. undanderer Mangrovebäume auf Puerto Rico. *Osterr. Bot. Z.* 112: 56–93.

Billings, W. D. 1957. Physiological ecology. *Ann. Rev. Plant Physiol.* 8:375–392.

Brown, J. M. A., H. A. Outred, and C. F. Hill. 1969. Respiratory metabolism in mangrove seedings. *Plant Physiol.* 44:287–294.

Bünning, E. 1952. Über die Ursachen der Blühreife und Blühperiodizität. *Z. Bot.* 40: 293–306.

Chouard, P., and E. Bilderling. 1972. *Phytotronique et Prospective Horticole.* Paris: Gauthier-Villars.

Coutinho, L. M. 1965. Algunas informações sôbre a capacidade ritmica diaria de fixaçao e acumulação de $CO_2$ no escuro em epifitas e erbaceas terrestre da mata pluvial. *Bol. Fac. Fil. Cien Let. Univ. Sao Paulo* 294:397–400.

Crawford, R. M. M. 1969. The physiological basis of flooding tolerance. *Ber. Deut. Bot. Ges.* 82:111–114.

Downs, R. J. 1964. Photocontrol of germination of seeds of the Bromeliaceae. *Phyton* 21:1–6.

Dueker, J., and J. Arditti. 1968. Photosynthetic $^{14}CO_2$-fixation by green Cymbidium (Orchidaceae) flowers. *Plant Physiol.* 43:130.

Dycus, A. M., and L. Knudson. 1958. The role of the velamen of the aerial roots of orchids. *Bot. Gaz.* 119:78–87.

Eckardt, F. E. (ed.). 1965. *Methodology of Plant Eco-physiology.* Paris: Arid Zone Research UNESCO.

Evans, L. T. (ed.). 1963. *Environmental Control of Plant Growth.* New York: Academic Press.

Ferri, M. G. 1960. Contributions to the knowledge of the ecology of the "Rio Negro Caatinga" (Amazon). *Bull Res. Council Israel Sect. D* 8:195–207.

———, and L. G. Labouriau. 1952. Water balance of the "caatinga." I. Transpiration of some of the most frequent species of the "caatinga" of Paulo Alfonso (Bahia) in the rainy season. *Rev. Brasil. Biol.* 12:301–312.

Foldats, E., and E. Rutkis. 1961. Suelo y agua como factores determinantes en la seleccion de algunas de las especies de arboles que en forma aislada acompañan nuestros pastizales. *Bol. Soc. Ven. Cienc. Nat.* 21:9–50.

Gates, D. M. 1968. Energy exchange in the biosphere. In *Functioning of Terrestrial Ecosystems at the Primary Production Level.* Natural Resources Research 5. Paris: UNESCO.

Gessner, F. 1960. Die Bluetenoeffnung der *Victoria regia* in ihrer Beziehung zum Licht. *Planta* 54:453–465.

———. 1965. Untersuchungen über den Gefäss-saft tropischer Lianen. *Planta* 64: 186–190.

———. 1968. Zur ökologischen Problematik der Ueberschwemmungswaelder des Amazonas. *Intern. Rev. Ges. Hydrobiol.* 53:525–547.

———, and L. Hammer. 1962. Ökologisch-physiologische Untersuchungen an den Podostemonaceen des Caroni *Inter. Rev. Ges. Hydrobiol.* 47:497–541.

Gill, A. 1971. Endogenous control of growth ring development in *Avicennia. Forest Sci.* 17:462.

Grisebach, A. H. R. 1884. Die Vegetation der Erde. Auf. 2, Leipzig.

Grubert, M. H. 1970. Untersuchungen ueber die Verankerung der Samen von Podostemaceen. *Intern. Rev. Ges. Hydrobiol.* 55:83–114.

Haberlandt, G. 1896. *Eine botanische Tropenreise.* Leipzig.

Härtel, O. 1971. Oekophysiologie und anthropogene Umweltsveraenderungen. *Ber. Schweiz. Bot. Ges.* 84:497–506.

Humboldt, A. V. 1806. Ideen zu einer Physionomik der Gewächse.

Kinzl, H. 1971. Biochemische Oekologie. Ergebnisse und Aufgaben. *Ber. Deut. Sch. Bot. Ges.* 84:391–404.

Klebs, G. 1915. Ueber Wachstum und Ruhe tropischer Baumarten. *Jahrb. Wiss. Bot.* 56:734.

Kozlowski, T. T. (Ed.). 1968. *Water Deficits and Plant Growth.* Vol. 1. New York-London: Academic Press.

Kreeb, K. 1971. Oekophysiologie natuerlicher Stresswirkungen. *Ber. Deut. Bot. Ges.* 84:485–496.

Lang, A. 1963. Achievements, challenges and limitations of phytotrons. In *Environmental Control of Plant Growth,* L. T. Evans, ed. New York-London: Academic Press.

Lundegardh, H. 1857. *Klima und Boden.* 5th edit. Jena: VEB Gustav Fischer Verlag.

Mägdefrau, K., and A. Wutz. 1961. Leichthoelzer und Tonnenstaemme in Schwarzwassergebieten und Dornbuschwaeldern des tropischen Suedamerikas. *Forstwiss. Zentr.* 80:17–28.

Medina, E. 1967. Intercambio gaseoso de arboles de las sabanas de *Trachypogon* en Venezuela. *Bol. Soc. Ven. Cienc. Nat.* 27:56–59.

———. 1969. Expedición Aso VAC al Alto Orinoco. *Acta Cientif. Venez.* 20:9–13.

———. 1972. Fijacion nocturna del $CO_2$, distribución ecológica y evolución de las Bromeliáceas. *Comun. 2nd Congr. Venez. Bot. Mérida.*

Pannier, F. 1960. Physiological responses of Podostemonaceae in their natural habitat. *Intern. Rev. Ges. Hydrobiol.* 45:347–354.

———. 1968. Transport problems in *Rhizophora Mangle* L. *Intern. Symp. Trans. Higher Plants.* Reinhardsbrunn. DDR. Vol. 2, pp. 125–129. Berlin: Abh. Deut. Akad. Wissensch.

———. 1969. Untersuchungen zur Keimung und Kultur von *Espeletia,* eines endemischen Megaphyten der alpinen Zone ("páramos") der venezolanischen-kolumbianischen Anden. *Ber. Deut. Bot. Ges.* 82:559–571.

———. 1971. Compensación del incremento respiratorio de raices de *Rhizophora* inducido por elevadas concentraciones salinas, por los paramentros luz y temperatura. *Com. IV. Simp. Latinoam. Fisiol. Veg. Lima, Peru.*

———. 1974. Physioecological problems of mangroves. INTECOL-Symp. The physioecology of plants and animals in extreme habitats (in press).

Rashid, M. 1947. Transpiracao e sistemas subterraneos de vegetacao dos Campos Cerrados de Emas. *Bol. Fac. Fil. Cienc. Letr. Univ. Sao Paulo.* Bot. 5.

Sander-Viehbahn, G. 1962. Der winterliche Kohlensäurehaushalt tropischer Gewächshauspflanzen. *Beitr. Biol. Pflazen* 37:13–53.

Salisbury, B., G. G. Spomer, M. Sobral, and R. T. Ward. 1969. *Bot. Gaz.* 129:16–32.

Schimper, F. 1935. *Pflanzengeographie auf physiologischer Grundlage.* 3. Aufl. bearb. v. F. C. v. Faber. 2 Bd. Jena., Gust Fischer Verlag. 1612 p.

Scholander, P. 1958. The rise of sap in lianas. In *The Physiology of Forest Trees,* K. V. Thimann, ed. New York: Ronald Press.

———. 1968. How mangroves desalinate seawater. *Physiol. Plant.* 21:251–261.

———, L. van Dam, and S. I. Scholander. 1955. Gas exchange in the roots of mangrove. *Am. J. Bot.* 42: 92–98.

———, H. T. Hammel, E. Hemmingsen, and W. Garey. 1962. Salt balance in mangrove. *Am. J. Bot.* 42:92–98.

Sesták, Z., J. Catsky, and P. G. Jarvis (eds.). 1971. *Plant Photosynthetic Production. Manual of Methods.* The Hague: W. Junk.

Simon, S. V. 1914. Studien ueber die Periodizitaet der Lebensprozesse der in dauernd feuchten Tropengebieten heimischen Bäume. *Jahrb. Wiss. Bot.* 54:71–187.

Sioli, H. 1954. Beitraege zur regionalen Limnologie des Amazonasgebiets. *Arch. Hydrobiol.* 49:441–447.

Stocker, O. 1931. Ueber die Asimilationsbedingungen im tropischen Regenwald. *Ber. Deut. Bot. Ges.* 49:267–299.

Trautner-Jaeger, E. 1962. La formación de zonas generatrices en plantas leñosas del limite selvático andino. *Acta Cientif. Venez.* 13:126–134.

Vaartaja, O. 1959. Evidence of photoperiodic ecotypes in trees. *Ecol. Monographs* 29:91–111.

Vareschi, V. 1954. Der Wasserhaushalt von Bäumen, welche zur Aufforstung entwaldeter Gebiete Venezuelas verwendet werden. *Angew. Pfl. Soziol. Festschr. A. Aichinger* 2:721–729.

———. 1960. Observaciones sobre la transpiración de arboles llaneros durante la época de sequía. *Bol. Soc. Ven. Cienc. Nat.* 96:128–134.

———, and F. Pannier. 1951. Ueber den Wasserhaushalt tropischer Loranthaceen am natuerlichen Standort. *Phyton (Austria)* 5:140–152.

Volkens, G. 1912. *Laubfall und Lauberneuerung in den Tropen.* Berlin.

Wallach, R. 1938/39. Beitraege zur Kenntnis der Wasseraufnahme durch Luftwurzeln tropischer Orchideen. *Z. Bot.* 33:433–468.

Wulff, R. and E. Medina. 1971. Germination of seed in *Hyptis suaveolens* Poit. *Plant Cell Physiol.* 12:567–579.

# Dynamics of Populations

Study of population dynamics requires information on density, frequency, movement, and birth and death rates. While the literature is replete with data from temperate, arctic, and montane terrestrial and aquatic environments, very few tropical animals or plants have been studied adequately from this point of view. Investigations of tropical pest species, insects (Owen, 1971; Ehrlich and Gilbert, 1973), and birds (Willis, 1967; Snow, 1962) suggest that selected tropical populations may be quite stable, have low reproductive rates, and be relatively sedentary. The examination of tropical populations in the context of Levins' (1968) and other theories warrants more attention from ecologists.

The following papers discuss several topics of interest to tropical population ecology. Rabinovich analyzes 18 populations of animals to test the MacArthur-Wilson model of colonization. In this model the concepts of *r* and *K* selection are used to evaluate the fate of populations established in a new environment. Rabinovich attempts to verify the model with actual data. Hartshorn's study of the dynamics of tree populations is the first of this type for tropical trees. Knight suggests a method of study of the entire population of trees when the age of the trees is unknown. de Kruijf's paper is an example of a study of the interaction of an environmental factor, in this case, rainfall, and population of mosquitoes.

## References


Ehrlich, P. R., and L. E. Gilbert. 1973. The population structure and dynamics of a tropical butterfly. *Heliconium ethilla. Biotropica* 5:69–82.

Levins, R. 1968. *Evolution in Changing Environments. Some Theoretical Explorations.* 120 pp. Princeton, N.J.: Princeton Univ. Press.

Owen, D. F. 1971. Tropical butterflies: The ecology and behavior of butterflies in the tropics with special reference to African species. 214 pp. Oxford: Clarendon Press.

Snow, D. W. 1962. A field study of the black-and-white manakin, *Manacus manacus,* Trinidad. *Zoologica* 47:65–104.

Willis, E. O. 1967. The behavior of the bicolored ant birds. *Univ. Calif. Publ. Zool.* 79:1–127.

# CHAPTER 3

## Demographic Strategies in Animal Populations: A Regression Analysis

JORGE E. RABINOVICH

The idea put forward by Dobzhansky (1950) that tropics and temperate zones are areas where selection operates differently generated fruitful lines of thinking and research. His contention was that in temperate areas mortality was essentially climatically determined, with little or no competition pressure, while in the tropics, where the environment is relatively more constant, mortality is the result of the effects of population size. In the first case, high fecundity and short developmental time are favored by selection, while in the latter selection would favor competitive ability through lower fecundity and longer developmental time, increasing overall fitness by allocating more energy into each individual but producing fewer offspring.

A similar type of analysis, but including a comparison of the mode of selection operating in different types of ecosystems, was presented by Margalef (1958), who showed that differences in mortality curves would be expected between species belonging to mature communities and those belonging to pioneer communities. However, this line of thinking remained relatively undeveloped until the appearance of MacArthur and Wilson's (1967) work, where the differences between the two types of selection were formally defined, and the resulting properties of these processes were clearly characterized; these authors coined the terms $r$ and $K$ *selection*.

MacArthur and Wilson (1967) analyzed the strategies of colonization by relating the basic properties of the life history of a species to its chances of successful colonization and, if it failed, to the time it would persist before becoming extinct. They derived a model that could be used to predict the mean survival time of populations, a parameter that the authors considered to be one of the most relevant to the success of colonization. This model contains three basic parameters: the per capita birth rate ($\lambda$), the per capita death rate ($\mu$), and the carrying capacity of the environment ($K$) (i.e., the number of individuals the environment can hold). The algebraic difference between $\lambda$ and $\mu$ constitutes the intrinsic rate of natural increase of a species ($r$).

There are two versions of this model: one in which births are made density dependent, and another in which deaths are made density dependent. These assumptions are one of the weaknesses of the model. If there is a density dependent change in the population parameters, this change will probably affect both the death rate and the birth rate simultaneously. In addition the density dependent controls are introduced in a very crude form. The populations are allowed to grow unchecked until they reach the value of the carrying capacity of the environment; then growth stops abruptly, either by reducing the birth rate or by increasing the death rate. However, even with these limitations the general conclusions of the model are valid and useful.

Figure 3–1 shows the results of the MacArthur and Wilson model. The extinction time $T$ has a 1 as a subscript because it represents the particular case of the estimate of the mean survival time of the population resulting from a colonizing propagule. Propagule is defined by the authors as "the minimal number of individuals capable of reproducing—ordinarily a gravid female, a seed, or an unmated female plus a male." A large extinction time (that is, a long mean survival time) can be achieved in two ways: (1) for a given carrying capacity ($K$) by large values of the ratio $\lambda/\mu$, and (2) for a given value of the ratio $\lambda/\mu$ by larger K values.

Large values of $T_1$ do not mean that all propagules will persist that long in their colonizing process. In general, when $\lambda$ is larger than $\mu$, the fate of a propagule is either rapid extinction or development into a population size near the K value, from which extinction takes very long. MacArthur and Wilson show that

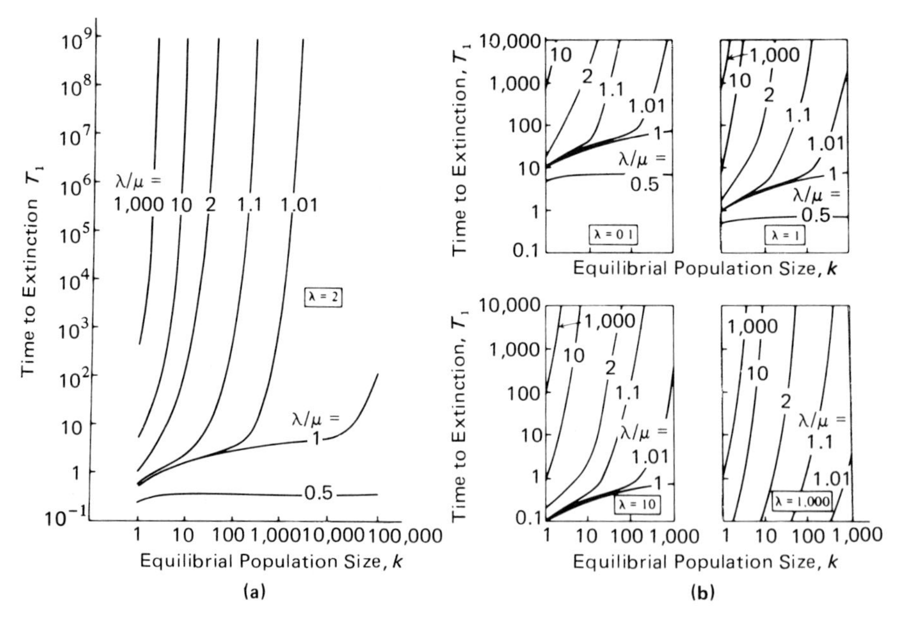


**Fig. 3–1.** Mean survival time of a population beginning with a single propagule, as a function of the per capita birth rate $\lambda$, per capita death rate $\mu$, and maximum number $K$ of individuals the environment can hold, under the conditions of density-dependent deaths. (After MacArthur and Wilson, 1967.)

$(\lambda - \mu)/\lambda$ or $r/\lambda$ is the fraction of the propagules of size 1 that will reach size $K$ and take time $T_K$ to go extinct; and $1 - [(\lambda - \mu)/\lambda]$ or $\mu/\lambda$ is the remaining fraction of $r/\lambda$ that will go extinct very rapidly. Also for relatively large values of $K$, the time to extinction of a propagule of size $K$ is approximately equal to $(r/\lambda)T_1$.

In summary, the model allows an estimate, based only on the values of the parameters $\lambda$, $\mu$, and $K$ of the rate at which immigrants must arrive if the species is to be maintained. The chances of a quick success of a colonizing propagule should be larger when the ratio $r/\lambda$ is large. How can a large $r/\lambda$ ratio be achieved? Obviously by increasing $r$ or decreasing $\lambda$; however, the way in which this happens is not obvious at all. Decreasing $\lambda$ may mean a decrease in $r$, and vice versa, depending upon how $\mu$ changes. Analysis of the possible simultaneous changes of $\lambda$ and $\mu$ for a given value of $K$ led MacArthur and Wilson to postulate a very important conclusion: *A good colonizer will have a large r, but this large r value will be the result of a low death rate and not the result of a high birth rate.* However, it must be kept in mind that it is the *ratio* $r/\lambda$ which is important in characterizing a good colonizer, and not the exclusive value of $r$.

MacArthur and Wilson (1967) introduced the concepts of $r$ and $K$ selection when they were trying to determine the fate of successful populations once they had established in a new environment. The arriving propagules encounter a situation with no crowding so that genotypes that harvest the largest amount of food, even wastefully, will leave more descendants. In other words, through the $r$ selection process evolution will favor productivity. In contrast, when the environment is crowded genotypes that can replace themselves with a low reproductive effort at the lowest food level will be favored; through $K$ selection evolution will favor efficiency of conversion of food into offspring and minimize waste.

Since 1967 many papers have been published on these problems. Some investigate the genetical aspect of the selection processes (Anderson and King, 1970; Levins, 1970; Charlesworth, 1971; King and Anderson, 1971; Roughgarden, 1971); others analyze the importance of life cycle parameters (Demetrius, 1969; Gadgil and Bossert, 1970; Pianka, 1970, 1972; Hairston et al., 1970). All refer to the theoretical aspects of the problem, either mathematically or conceptually (for example, Force, 1972). One important exception is the paper by Gadgil and Solbrig (1972), where actual data are presented to verify predictions made by the theory.

This paper will present the analysis of a number of animal species to verify some of the predictions of the MacArthur and Wilson model. This study was restricted to only 18 species for two reasons: (1) numerical evaluation of $\lambda$ and $\mu$ is essential but in many reports the intrinsic rate of natural increase is published but not the per capita birth and/or death rates; and (2) the data were required to be estimated from laboratory (supposedly optimal) situations. The latter restriction needs some justification.

The intrinsic rate of natural increase has been defined as the actual rate of increase of a population with a stable age distribution (Andrewartha and Birch, 1954). This parameter, as well as its components, the instantaneous birth and death rates, depend upon the age distribution of the population. Thus, *only* when calculated under conditions of stable age distributions will our estimate be time-invariant. This is true because age distributions of a population change constantly unless it has reached the stable age distribution, even if the birth and death age schedule ($l_x$ and $m_x$) remain constant through time. For this reason, the intrinsic rate of natural increase is the only statistic that adequately summarizes the functional qualities of an animal that are related to its capacity of increase, thus useful to compare species (Andrewartha and Birch, 1954). Models of colonization (MacArthur and Wilson, 1967; Richter-Dyn and Goel, 1972) use the intrinsic rate of natural increase and its components (the instantaneous birth and death rates), to characterize the colonizing ability of a species, and are thus consistent with the use made of these statistics in this paper.

However, because field populations rarely, if ever, exhibit a stable age distribution, any evaluation under field conditions will not constitute adequate estimation of these statistics. One of the best methods of estimating the instantaneous parameters of a population is by the analysis of cohort data obtained from laboratory experiments, where density, food, and space are kept near an optimum level for the species and the physical conditions of the environment, usually light, temperature and relative humidity, are also maintained near to what the investigator believes to be the optimum for the species.

In the present analysis when parameter estimates were available for several conditions of temperature and/or relative humidity, only those conditions that resulted in the highest estimates of the intrinsic rate of natural increase were included.

Table 3–1 lists the 18 species and gives the numerical values of the population parameters used in this analysis. In some instances, when the birth or death rates were not provided by the source, these were calculated from the author's original data either by using the original survivorship and fecundity tables, if available, or by reading from the survivorship and fecundity curves if the tables were not available. In the latter case, the estimated values were considered acceptable only if the estimation of the intrinsic rate of natural increase ($r$) was close to the third decimal when compared with the published value.

Table 3–2 shows the ratios $r/\lambda$ and $\lambda/\mu$ and the estimates of the extinction times calculated with MacArthur and Wilson's formula (4–5, p. 73). Extinction time was restricted to the values of a propagule of size 1 when $K = 1$ $\{T1(1)\}$, both in units of days and of generations, to the values of a propagule of size $K$ when $K = 1$ and $K = 5$ in units of generations, and for the proportional increase in the extinction time when $K$ increases from 1 to 5 $[\{TK(5) - TK(1)\}/TK(1)]$.

Figure 3–2 shows the extinction times for five selected species as a function of the carrying capacity of the environment. For small values of $K$ the extinction time does not correlate well with the expected $r/\lambda$ values. However, the use of an absolute time unit is not very meaningful in comparing species having different generation times. When the extinction times are expressed in generations (Fig. 3–3), the trend of increased extinction time is consistent for any value of $K$, and now the position of the curves for all five species with respect to the

**Table 3–1.** Identification and population parameters[a] of the species used in the analysis.

| | Species and Number | $r$ | $\lambda$ | $\mu$ | $T$ | $R_o$ | Source |
|---|---|---|---|---|---|---|---|
| 1 | *Triatoma infestans* (Reduviidae) | 0.01457 | 0.02257 | 0.00799 | 216.1 | 25.0 | Rabinovich (1972a) |
| 2 | *Rhodnius prolixus* (Reduviidae) | 0.02367 | 0.03218 | 0.00851 | 197.7 | 49.3 | Rabinovich (1972b) |
| 3 | *Phyllopertha horticola* (Rutelidae) | 0.00293 | 0.0061 | 0.00316 | 367.4 | 2.9 | Laughlin (1965)[c] |
| 4 | *Lasioderma serricorne* (Anobiidae) | 0.07270 | 0.11911 | 0.04641 | 35.6 | 37.2 | Lefkovitch (1963)[c] |
| 5 | *Daphnia pulex* (Daphnidae) | 0.33354 | 0.39511 | 0.06156 | 10.6 | 21.4 | Marshall (1962)[c] |
| 6 | *Nasonia vitripennis* (Pteromalidae) | 0.30071 | 0.38460 | 0.08389 | 20.7 | 334.9 | Rabinovich (1968) |
| 7 | *Pediculus humanus* (Pediculidae) | 0.10737 | 0.12849 | 0.02111 | 34.6 | 30.9 | Evans and Smith (1952)[c] |
| 8 | *Culex pipiens fatigans* (Culicidae) | 0.14535 | 0.49546 | 0.35011 | 41.8 | 77.5 | Gómez (1972)[c] |
| 9 | *Tribolium castaneum* (Tenebrionidae) | 0.101 | 0.154 | 0.053 | 55.6 | 275.0 | Leslie and Park (1949)[b] |
| 10 | *Drosophila melanogaster* (Drosophilidae) | 0.37566 | 0.44476 | 0.06910 | 16.7 | 198.9 | Siddiqui and Barlow (1972)[c] |
| 11 | *Rattus norvegicus* (Muridae) | 0.01560 | 0.02140 | 0.00580 | 261.1 | 58.6 | Leslie, Venables, and Venables (1952)[b] |
| 12 | *Microtus agrestis* (Cricetidae) | 0.0125 | 0.0161 | 0.0036 | 141.8 | 5.9 | Leslie and Ranson (1940)[b] |
| 13 | *Peromyscus maniculatus* (Cricetidae) | 0.0104 | 0.0130 | 0.0025 | 407.9 | 69.6 | French and Kaaz (1968)[c] |
| 14 | *Synthesiomyia nudiseta* (Muscidae) | 0.08205 | 0.14214 | 0.06008 | 51.3 | 47.1 | Rabinovich (1970) |
| 15 | *Ovis aries* (Bovidae) | 0.0000548 | 0.000849 | 0.000301 | 1862.6 | 2.5 | Caughley (1967)[c] |
| 16 | *Oncopeltus fasciatus* (Lygaeidae) | 0.04120 | 0.0793 | 0.03809 | 123.9 | 107.6 | Landahl and Root (1969)[d] |
| 17 | *Oncopeltus unifasciatellus* (Lygaeidae) | 0.03353 | 0.06570 | 0.03217 | 143.7 | 67.4 | Landahl and Root (1969)[d] |
| 18 | *Stagmatoptera biocellata* (Mantidae) | 0.00651 | 0.01814 | 0.01753 | 301.4 | 7.1 | Rabinovich and Maldonado (unpubl.) |

[a] All parameter units in days. $r$, intrinsic rate of natural increase; $\lambda$, per capita birth rate; $\mu$, per capita death rate; $T$, generation time; $R_o$, net reproduction rate.

[b] All parameters given by original sources.

[c] Parameters calculated from source's tables.

[d] Parameters calculated from source's curves.

**Table 3–2.** Colonizing ability ratios and extinction times.[a]

| | Species and Number | $r/\lambda$ | $\lambda/\mu$ | $T1(1)$ (days) | $T1(1)$ (Gen.) | $T1(1)$ (Gen.) | ($TK(5)$ (Gen.) | $\frac{TK(5)-TK(1)}{Tk(1)}$ |
|---|---|---|---|---|---|---|---|---|
| 1 | *Triatoma infestans* | 0.6392 | 2.813 | 125 | 0.577 | 0.895 | 20.9 | 22.4 |
| 2 | *Rhodnius prolixus* | 0.7356 | 3.782 | 117 | 0.610 | 0.829 | 51.5 | 61.1 |
| 3 | *Phyllopertha horticola* | 0.4809 | 1.925 | 315 | 0.859 | 1.780 | 13.8 | 6.8 |
| 4 | *Lasioderma serricorne* | 0.6104 | 2.566 | 22 | 0.605 | 0.992 | 17.2 | 16.3 |
| 5 | *Daphnia pulex* | 0.8442 | 6.418 | 16 | 1.520 | 1.800 | 766.0 | 424.6 |
| 6 | *Nasonia vitripennis* | 0.7819 | 4.584 | 12 | 0.574 | 0.735 | 90.2 | 121.7 |
| 7 | *Pediculus humanus* | 0.8357 | 6.086 | 47 | 1.370 | 1.640 | 569.0 | 346.0 |
| 8 | *Culex pipiens fatigans* | 0.2934 | 1.415 | 3 | 0.068 | 0.233 | 0.9 | 2.9 |
| 9 | *Tribolium castaneum* | 0.6560 | 2.906 | 19 | 0.339 | 0.517 | 13.2 | 24.5 |
| 10 | *Drosophila melanogaster* | 0.8446 | 6.437 | 14 | 0.868 | 1.020 | 440.0 | 430.0 |
| 11 | *Rattus norvegicus* | 0.7294 | 3.696 | 172 | 0.661 | 0.907 | 18.1 | 19.0 |
| 12 | *Microtus agrestis* | 0.7781 | 4.508 | 280 | 1.970 | 2.530 | 83.5 | 32.0 |
| 13 | *Peromyscus maniculatus* | 0.8067 | 5.173 | 399 | 0.979 | 1.210 | 57.1 | 46.2 |
| 14 | *Synthesiomya nudiseta* | 0.5773 | 2.366 | 17 | 0.324 | 0.562 | 7.7 | 12.6 |
| 15 | *Ovis aries* | 0.6456 | 2.822 | 3320 | 1.780 | 2.760 | 64.6 | 22.4 |
| 16 | *Oncopeltus fasciatus* | 0.5196 | 2.082 | 26 | 0.211 | 0.407 | 3.9 | 8.5 |
| 17 | *Oncopeltus unifasciatellus* | 0.5104 | 2.043 | 31 | 0.216 | 0.423 | 3.8 | 8.0 |
| 18 | *Stagmatoptera biocellata* | 0.3588 | 1.560 | 86 | 0.285 | 0.794 | 3.8 | 3.7 |

[a] $T1(1)$, extinction time of a propagule for $K = 1$; $TK(1)$, extinction time of a population size $K$ when $K = 1$; $TK(5)$ = extinction time of a population size $K$ *when* $K = 5$; $[TK(5) - TK(1)]/TK(1)$, proportional increase in the extinction time when $K$ increases from 1 to 5.

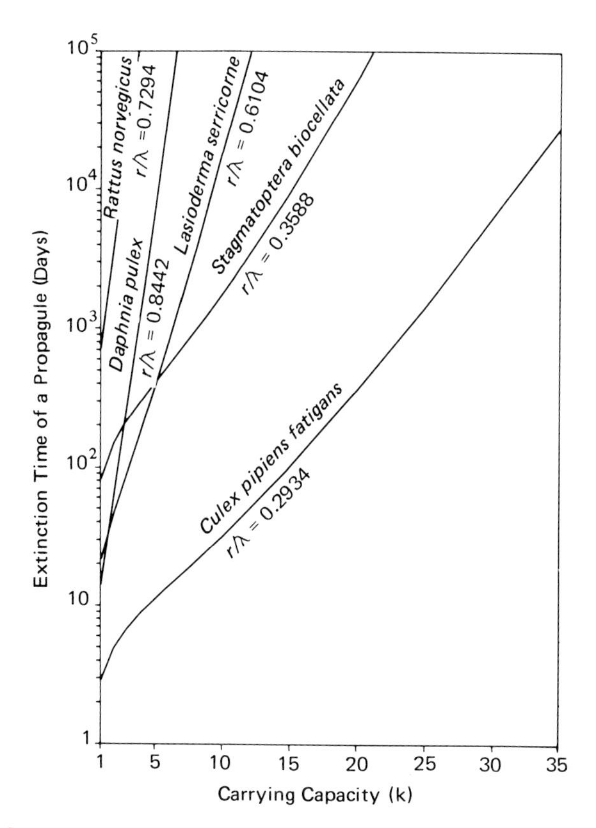


**Fig. 3–2.** Mean number of days that a population will survive beginning with a single propagule for five selected species, as a function of $\lambda$, $\mu$, and $K$.

value of the ratio $r/\lambda$ is consistent with the predictions of the model. The larger the ratio $r/\lambda$, the larger the mean survival time of the population, indicating a better colonizing species.

Figure 3–4 shows a statistically significant relationship between $T1(1)$ and the generation time ($T$) and the net reproduction rate ($R_o$). Here we find one of the first counterintuitive results: extinction times become longer with increasing generation times and shorter for larger net reproduction rates. Rather, if, as MacArthur and Wilson have claimed, extinction times are one of the best indicators of colonizing ability, one would expect that they correlate well with *short* generation times and with high net reproduction rates. A similar type of difficulty is found when analyzing the extinction times in relation to the intrinsic rate of natural increase and the per capita birth rate (Figures 3–5 and 3–6 respectively): large extinction times correlate well with small $r$ and $\lambda$ values while the opposite would have been expected. Finally, there is no significant relationship between extinction time and the colonizer indices $r/\lambda$ and $\lambda/\mu$.

It was suspected that a spurious effect was being introduced by the fact that the species being compared had completely different generation time values. Thus, a similar set of plots of extinction time in generation units versus popu-

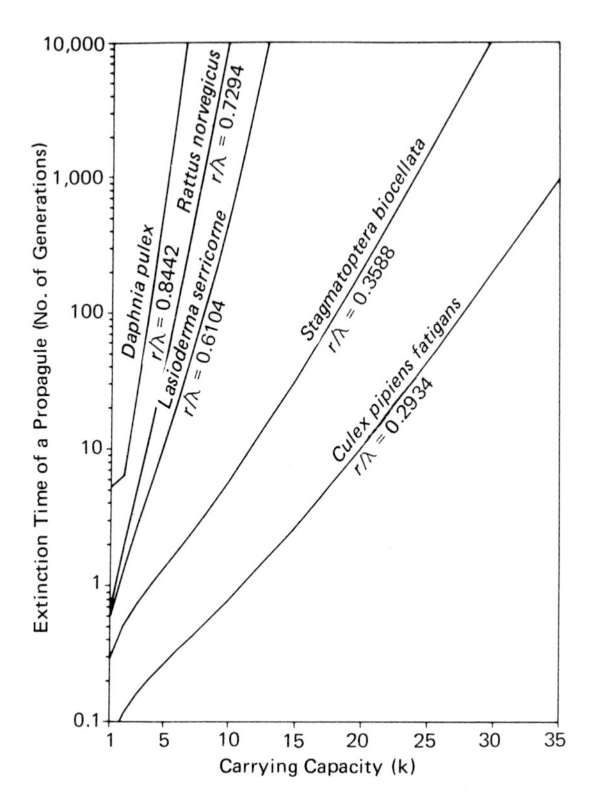


**Fig. 3–3.** Mean number of generations that a population will survive beginning with a single propagule for the same selected species of Figure 3–2, as a function of $\lambda$, $\mu$, and $K$.

lation parameters and colonizing ratios was prepared (Figures 3–7 to 3–9). We see now that no *single* population parameters shows any clear relationship with extinction time. However, the relationship between the extinction time and the ratios $r/\lambda$ and $\lambda/\mu$ shows a statistically significant correlation in the direction predicted by the model: the larger the values of the ratios the longer the extinction time (Figures 3–8 and 3–9).

The possible relationship between the population parameters and the colonizing ratios $r/\lambda$ and $\lambda/\mu$ and the proportional increase in the extinction times when carrying capacity is increased from 1 to 5 was also considered. Again, neither the generation time, the net reproduction rate, nor the intrinsic rate of natural increase show a particular pattern when related to the proportional increase in the extinction times. However, there is a strong relationship between the proportional increase and both the ratios $r/\lambda$ (correlation coefficient = 0.9929) and $\lambda/\mu$ (correlation coefficient = 0.9504).

The analysis of Figure 3–10 is of particular interest because it was found that the best way to linearize this relationship was by means of the inverse transformation. Both the equation and the plot of data show, at least to the eye, three

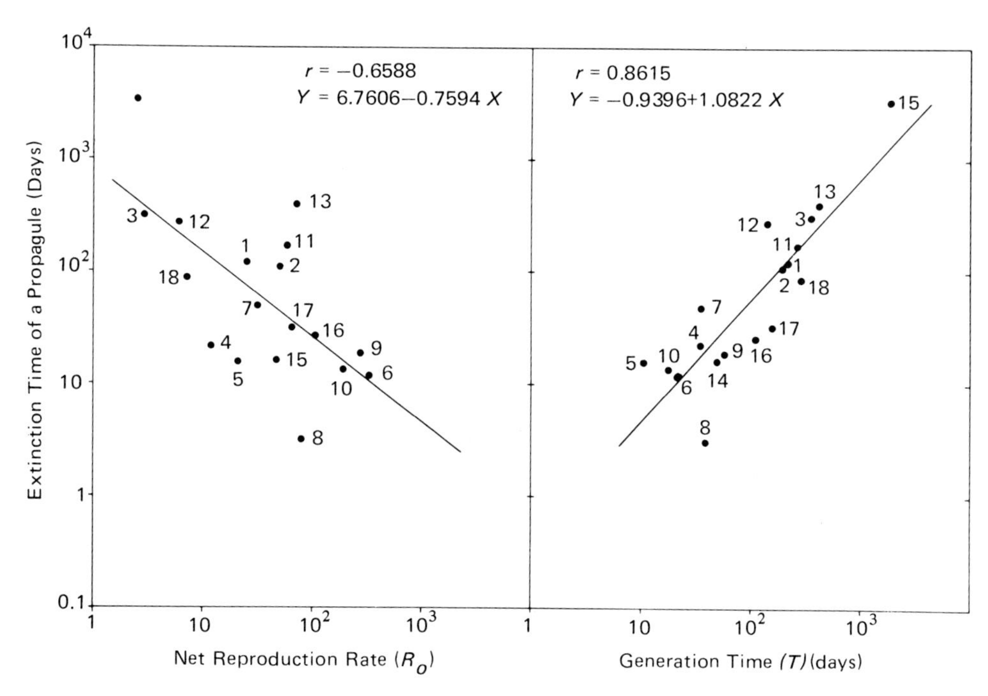


**Fig. 3–4.** Relationships between the extinction time (in days) of a population originated from a single propagule and its net reproduction rate and generation time.

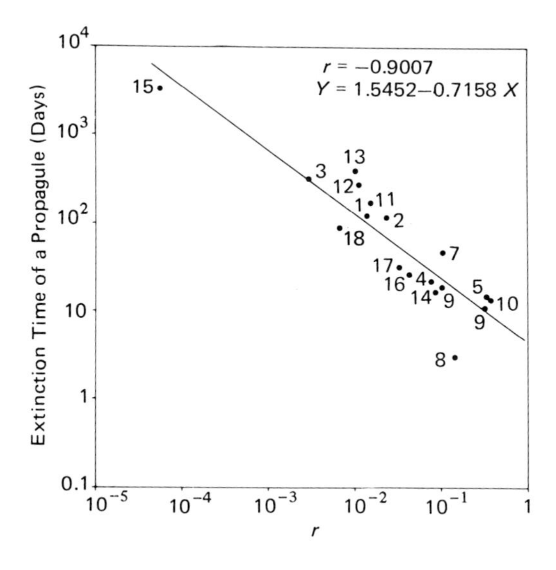


**Fig. 3–5.** Relationship between the extinction time (in days) of a population originated from a single propagule and its intrinsic rate of natural increase.

different types of species: Type I, on the lower part of the curve, are the species almost insensitive to an increase in the carrying capacity of the environment; type II are the species with intermediate or medium sensitive to an increase in the carrying capacity of the environment; and type III are species most sensitive to an increase in the carrying capacity of the environment. Of course this division is arbitrary, particularly the inclusion of species 3 in type I or II.

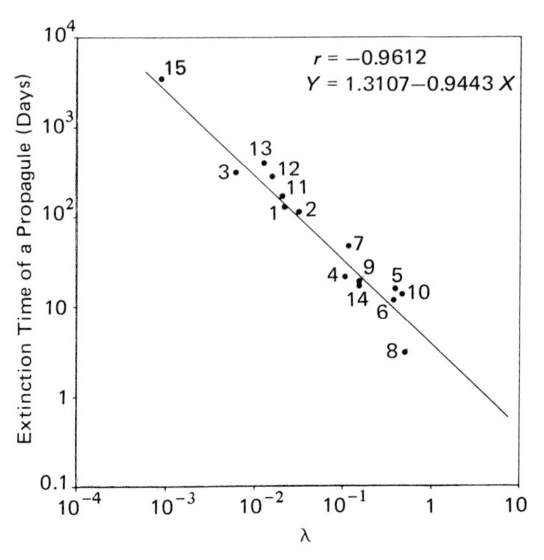


**Fig. 3–6.** Relationship between the extinction time (in days) of a population originated from a single propagule and its instantaneous birth rate.

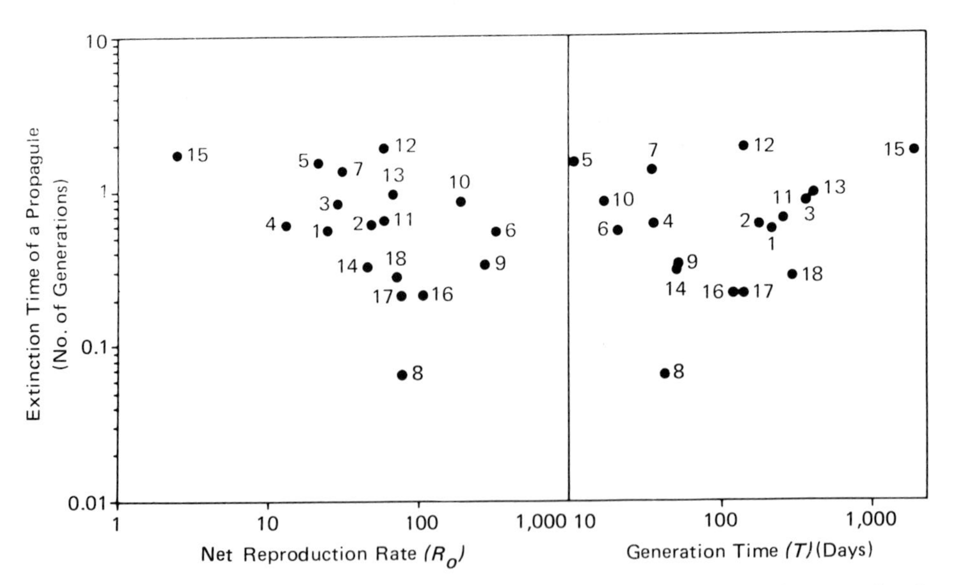


**Fig. 3–7.** Relationship between the extinction time (in number of generations) of a population originated from a single propagule and its net reproduction rate and generation time.

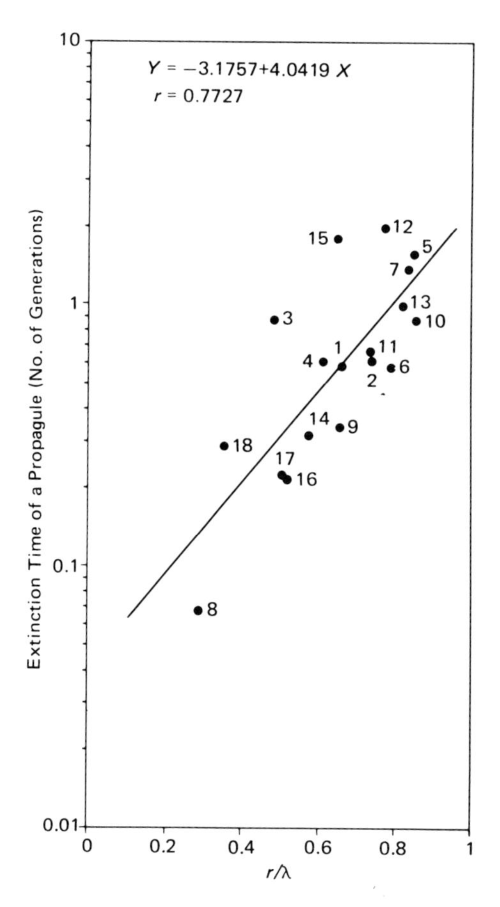


**Fig. 3–8.** Relationship between the extinction time (in number of generations) of a population originated from a single propagule and the demographic ratio intrinsic rate of natural increase over instantaneous birth rate.

Finally, I also investigated the relative values of the per capita birth and death rates after ranking the species in increasing order of colonizing ability (Figure 3–11). Contrary to the theory, there seems to be a general trend by which the "best" colonizers have increased their $r/\lambda$ ratios by increasing the per capita birth rate. This is confirmed by the plot of ratio $\lambda/\mu$ that increases steadily with the species ranked after $r/\lambda$.

Richter-Dyn and Goel (1972) have presented a stochastic model of the process of colonization based on a more realistic dependence of $\lambda$ and $\mu$ in the population size. For a population with a birth rate reasonably larger than the death rate, there exists a critical value $n_c$ such that for all $m > n_c$ the colonization is practically a certain event; thus the colonizing group should consist of $m > n_c$ individual where $n_c \simeq$ 10–15. When this is not possible, the best strategy is to increase the probability of reaching $n_c$ without going extinct. For an initial population size $m$ this probability is $\simeq 1 - (\mu/\lambda)^m$, which indicates a possible strategy: if $m$ is not increased, then the probability of reaching $n_c$ without going extinct increases with decreasing $\mu/\lambda$.

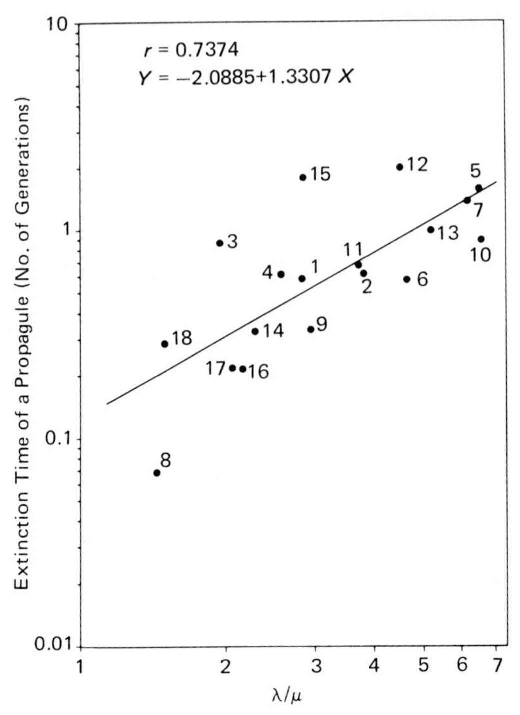


**Fig. 3–9.** Relationship between the extinction time (in number of generations) of a population originated from a single propagule and the demographic ratio of birth over death instantaneous rates.

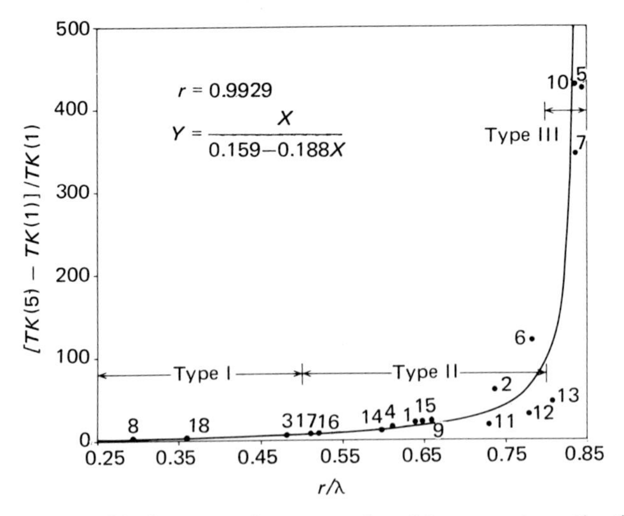


**Fig. 3–10.** Relationship between the proportional increase in extinction time (in number of generations) after an increase in $K$ from 1 to 5 and the demographic ratio intrinsic rate of natural increase over instantaneous birth rate. Types I, II, and III represent poor, intermediate, and good colonizers, respectively.

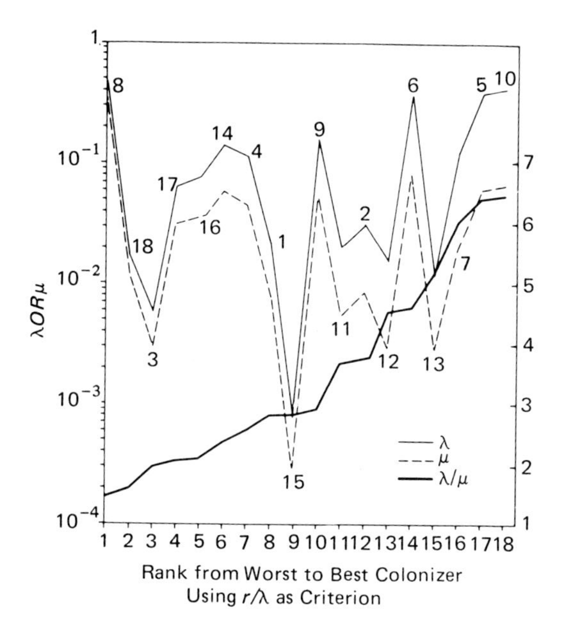


**Fig. 3–11.** Changes in the instantaneous birth and death rates and in their ratio after ranking all species using the colonizer index $r/\lambda$ as a criterion.

Figure 3–12 shows how the 18 species are arranged when the size of the colonizing population ($m$) is expressed as a function of the colonization index ($r/\lambda$) for a probability 0.5 of reaching the critical size $n_c$ before going extinct (line A), and for a value of 0.9 for this probability (line B). For 50 percent of the propagules to reach the safe value $n_c$, the size of the propagules must be larger than 1 only for three species (species number 3, 8, and 18); on the other hand, for 90 percent of the propagules to reach the safe value $n_c$, the size of the propagules must be larger than 1 for all species (species 3, 8, and 18 must have a colonizing population of approximately 4, 7, and 5 individuals respectively).

Another advantage of the Richter-Dyn and Goel model is that a comparison can be made between strategies of colonization for a single population and ones with several subpopulations. A critical number of subpopulations exists above which the ensemble is likely to persist for a long time; if a "good" strategy of migration is achieved the population can increase its prospects of persisting when divided into a number of subpopulations. Table 3–3 shows the results of the estimations of the possible rates of immigration that would "justify" a strategy of an ensemble of subpopulations. It is readily seen that in none of the 18 species analyzed is the rate of immigration to avoid extinction larger than the rate of immigration for which the strategy of an ensemble of subpopulations would be "justified." The reason is that for no species is the conditions $\tilde{\lambda}\ (1 + \mu/2\lambda)(1 - \mu/\lambda) > \lambda$ (eq (3.17), p. 422) satisfied, where $\tilde{\lambda}$ is the rate of immigration characterizing successful subpopulations.

As $\tilde{\lambda}$ is the minimum rate of immigration to avoid extinction, its inverse is

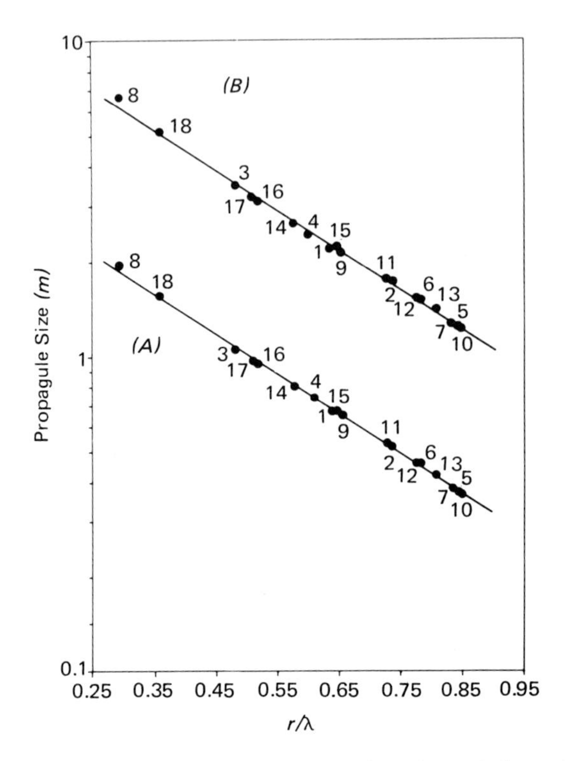


**Fig. 3–12.** Size of colonizing propagule as a function of the colonizer ratio $r/\lambda$ to assure 0.5 (A) and 0.9 (B) probability of successful colonization. (From Richter-Dyn and Goel's model.)

the extinction time of a colonizing population if no new arrivals occur. This is shown in the third column of Table 3–3 (expressed in units of days); it can be compared with the column labeled $T1(1)$ (days) of Table 3–2 that gives the extinction time of a propagule after the MacArthur and Wilson model. As expected, the values for the extinction time of the Richter-Dyn and Goel model are at least equal but in general lower than the corresponding values of the MacArthur and Wilson model. As mentioned before, a reason for the direction of these differences is the simple mechanism of density dependence used in the MacArthur and Wilson model that leads to an overestimation of the extinction times.

Our conclusion is that for all species analyzed the best strategy is colonization with a single subpopulation. Richter-Dyn and Goel's model gives an expression to estimate the probability of extinction per unit time of a single subpopulation ($\nu$). This probability is shown in the fourth column of Table 3–3. Figures 3–13 to 3–15 evaluate the relationships between this probability and the intrinsic rate of natural increase, generation time, and the colonizing ratio $r/\lambda$. The latter does not show any particular influence on the probability of extinction of a single population, while this probability is larger for short generation times and for large $r$ values, as expected.

**Table 3–3.** Rates of immigration, extinction times, and extinction probabilities estimated with the Richter-Dyn and Goel model.

| Species | Minimum Rates of Immigration Such That the Species Does not Go Extinct ($\bar{\lambda}$) | Minimum Rate of Immigration Such That the Prospect of the Species to Persist Is Better Than Its Prospect of Existence as One Subpopulation ($\bar{\lambda}_s$) | Days to Extinction if No Arrivals Occur or Number of Propagules That Have to Arrive per Day for the Species to Persist ($1/\bar{\lambda}$) | Probability of Extinction per Day of a Single Subpopulation ($\nu$) |
|---|---|---|---|---|
| 1 | 0.0092 | 0.0296 | 109 | 0.0068 |
| 2 | 0.0092 | 0.0386 | 109 | 0.0075 |
| 3 | 0.0043 | 0.0101 | 231 | 0.0025 |
| 4 | 0.0547 | 0.1633 | 18 | 0.0388 |
| 5 | 0.0631 | 0.4342 | 16 | 0.0571 |
| 6 | 0.0881 | 0.4435 | 11 | 0.0756 |
| 7 | 0.0217 | 0.1421 | 46 | 0.0195 |
| 8 | 0.6992 | 1.2478 | 1 | 0.2587 |
| 9 | 0.0601 | 0.2003 | 17 | 0.0452 |
| 10 | 0.0708 | 0.4886 | 14 | 0.0641 |
| 11 | 0.0063 | 0.0258 | 158 | 0.0051 |
| 12 | 0.0038 | 0.0186 | 263 | 0.0032 |
| 13 | 0.0026 | 0.0147 | 384 | 0.0023 |
| 14 | 0.0732 | 0.2033 | 14 | 0.0496 |
| 15 | 0.0003 | 0.0011 | 2907 | 0.0003 |
| 16 | 0.0495 | 0.1231 | 20 | 0.0307 |
| 17 | 0.0423 | 0.1034 | 24 | 0.0258 |
| 18 | 0.0198 | 0.0383 | 51 | 0.0088 |

## Discussion

These results are one of the first attempts to verify the applicability of some colonization models with actual data from animal populations. In Table 3–1 the maximum potential of each species is given. This is a very important criterion in performing the analysis presented here. We have been dealing with models in which the population parameters are subjected to selection of two different types, $r$ and $K$ selection. Thus if we want to check their applicability, the parameter estimates should reflect the end results of these selection processes. To achieve this, local or temporary factors that reflect ecological conditions affecting the *complete* expression of a given trait should be eliminated as much as possible.

Most of the general features of the different regressions have been discussed. It is interesting to consider here the relative positions of some species as resulting

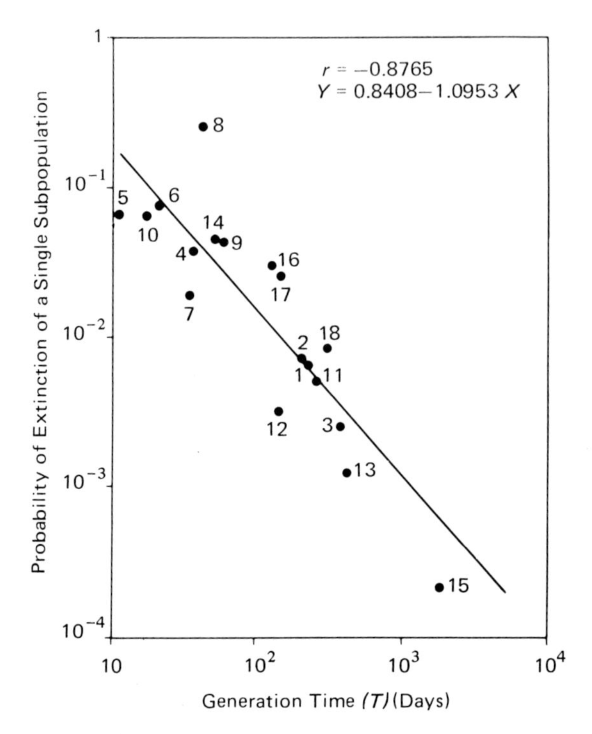


**Fig. 3–13.** Probability of extinction of a single subpopulation and the relationship with its generation time. (From Richter-Dyn and Goel's model.)

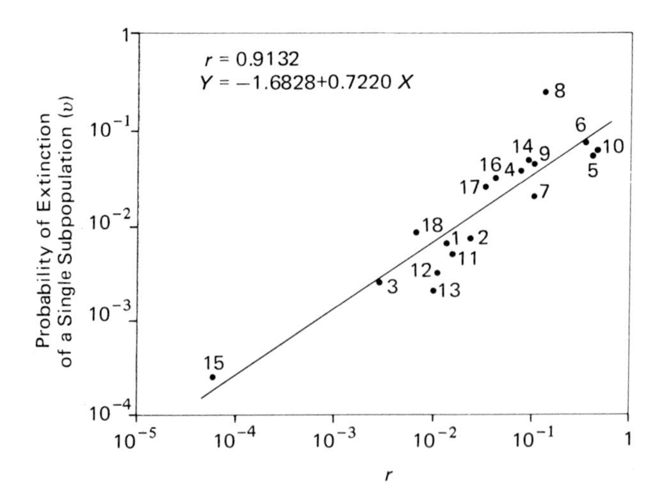


**Fig. 3–14.** Relationship between the probability of extinction of a single subpopulation and the intrinsic rate of natural increase.

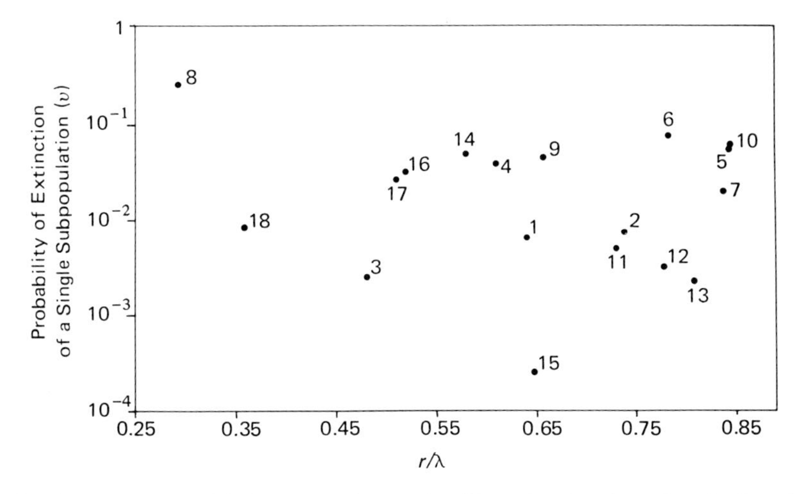


**Fig. 3–15.** Relationship between the probabiilty of extinction of a single subpopulation and the demographic ratio intrinsic rate of natural increase over instantaneous birth rate.

from the regression analyses. We will restrict the discussion to extinction times expressed in generation units.

If we consider population extinction time as one of the best indicators of colonizing success, as proposed by MacArthur and Wilson (1967), some species (such as *Culex pipiens fatigans*) which are considered "good" colonizers, have a short extinction time, while a species like the domestic sheep (*Ovis aries*) has a long extinction time. A similar counterintuitive case is found comparing species such as *Drosophila melanogaster, Daphnia pulex,* and *Pediculus humanus* with the rodent species *Microtus agrestis* and *Peromyscus maniculatus*, one would expect the former to show a larger colonizing ability than the latter; however, all five show approximately the same extinction time values.

These differences disappear in part if we analyze the capacity of the species to increase their extinction time with an increase in the carrying capacity of the environment: species like *D. melanogaster, D. pulex,* and *P. humanus* show a much higher value than the rodents; however, *C. p. fatigans* still keeps the lowest value for all species.

It is here that the results of this regression analysis suggest a revision of the concept of colonizing species within the framework of ecosystem maturity and stability, as proposed by Margalef (1958, 1968). Gadgil and Solbrig (1972) claim that in habitats temporary in space and time, an *r* strategy may be expected but not a colonizing ability. This implies that *r* strategists are not restricted to "good" colonizers only, but is a broader category including fugitive or opportunistic species as well.

In Figure 3–10 we had recognized poor, intermediate, and good colonizers. Let us analyze the first group. In type I three species were included that have obvious differences among them. *Culex pipiens fatigans* is a short-lived mosquito, typical of ephimerous habitats; *Stagmatoptera biocellata* is a long-lived mantid with a fairly sophisticated behavior and inhabiting typically stable environments; and the garden chafer *Phyllopertha horticola* is a Rutelid beetle that inhabits light, well-drained soils under permanent pasture in the northwest, west, and south of England, and goes through a winter diapause (Laughlin, 1965). How can we sort out the fugitive or opportunistic species, that is, the *r* strategists, from the *K* strategists, among the "poor" colonizers?

First consider the intrinsic rate of natural increase. The *r* values of these three species are 0.1454, 0.0065, and 0.0029 respectively (Table 3–1), indicating a trend towards *r* strategy in *C. p. fatigans*, and a trend towards *K* strategy for *S. biocellata* and *P. horticola.* Also, if we assume that the immigration rates (Table 3–3) indicate dispersibility, we could use them to check for a trend. The $\tilde{\Lambda}$ values are 1.2478, 0.0383, and 0.0101 for species 8, 18, and 3 respectively. Thus, we confirm that *C. p. fatigans* is an *r* strategist, while *S. biocellata* and *P. horticola* are *K* strategists, for it is accepted that a high dispersal ability is a characteristic feature of *r* strategists (Pianka, 1970).

In Table 3–4 the 18 species have been ranked by their $\tilde{\Lambda}$ values, and their colonizing type (from Figure 3–10) is listed for comparison. This table shows that there is a clear discontinuity between the first 10 species and the last 8; the $\tilde{\Lambda}$ value jumps approximately 2.7 times between species 17 and 2. If we consider the first 10 species as having a high dispersal ability (group A) and the last 8 species as having a low dispersibility (group B), then (1) all species of type III belong to group A; (2) of the species of type I, only the *r* strategist (species 8) belongs to group A, while the *K* strategists (species 18 and 3) belong to group B; and (3) species of type II are distributed in both groups A and B.

**Table 3–4.** Ranking of species after their immigration rate ($\tilde{\Lambda}$) (see Table 3–3), their classification in species types as in Figure 3–17, and ratio $\lambda/\overset{*}{\Lambda}$.

| Species | Rank | $\tilde{\Lambda}$ | Type | $\lambda/\overset{*}{\Lambda}$ |
|---|---|---|---|---|
| 8 | 1 | 1.2478 | I | 1.79 |
| 10 | 2 | 0.4886 | III | 6.91 |
| 6 | 3 | 0.4435 | II | 5.03 |
| 5 | 4 | 0.4342 | III | 6.88 |
| 14 | 5 | 0.2033 | II | 2.78 |
| 9 | 6 | 0.2003 | II | 3.33 |
| 4 | 7 | 0.1633 | II | 2.99 |
| 7 | 8 | 0.1421 | III | 6.56 |
| 16 | 9 | 0.1231 | II | 2.48 |
| 17 | 10 | 0.1034 | II | 2.43 |
| 2 | 11 | 0.0386 | II | 4.22 |
| 18 | 12 | 0.0383 | I | 1.93 |
| 1 | 13 | 0.0296 | II | 3.24 |
| 11 | 14 | 0.0258 | II | 4.12 |
| 12 | 15 | 0.0186 | II | 4.88 |
| 13 | 16 | 0.0147 | II | 5.65 |
| 3 | 17 | 0.0101 | I | 2.33 |
| 15 | 18 | 0.0011 | II | 32.65 |

Table 3–5 summarizes this discussion, grouping the strategies characterized by a given colonizing ability and a given dispersal ability with the ratio $r/\lambda$ and $\lambda/\mu$, and the statistic $\tilde{\Lambda}$.

**Table 3–5.** Colonizer ratios and immigration rates that characterize demographic strategies for given colonizing and dispersal abilities. (Species type as in Figure 3–16.)

| Species Type | $r/\lambda$ | $\lambda/\mu$ | Colonizing Ability | Dispersibility: High ($\tilde{\Lambda} > 0.1$) | Dispersibility: Low ($\tilde{\Lambda} < 0.1$) |
|---|---|---|---|---|---|
| I | $\rightarrow 0.5$ | $\rightarrow 2$ | Poor | $r$ strategist (opportunistic) | $K$. strategist |
| II | $0.5 \rightarrow 0.8$ | $2 \rightarrow 6$ | Intermediate | Relatively $r$ strategist | Relatively $K$. strategist |
| III | $0.8 \rightarrow$ | $6 \rightarrow$ | Good | $r$ strategist ? | —— |

A question mark goes with the $r$ strategists that are good colonizers for the following reasons. In the MacArthur and Wilson model these would be the species that are in their first postcolonization phase; that is to say, species that are supposed to switch to a $K$ strategy if the environment remains relatively constant in time and space. However, the only species that fall under this class are *D. melanogaster, D. pulex,* and *P. humanus*. It is not clear at all why these species should be considered pioneer colonizers. These three might be the type that are good colonizers through the strategy of an ensemble of subpopulations. However, the last column of Table 3–4 shows this is not the case. The symbol $\overset{*}{\Lambda}$ is used to denote the product $\tilde{\Lambda}(1 + \mu/2\lambda)(1 - \mu/\lambda)$, which has to be larger than $\lambda$ if the strategy of an ensemble of subpopulations is to be advantageous. We have shown that none of the 18 species fulfills this requirement. However, as $\tilde{\Lambda}$ has been estimated as the *minimum* immigration rate for successful colonization, it is probably an *underestimate* of the real immigration rate of the species. Thus the ratio $\lambda/\overset{*}{\Lambda}$ can be used to check which species are furthest away from fulfilling the condition $\overset{*}{\Lambda} > \lambda$. The lowest value of $\lambda/\overset{*}{\Lambda}$ is found for *Culex pipiens fatigans*, the only species that can be clearly classified as a fugitive species. *C. p. fatigans* is the most likely species to justify a strategy of an ensemble of subpopulations, a conclusion that is consistent with its opportunistic characteristics and high dispersibility. However, *D. melanogaster, D. pulex,* and *P. humanus* have a high $\lambda/\overset{*}{\Lambda}$ ratio, suggesting that this possibility is not an appropriate explanation of their properties of good colonizers with an $r$ strategy.

Figure 3–16 is a summary of this discussion; it shows the different strategies in relation to habitat stability, to yield successful colonization. It is an attempt to avoid confusion between demographic strategies and colonization abilities. This figure suggests a very simple testable idea. Since selection in relation to crowding may operate rapidly (in the order of five to six generations in *Drosophila pseudoobscura*) (Birch, 1955), the population parameters of a $K$ strategist species subjected to a competition-free environment (colonization), may be estimated after a few generations, thus verifying Figure 3–16. Furthermore, if

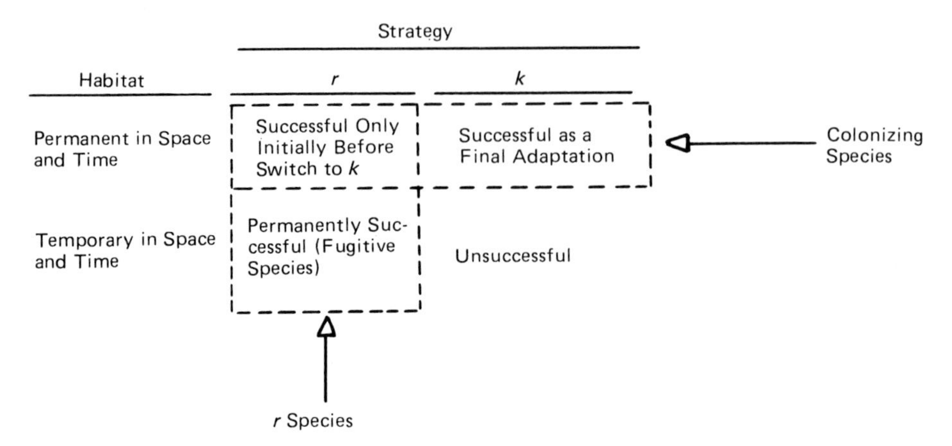


**Fig. 3–16.** Demographic strategies in relation to habitat stability to yield successful colonization.

the competition-free environment can be physically manipulated (as in a programable climatic chamber, for example) to simulate a periodically fluctuating environment, then changes in the life table components predicted by MacArthur (1968) may be verified.

## References

Anderson, W. W., and C. E. King. 1970. Age specific selection. *Proc. Natl. Acad. Sci.* 66(3):780–786.

Andrewartha, H. G., and L. C. Birch. 1954. *The Distribution and Abundance of Animals.* Chicago: Univ. Chicago Press.

Birch, L. C. 1955. Selection in *Drosophila pseudoobscura* in relation to crowding. *Evolution* 9:389–399.

Caughley, G. 1967. Parameters for seasonally breeding populations. *Ecology* 48(5): 834–839.

Charlesworth, B. 1971. Selection in density-regulated populations. *Ecology* 52(3): 469–474.

Demetrius, L. 1969. The sensitivity of population growth rate to perturbations in the life cycle components. *Math. Biosci.* 4(1/2):129–136.

Dobzhansky, T. 1950. Evolution in the tropics. *Am. Sci.* 38:209–221.

Evans, R. C., and R. E. Smith. 1952. The intrinsic rate of natural increase for the human louse *Pediculus humanus* L. *Am. Nat.* 86:299–310.

Force, D. C. 1972. r- and K-strategists in endemic host-parasitoids communities. *Bull. Entomol. Soc. Am.* 18(3):135–137.

French, N. R., and H. W. Kaaz. 1968. The intrinsic rate of natural increase of irradiated *Peromyscus* in the laboratory. *Ecology* 49:1172–1179.

Gadgil, M., and W. H. Bossert. 1970. Life historical consequences of natural selection. *Am. Nat.* 104(935):1–24.

Gadgil, M., and O. T. Solbrig. 1972. The concept of r- and K-selection: Evidence from wild flowers and some theoretical considerations. *Am. Nat.* 106(947): 14–31.

Gomez, C. 1972. Contribución al estudio de la dinámica poblacional de *Culex pipiens fatigans* Wiedeman (Diptera: Culicidae). Thesis, Fac. Sci. Central Univ. Venezuela.

Hairston, N. G., D. W. Tinkle, and H. M. Wilbur. 1970. Natural selection and the parameters of population growth. *J. Wildlife Mgmt.* 34(4):681–690.

King, C. E., and W. W. Anderson. 1971. Age specific selection. II. The interaction between r and K during population growth. *Am. Nat.* 105(94):137–156.

Landahl, J. T., and R. B. Root. 1969. Differences in the Life Tables of tropical and temperate milkweed bugs, genus *Oncopeltus* (Hemiptera: Lygaeidae). *Ecology* 50(4):734–737.

Laughlin, R. 1965. Capacity for increase: A useful population statistic. *J. Animal Ecol.* 34:77–91.

Lefkovitch, L. P. 1963. Census studies on unrestricted populations of *Lasioderma serricorne* (F.) (Coleoptera: Anobiidae). *J. Animal Ecol.* 32:221–231.

Leslie, P. H., and T. Park. 1949. The intrinsic rate of natural increase of *Tribolium castaneum* Herbst. *Ecology* 30:469–477.

Leslie, P. H., and R. M. Ranson. 1940. The mortality, fertility, and rate of natural increase of the vole (*Microtus agrestis*) as observed in the laboratory. *J. Animal Ecol.* 9:27–52.

Leslie, P. H., U. M. Venables, and L. S. V. Venables. 1952. The fertility and population structure of the brown rat (*Rattus norvegious*) in corn-ricks and some other habitats. *Proc. Zool. Soc. London* 122:187–238.

Levins, R. 1970. Fitness and optimization. *Mathematical Topics in Population Genetics*; In K. Kojima, ed. 400 pp. New York: Springer-Verlag.

MacArthur, R. H. 1968. Selection for life tables in periodic environments. *Am. Nat.* 102(926):381–383.

MacArthur, R. H., and E. O. Wilson. 1967. *The Theory of Island Biogeography. Monographs in Population Biology.* No. 1, 203 pp. Princeton, N.J.: Princeton Univ. Press.

Margalef, R. 1958. Mode of evolution of species in relation to their places in ecological succession. *XV Intern. Congr. Zool. (London).* Sect. X, paper 17.

———. 1968. *Perspectives in Ecological Theory.* 111 pp. Chicago: Univ. Chicago Press.

Marshall, J. S. 1962. The effects of continuous gamma radiation on the intrinsic rate of natural increase of *Daphnia pulex. Ecology* 43:598–607.

Pianka, E. R. 1970. On r- and K-selection. *Am. Nat.* 104(940):592–597.

———. 1972. r and K selection or b and d selection? *Am. Nat.* 106(951):581–588.

Rabinovich, J. E. 1968. Contribución al estudio de la dinámica de poblaciones. I. Análisis poblacional de *Nasonia vitripennis* Walk. (Hymenoptera: Pteromalidae) *Acta Biol. Venez.* 6:68–81.

———. 1970. Vital statistics of *Synthesiomyia nudiseta* (Diptera: Muscidae). *Ann. Ent. Soc. Am.* 63:749–752.

———. 1972a. Vital statistics of triatominae (Hemiptera: Reduviidae) under laboratory conditions. I. *Triatoma infestans* Klug. J. Mèd. Ent. 9(4):351–370.

———. 1972b. Vital statistics of triatominae (Hemiptera: Reduviidae) under laboratory conditions. II. *Rhodnius prolixus* Staal. *Abst. 14th Intern. Congr. Entomol.* Canberra 22–30 August 1972. p. 291.

Richter-Dyn, N., and N. S. Goel. 1972. On the extinction of a colonizing species. *Theor. Pop. Biol.* 3:406–433.

Roughgarden, J. 1971. Density-dependent natural selection. *Ecology* 52(3):453–468.

Sidiqqui, W. H., and C. A. Barlow. 1972. Population growth of *Drosophila melanogaster* (Diptera: Drosophilidae) at constant and alternating temperatures. *Ann. Ent. Soc. Am.* 65(5):993–1001.

# CHAPTER 4

# A Matrix Model of Tree Population Dynamics

Gary S. Hartshorn

## Introduction

The purpose of this short paper is to explain how a projection matrix model has been used to describe the population dynamics of a tropical tree species (Hartshorn, 1972). A detailed description of model development and some potential uses of the model will be presented. The tree *Pentaclethra macroloba* (Willd.) Ktze. (Mimosaceae) is a large canopy species dominating areas of tropical wet forest in the Atlantic lowlands of Costa Rica.

Lefkovitch (1965) described a population model for unequal sized stages in an organism's life cycle. For a species there are $s$ distinct stages and $n_{i,t}$ individuals in stage $i$ ($i = 1, 2, \ldots, s$) at time $t$. The total population at time $t$ is

$$n_{1,t} + n_{2,t} + \ldots + n_{s,t} = N_t$$

The numbers of individuals in each stage at time $t+1$ are functions of the numbers at time $t$ and the mortality, growth, fecundity, immigration, or emigration of each stage during the time interval $(t, t+1)$. The relation of the numbers in stage $p$ at time $t+1$ to those in all other stages at time $t$ are described deterministically by the linear equation

$$n_{1,t,}m_{p,1} + n_{2,t}m_{p,2} + \ldots + n_{p,t}m_{p,p} + \ldots + n_{s,t}m_{p,s} = n_{p,t+1}$$

The constants $m_{p,j} \geqq 0$ ($j = 1, 2, \ldots, s$) represent the biological dependence of the $p$th stage at time $t+1$ on the $j$th stage at time $t$. The series of equations for all $s$ stages at time $t$ and $t+1$ are expressed in matrix notation as

$$\begin{bmatrix} m_{1,1} & m_{1,2} & \ldots & m_{1,s} \\ m_{2,1} & m_{2,2} & \ldots & m_{2,s} \\ \cdot & \cdot & & \cdot \\ \cdot & \cdot & & \cdot \\ \cdot & \cdot & & \cdot \\ m_{s,1} & m_{s,2} & \ldots & m_{s,s} \end{bmatrix} \begin{bmatrix} n_1 \\ n_2 \\ \cdot \\ \cdot \\ \cdot \\ n_s \end{bmatrix}_t = \begin{bmatrix} n_1 \\ n_2 \\ \cdot \\ \cdot \\ \cdot \\ n_s \end{bmatrix}_{t+1}$$

or as $M_t n_t = n_{t+1}$ where $M_t$ is a matrix of the coefficients $m_{p,j}$ and $n_t$ and $n_{t+1}$ are vectors representing the stage compositions of the populations $N_t$ and $N_{t+1}$. The square ($s \times s$) matrix $M_t$ has $s$ latent roots $\lambda_i$, each with its appropriate stable vector $v_i$, such that $M_t v_i = \lambda_i v_i$ ($i = 1, 2, \ldots, s$).

The Lefkovitch matrix is classified mathematically as an irreducible, nonnegative matrix. It is nonnegative because all the coefficients are nonnegative, that is, either zero or positive. An irreducible matrix always has one positive real root $\lambda_1$ (or eigenvalue) and the absolute values of all other roots are less than or equal to $\lambda_1$ (Gantmacher, 1959).

Working with the cigarette beetle *Lasioderma serricorne*, under experimental conditions that allowed unrestricted growth of all stages, Lefkovitch (1965) showed that a dominant, positive latent root existed and that each coefficient in the stable vector was real and of the same sign. Thus the Lefkovitch matrix $M$ is analogous to $A$, the age specific matrix of Leslie (1945). $M$ is identical with $A$ when each life history stage lasts one unit of time and the individuals show no variation (and assuming that the stages are taken in their temporal sequence); consequently Leslie's $A$ is a special case of the more general Lefkovitch matrix $M$ (Lefkovitch, 1965). When the various stages have different durations, it may be possible to derive $M$ from $A$ (if known), but not $A$ from $M$.

Leslie (1945) showed that the dominant latent root of $A$ is equal to $e^r$, where $r$ is the intrinsic rate of natural increase in the equation $N_t = N_0 e^{rt}$. If $\lambda_1 = e^r = 1$, then $r = 0$ and the population neither increases nor decreases since $r$ equals births minus deaths. Lefkovitch (1965) illustrated that regardless of how the population is classified, its rate of increase in the stable state is unchanged; thus the dominant latent root of $A$ is equal to that of $M$, both of which are equal to $e^r$. If the appropriate environmental conditions are simulated for a population, then a study of the numbers of each distinguishable stage will give an estimate of $e^r$. This is true no matter how the population changes or whether the change is considered in terms of age or stage groups. The Lefkovitch projection matrix is especially applicable when distinct morphological stages exist in the life cycle.

## Life Table

Foresters have long been interested in the size structure of forest stands and the recruitment of trees into the next larger size class. The use of size classes is actually preferred to age classes when the stages of a population are easily recognized, for example, for insects (Lefkovitch, 1965) and trees (Usher, 1966, 1969). In mature forests the range of ages in a size class may be extreme due to differential growth responses resulting from succession, suppression, competition, and the like. The present difficulty of determining the ages of tropical wet forest trees from increment cores clearly prevents the use of age classes. However, most trees can be assigned to stages with relative ease. Indeed, Pelton

(1953) proposed that seed, seedling, juvenile, reproductive, and senescent stages be used for life cycle studies of plants.

Regardless of the number of stages, it is essential to have the following information for each stage over a given time period: (1) percentage of individuals that survive, (2) percentage of individuals that move to the next larger stage class in the chosen time interval, and (3) number of progeny produced per individual that survive the first time period. Thus an individual can do one of three things—die, remain in that stage, or move into the next larger stage. These rates can then be used to calculate the coefficients of the Lefkovitch projection matrix.

A life table (Table 4–1) is an efficient means of tabulating the data necessary for calculating the coefficients of the Lefkovitch projection matrix. Fifteen unequal stages ranging from seeds to trees >100 cm dbh are used to represent the life cycle of *Pentaclethra macroloba*. The size class limits were arbitrarily chosen to take advantage of the large amount of data on the seedling and sapling stages. For size classes 2 through 15, the total number of individuals, the number of live individuals as of 15 August 1971, and the mortality rates are all based on a study of marked individuals in a 4-ha plot of mature, undisturbed forest. For use in the model, it is assumed that the proportion of individuals moving to the next class consists of those individuals that are promoted into the next larger class when the average annual growth increment for their size class is added to their actual height or diameter. For example, of the 88 individuals in size class 4, only 12 are between 141 and 150 cm tall, thus the proportion of 0.14 moving to class 5. The proportion remaining is simply the remainder since no mortality enters into these calculations. *Pentaclethra macroloba* seeds do not have a dormancy mechanism, so no seeds remain in class 1 more than one year. The proportion of total seed input for each reproductive class is a crude estimate based on a minimum reproductive size of 35 cm dbh, the number of individuals in each reproductive class and my assumption that annual reproductive output increases with increasing size. The proportions in line 7 are multiplied by the total seed input to obtain the number of seeds produced by each adult in each reproductive class.

## Initial Matrix

The 15 stages in the life cycle of *P. macroloba* are represented in the model by the matrix

$$\begin{bmatrix} m_{1,1} & m_{1,2} & m_{1,3} & \dots & m_{1,15} \\ m_{2,1} & m_{2,2} & m_{2,3} & \dots & m_{2,15} \\ m_{3,1} & m_{3,2} & m_{3,3} & \dots & m_{3,15} \\ \cdot & \cdot & \cdot & & \cdot \\ \cdot & \cdot & \cdot & & \cdot \\ \cdot & \cdot & \cdot & & \cdot \\ m_{15,1} & m_{15,2} & m_{15,3} & \dots & m_{15,15} \end{bmatrix}$$

**Table 4–1.** Life table for *Pentaclethra macroloba* on intensive study plot (4 ha), La Selva, based on the 12-month period ending 15 August 1971.[a]

| Stage Number (Size Classes) | 1 | 2 | 3 | 4 | 5 | 6 | 7 | 8 | 9 | 10 | 11 | 12 | 13 | 14 | 15 |
|---|---|---|---|---|---|---|---|---|---|---|---|---|---|---|---|
| Height (cm) | | 0–50 | 50–100 | 100–150 | 150–200 | 200–250 | 250–300 | >300 | | | | | | | |
| Diameter (cm) | Seeds | | | <1.0 | 1.0<1.2 | 1.2<1.5 | 1.5<2.0 | 2.0<5.0 | 5<10 | 10<20 | 20<40 | 40<60 | 60<80 | 80<100 | >100 |
| Total individuals | 21,400 | 2,292 | 137 | 88 | 57 | 36 | 37 | 38 | 19 | 37 | 41 | 54 | 44 | 21 | 4 |
| Survival rate | 0.4294 | 0.4800 | 0.7226 | 0.8863 | 0.8596 | 0.9722 | 0.9189 | 0.9736 | 0.9999 | 0.9730 | 0.9757 | 0.9815 | 0.9773 | 0.9048 | 0.7500 |
| Number of live individuals | 9202 | 1100 | 99 | 78 | 49 | 35 | 34 | 37 | 19 | 30 | 40 | 53 | 43 | 19 | 3 |
| Growth rate (cm) | — | 3.2 | 4.6 | 9.0 | 13.9 | 24.0 | 30.2 | 0.09 | 0.20 | 0.30 | 0.37 | 0.37 | 0.32 | 0.67 | 1.03 |
| Proportion moving up to next class | 1.00 | 0.10 | 0.11 | 0.14 | 0.23 | 0.39 | 0.49 | 0.01 | 0.05 | 0.03 | 0.10 | 0.06 | 0.02 | 0.01 | 0 |
| Proportion remaining in size class | 0 | 0.90 | 0.89 | 0.86 | 0.77 | 0.61 | 0.51 | 0.99 | 0.95 | 0.97 | 0.90 | 0.94 | 0.98 | 0.99 | 1.00 |
| Proportion of total seed input | 0 | 0 | 0 | 0 | 0 | 0 | 0 | 0 | 0 | 0 | 0.15 | 0.30 | 0.30 | 0.20 | 0.05 |
| Number of seeds per adult | 0 | 0 | 0 | 0 | 0 | 0 | 0 | 0 | 0 | 0 | 80 | 121 | 149 | 225 | 357 |

[a] ∧ indicates interpolated values.

where $m_{p,j}$ is the contribution of stage $j$ ($j = 1, 2, \ldots, 15$) to stage $p$ ($p = 1, 2, \ldots, 15$). Data from the life table (Table 4–1) are used to calculate the coefficients of the initial matrix for *Pentaclethra macroloba* (Table 4–2). The $m_{1,1}$ coefficient comes from two sources, both of which equal zero: (1) the number of seeds produced by an individual in stage 1 (for example, seeds do not produce seeds), and (2) the proportion of individuals that remain in stage 1. Stages 2 through 10 do not produce seeds, so coefficients $m_{1,2}$ through $m_{1,10}$ are also zero. Coefficients $m_{1,11}$ to $m_{1,15}$ are the number of seeds produced by each adult in the respective stages (Table 4–1, line 8). The diagonal coefficients (where $p = j$; for example, $m_{2,2}, m_{3,3}, \ldots, m_{15,15}$) are the proportion of individuals remaining in stage $j$ (Table 4–1, line 6) multiplied by the survival rate of stage $j$ (Table 4–1, line 2). The subdiagonal coefficients (where $p = j+1$; for example, $m_{2,1}, m_{3,2}, \ldots, m_{15,14}$) are the proportion of individuals that move up into stage $j+1$ from stage $j$ (Table 4–1, line 5) multiplied by the survival rate of stage $j$ (Table 4–1, line 2). Except on the diagonal and subdiagonal, all the coefficients in rows 2 to 15 of the initial matrix are zero because a *P. macroloba* individual does not regress a stage (for example, $m_{2,3}$ indicates an individual goes from stage 3 to stage 2) nor pass through two stages in one year (for example, $m_{3,1}$ indicates an individual moves from stage 1 to stage 3).

**Dominant Latent Root and Stable Column Vector**

Of the several methods available, exponentiation of the initial matrix was used to determine the dominant latent root (Keyfitz, 1968). If an initial matrix, $X$, is raised to a sufficiently high power, all the coefficients become positive and stable, regardless of the initial stage (or age) distribution. After raising the initial matrix to the $k$th power of 2, the $X^{2k}$ or $Y$ matrix is multiplied by the initial matrix $X$ to give the $X^{(2k)+1}$ or $Z$ matrix. Each coefficient of the $Z$ matrix is divided by its corresponding coefficient in the $Y$ matrix to yield a latent root for each division of coefficients. All the numbers in the final matrix of latent roots approach equality if the initial matrix has been raised to a sufficiently high power. Thus the repetitive number occurring in the matrix of latent roots is the dominant latent root of the initial matrix. The matrix of latent roots was identical to two decimal places when the initial matrix was raised to the 256th power. These operations were carried out on the College of Forest Resources' (University of Washington) NOVA computer; the BASIC program can be obtained from the author.

The dominant latent root for the initial matrix of *Pentaclethra macroloba* is 1.002, which is remarkably close to the theoretical value of 1.0 for population stability. This is the first valid determination of $r$ or $\lambda_1$ for a natural plant population. Two other estimates of $\lambda_1$ for plant populations have been published: Bosch's (1971) erroneous value of 12.87 for redwoods, and a $\lambda_1$ value of 1.204 given by Usher (1966) for a managed Scot's pine forest in England.

The dominant latent root indicates whether a population is decreasing, remaining the same, or increasing in size (numbers), but it gives no information

**Table 4–2.** Initial matrix for *Pentaclethra macroloba*.

| | | | | | | | | | | | | | | |
|---|---|---|---|---|---|---|---|---|---|---|---|---|---|---|
| 0 | 0 | 0 | 0 | 0 | 0 | 0 | 0 | 0 | 0 | 80 | 121 | 149 | 225 | 357 |
| 0.43 | 0.43 | 0 | 0 | 0 | 0 | 0 | 0 | 0 | 0 | 0 | 0 | 0 | 0 | 0 |
| 0 | 0.05 | 0.64 | 0 | 0 | 0 | 0 | 0 | 0 | 0 | 0 | 0 | 0 | 0 | 0 |
| 0 | 0 | 0.08 | 0.76 | 0 | 0 | 0 | 0 | 0 | 0 | 0 | 0 | 0 | 0 | 0 |
| 0 | 0 | 0 | 0.12 | 0.66 | 0 | 0 | 0 | 0 | 0 | 0 | 0 | 0 | 0 | 0 |
| 0 | 0 | 0 | 0 | 0.20 | 0.59 | 0 | 0 | 0 | 0 | 0 | 0 | 0 | 0 | 0 |
| 0 | 0 | 0 | 0 | 0 | 0.38 | 0.47 | 0 | 0 | 0 | 0 | 0 | 0 | 0 | 0 |
| 0 | 0 | 0 | 0 | 0 | 0 | 0.45 | 0.96 | 0 | 0 | 0 | 0 | 0 | 0 | 0 |
| 0 | 0 | 0 | 0 | 0 | 0 | 0 | 0.01 | 0.95 | 0 | 0 | 0 | 0 | 0 | 0 |
| 0 | 0 | 0 | 0 | 0 | 0 | 0 | 0 | 0.05 | 0.94 | 0 | 0 | 0 | 0 | 0 |
| 0 | 0 | 0 | 0 | 0 | 0 | 0 | 0 | 0 | 0.03 | 0.88 | 0 | 0 | 0 | 0 |
| 0 | 0 | 0 | 0 | 0 | 0 | 0 | 0 | 0 | 0 | 0.10 | 0.92 | 0 | 0 | 0 |
| 0 | 0 | 0 | 0 | 0 | 0 | 0 | 0 | 0 | 0 | 0 | 0.06 | 0.96 | 0 | 0 |
| 0 | 0 | 0 | 0 | 0 | 0 | 0 | 0 | 0 | 0 | 0 | 0 | 0.02 | 0.90 | 0 |
| 0 | 0 | 0 | 0 | 0 | 0 | 0 | 0 | 0 | 0 | 0 | 0 | 0 | 0.01 | 0.75 |

about the distribution or stability of individuals in the defined size classes. By multiplying the stable matrix $Y$ by the initial stage distribution (Table 4–1, line 3), we obtain a stable column vector (or stable stage distribution). The relative distributions of individuals for actual and for predicted populations are very similar (Table 4–3). Both the dominant latent root and stable stage distribution are very strong evidence of the stable climax status for this *Pentaclethra macroloba* population [see Hartshorn (1972) for more detailed arguments].

**Table 4–3.** Comparisons of the initial stage distribution with the stable stage distribution for *Pentaclethra macroloba.*

| Stage Number | Initial Stage | Distribution | Stable Stage | Distribution |
|---|---|---|---|---|
| 1 | 21,400 | 88.05 | 22,523 | 67.50 |
| 2 | 2,292 | 9.43 | 10,188 | 30.53 |
| 3 | 137 | 0.56 | 202 | 0.61 |
| 4 | 88 | 0.36 | 78 | 0.23 |
| 5 | 57 | 0.23 | 48 | 0.14 |
| 6 | 36 | 0.15 | 33 | 0.10 |
| 7 | 37 | 0.15 | 31 | 0.09 |
| 8 | 38 | 0.16 | 53 | 0.16 |
| 9 | 19 | 0.08 | 18 | 0.06 |
| 10 | 37 | 0.15 | 36 | 0.11 |
| 11 | 41 | 0.17 | 37 | 0.11 |
| 12 | 54 | 0.22 | 54 | 0.16 |
| 13 | 44 | 0.18 | 45 | 0.14 |
| 14 | 21 | 0.09 | 20 | 0.06 |
| 15 | 4 | 0.02 | 3 | 0.01 |
| Total | 24,305 | 100.00 | 33,369 | 100.00 |

## Population Simulation

The effect of rate changes in fecundity, growth, or survival on the population can be simulated quite easily with the model. Simulation simply involves changing one or more coefficients of the initial matrix and calculating the dominant latent root for the new initial matrix. The dominant latent root from the simulation run can then be compared to the dominant latent root of the original initial matrix. If the same change is simulated for each stage, one is able to compare the effect on the different stages, or, in other words, do a sensitivity analysis. Comparisons of this sort are very useful for testing hypotheses and making predictions.

Janzen (1970, p. 523) predicted that the introduction of an insect host specific on *Pentaclethra macroloba* seeds would drastically reduce the population of adults. It is instructive to simulate decreased seed survival as a test of Janzen's prediction, and also to compare a reduction in survival in the other stages (Table 4–4). Seed survival reductions of 10 and 50 percent gave dominant latent roots

**Table 4–4.** Dominant latent roots resulting from simulated changes in survival of *Pentaclethra macroloba* population.

| Change Made in the Initial Matrix | | | Dominant Latent Root (for x 256) |
|---|---|---|---|
| Stage Number | Coefficient | Old → New Values *P. macroloba* | *P. macroloba* |
| None | Base run | | 1.002 |
| 1 | $m_{2,1}$ | 0.43 → 0.39 | 1.001 |
| 1 | $m_{2,1}$ | 0.43 → 0.215 | 0.996 |
| 1 | $m_{2,1}$ | 0.43 → 0.05 | 0.986 |
| 1 | $m_{2,1}$ | 0.43 → 0.01 | 0.978 |
| 1 | $m_{2,1}$ | 0.43 → 0.0001 | 0.972 |
| 2 | $m_{2,2}$<br>$m_{3,2}$ | 0.43 → 0.39<br>0.05 → 0.04 | 0.999 |
| 2 | $m_{2,2}$<br>$m_{3,2}$ | 0.43 → 0.215<br>0.05 → 0.025 | 0.994 |
| 3 | $m_{3,3}$<br>$m_{4,3}$ | 0.64 → 0.32<br>0.08 → 0.04 | 0.991 |
| 4 | $m_{4,4}$<br>$m_{5,4}$ | 0.76 → 0.38<br>0.12 → 0.06 | 0.989 |
| 5 | $m_{5,5}$<br>$m_{6,5}$ | 0.66 → 0.33<br>0.20 → 0.10 | 0.991 |
| 6 | $m_{6,6}$<br>$m_{7,6}$ | 0.59 → 0.295<br>0.38 → 0.19 | 0.992 |
| 7 | $m_{7,7}$<br>$m_{8,7}$ | 0.47 → 0.23<br>0.45 → 0.23 | 0.993 |
| 8 | $m_{8,8}$<br>$m_{9,8}$ | 0.96 → 0.49<br>0.01 → 0.005 | 0.976 |
| 9 | $m_{9,9}$<br>$m_{10,9}$ | 0.95 → 0.47<br>0.05 → 0.01 | 0.973 |
| 10 | $m_{10,10}$<br>$m_{11,10}$ | 0.94 → 0.47<br>0.03 → 0.01 | 0.978 |
| 11 | $m_{11,11}$<br>$m_{12,11}$ | 0.88 → 0.44<br>0.10 → 0.05 | 0.985 |
| 12 | $m_{12,12}$<br>$m_{13,12}$ | 0.92 → 0.46<br>0.06 → 0.03 | 0.987 |
| 11–15 | | 50% reduction of all coefficients | 0.972 |
| 1–15 | | 10% reduction of all coefficients | 0.903 |
| 1–15 | | 50% reduction of all coefficients | $\leqq$0.0 |

very close to 1.0. However, it is likely that a host-specific insect predator on *P. macroloba* seeds would cause appreciably greater seed survival reductions than 50 percent. The effects of extremely low seed survival were simulated with $m_{2,1}$ coefficients of 0.05, 0.01, and 0.0001, which gave dominant latent roots of 0.986, 0.978, and 0.972, respectively. These simulations indicate that very low seed survival rates will cause some reduction in the *P. macroloba* population, but will not have the drastic effect on the adult population that Janzen predicted.

It is also useful to look at the sensitivity of each stage to a 50 percent reduction in survival (Table 4–4). This was done by reducing survival in one stage, while holding all the other stages at their original value, and then repeating the procedure for a different stage. The sensitivity analysis showed that the seed stage is least sensitive to a 50 percent reduction in survival; that is, the smallest reduction of the dominant latent root occurred in stage 1. The most sensitive stage of *P. macroloba* to a 50 percent reduction in survival is stage 9 ($\lambda_1 =$ 0.973). Stages 8 ($\lambda_1 = 0.976$) and 10 ($\lambda_1 = 0.978$) are also quite sensitive to the same 50 percent reduction in survival. Stage 10 is the last prereproductive stage, and it is reasonable that a lowering of its survival has a strong effect on the population. The maximum reproductive potential of an organism occurs just as it enters into reproduction and reduced survival at this stage of the life cycle should have a strong effect on reproductive potential, especially in a species with a long prereproductive period. The sensitivity of stages 8 to 10 to reduced survival is probably accentuated by the very slow movement of individuals through these stages. Note that a comparable reduction in survival rate of reproductive stages 11 ($\lambda_1 = 0.985$) and 12 ($\lambda_1 = 0.987$) does not lower the dominant latent root nearly as much as in the late prereproductive stages. Even if the survival of all reproductive stages (11 to 15) is reduced 50 percent, the dominant latent root is only lowered to 0.972. The greatest reductions occurred when survival in all stages (1 to 15) was reduced by 10 percent and 50 percent; the respective dominant latent roots were 0.903 and 0 (suggesting population extinction).

## Discussion

The matrix model described in this paper has been successfully used to describe the population dynamics of two tropical wet forest tree species (Hartshorn, 1972). Through the use of unequal stage groupings, the matrix model should be applicable essentially to any plant population. It would be extremely interesting to test the generality and utility of this type of model with such diverse groups as epiphytes, marine algae, weedy annuals, or even temperate trees. It also would be useful to obtain the necessary input data from a population that can be aged as well as "sized," so that comparisons could be made between Leslie's age-specific matrix and Lefkovitch's unequal stage group matrix. The high degree of plasticity shown by *r* strategists (Harper, 1967; Gadgil and Solbrig, 1972) may cause some difficulty in interpreting $\lambda_1$ or $r$ values for such species.

This model should be very useful for determining the successional or sociologic status of a population. Questions and controversies about the stability or instability of a plant population or whether or not there is adequate regeneration to maintain the population are common in the ecological literature. Recent discussions concern the saguaro (*Cereus giganteus*) population (Shreve, 1910; Niering et al., 1963; Turner et al., 1969) and the *Araucaria* populations in New Guinea (Womersley, 1958; Havel, 1971). With this type of model we can determine how far a population is from stability and how many seeds or seedlings must survive to maintain the population.

An interesting application of the model would be to analyze the temporal pattern of $\lambda_1$ or $r$ for a short-lived $r$ strategist during a successional sequence. Ideally, it would be a colonizing species that is able to reach maturity before being outcompeted in a later successional stage. In such a situation, I would predict an initially high $\lambda_1$, considerably greater than 1.0, that would then decrease to less than 1.0 as succession progresses.

Another potential use of the matrix model is in forest management. Usher (1966, 1969) suggested that the rate of increase larger than 1.0 represents the harvestable yield in a selection forest. Although biologically and theoretically interesting, Usher's model is only applicable to a single species forest. In mixed forests, such a model would have to incorporate species interactions before the model could be used to predict allowable harvest accurately.

## Acknowledgment

The research reported here was supported by NSF grant GB25292 through the Organization for Tropical Studies. I thank W. H. Hatheway and C. Fowler for their helpful comments.

## References

Bosch, C. A., 1971. Redwoods: A population model. *Science* 172:345–349.

Gadgil, M., and O. T. Solbrig. 1972. The concept of r- and K-selection: Evidence from wild flowers and some theoretical considerations. *Am. Nat.* 106(947): 14–31.

Gantmacher, F. R. 1959. *The Theory of Matrices*, K. A. Hirsch, transl. Vol. II. New York: Chelsea.

Harper, J. L. 1967. A Darwinian approach to plant ecology. *J. Ecol.* 55(2):247–270.

Hartshorn, G. S. 1972. The ecological life history and population dynamics of *Pentaclethra macroloba*, a tropical wet forest dominant and *Stryphnodendron excelsum*, an occasional associate. Ph.D. dissertation, Univ. Washington, Seattle.

Havel, J. J. 1971. The Araucaria forests of New Guinea and their regenerative capacity. *J. Ecol.* 59(1):203–214.

Janzen, D. H. 1970. Herbivores and the number of tree species in tropical forests. *Am. Nat.* 104(940):501–528.

Keyfitz, N. 1968. *Introduction to the Mathematics of Populations*. Reading, Mass.: Addison-Wesley.

Lefkovitch, L. P. 1965. The study of population growth in organisms grouped by stages. *Biometrics* 21:1–18.

Leslie, P. H. 1945. On the use of matrices in certain population mathematics. *Biometrika* 33:183–212.

Niering, W. A., R. H. Whittaker, and C. H. Lowe. 1963. The saguaro: A population in relation to environment. *Science* 142:15–23.

Pelton, J. 1953. Ecological life cycle of seed plants. *Ecology* 34(3):619–628.
Shreve, F. 1910. The rate of establishment of the giant cactus. *Plant World* 13:235–240.
Turner, R. M., S. M. Alcorn, and G. Olin. 1969. Mortality of transplanted saguaro seedlings. *Ecology* 50(5):835–844.
Usher, M. B. 1966. A matrix approach to the management of renewable resources, with special reference to selection forests. *J. Appl. Ecol.* 3:355–367.
———. 1969. A matrix model for forest management. *Biometrics* 25:309–315.
Womersley, J. S. 1958. The Araucaria forests of New Guinea. In *Proc. Symp. Humid Tropics Vegetation.* pp. 252–257. Tjiawi, Indonesia: UNESCO.

# CHAPTER 5

# An Analysis of Late Secondary Succession in Species-Rich Tropical Forest

DENNIS H. KNIGHT

There have been few studies of succession in species-rich tropical forest, and most of these have focused on the early pioneer stages. This report is an analysis of late secondary succession in the forest on Barro Colorado Island (BCI), a 15Km$^2$ island in Gatun Lake, Panama Canal Zone.

The most effective way to study succession is by direct observation (Williams et al., 1969), but this approach is often impossible because of the time required. An alternative method is to examine adjacent stands of vegetation in the same habitat that have been disturbed at different times, under the assumption that they represent a time series. This comparative approach, using data on population structure, was used in my investigation on BCI.

The forests on BCI can be classified as semievergreen seasonal forest (*sensu* Beard, 1944), and have been undisturbed by man for the last 50 years. Approximately half of the island is covered by forest about 60 years old and past the early pioneer stage dominated by *Cecropia* spp., *Ochroma pyramidale*, and *Didymopanax morototoni*. The other half is dominated by older forest (probably more than 150 years old) that is believed by some to be the climax vegetation of the area. Tree species diversity ($\geqq$2.5 cm dbh) is high with an average Shannon-Wiener index of 4.8 (log base 2) and about 133 species per 500 trees. Total annual rainfall averages 267 cm, with a dry season from January to April.

## Methods

Nine upland stands were sampled for the successional analysis, five in the younger forest and four in the older forest. Each stand was three to five acres and appeared relatively homogeneous for vegetation structure and topography. An attempt was made to select stands that were homogeneous for species distribution, but Lang et al. (1971) concluded that this criterion was not very useful for stand selection on BCI. The physical and chemical characteristics of the soils in each stand did not differ greatly, and the stands were all within 3 km of each other. Thus it seemed reasonable to assume that the stands represented different stages of the sere for the habitat.

Each stand was sampled with 10 by 20 m quadrats distributed objectively throughout the stand. This quadrat size proved more efficient than larger plots (Lang et al., 1971). Approximately 30 to 50 percent of each stand was sampled for all trees greater than 2.5 cm diameter at breast height (dbh). This level of sampling intensity probably provided an adequate density estimate for only the commonest species (Lang et al., 1971). Tree ages could rarely be determined, so the data for each species were separated into seven diameter at breast height classes. Species were identified using vegetative features (Knight, 1970).

## Results and Discussion

The utility of species composition data for interpreting forest succession depends in part on knowledge about the population structure of the major species. Thus, in temperate regions a forest is labeled *climax* if the dominant canopy species comprise a large proportion of the individuals in the sapling size-class as well. However, since all species on BCI are relatively uncommon (Figure 5–1), it is not possible to focus on the population structure of a few

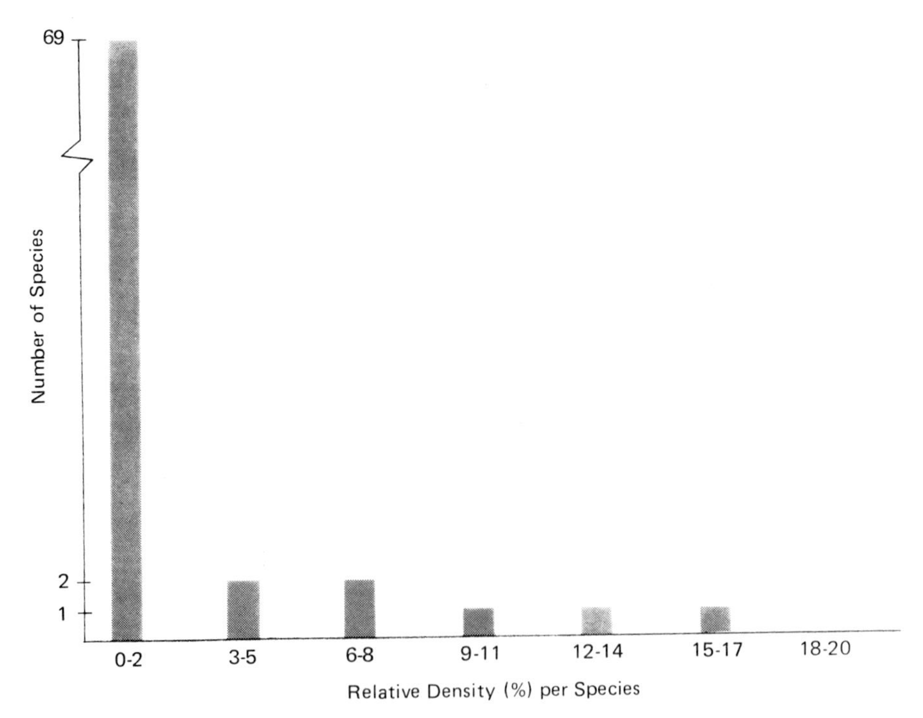


**Fig. 5–1.** The number of species ($\geqq$2.5 cm dbh) in seven relative density classes.

species as is often done in temperate vegetation, and an analysis that attempts to consider each species becomes quite confusing. To avoid this problem the species can be placed profitably into *ecological groups* based on population structure, each group including species with a similar structure. The analysis is then made on the basis of the abundance of one or another of the groups and is not dependent on the presence of any particular species, an especially attractive advantage where the "minimal area" cannot be entirely counted for species composition.

While at least five patterns of population structure were observed, two seem meaningful at this time for a successional interpretation (Figure 5–2). Pattern A in Figure 5–2, with about 60 species, is represented by trees with frequent reproduction; pattern B suggests little or no reproduction in the forest understory and is represented by about 45 species. Many of the pattern B species will probably persist on BCI by invading gaps in the forest canopy; they can be referred to as either pioneer or light-gap species.

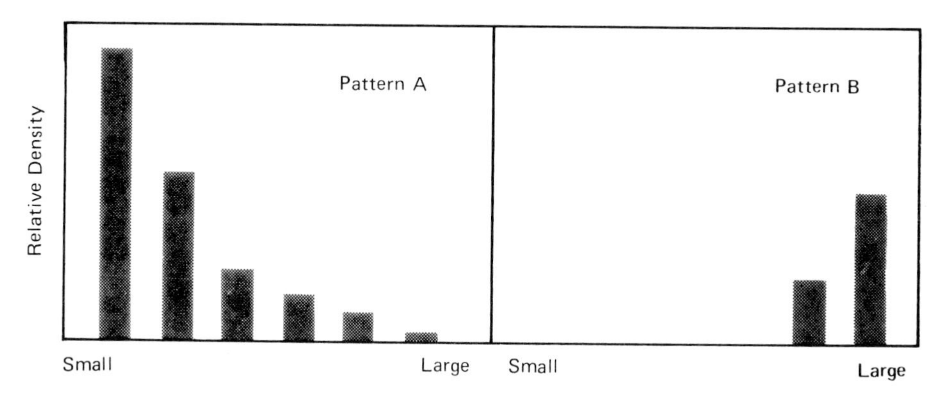


**Fig. 5–2.** Two patterns of population structure that were used to interpret successional status of the forests on BCI. Each bar is a different size class, ranging from the smallest on the left (2.5 to 6 cm dbh) to the largest on the right (>60 cm dbh).

With this classification, next the abundance of species with either pattern A or pattern B was determined in each of the nine stands. Group importance was evaluated first by calculating the percent of the total number of species in a stand belonging to each group, and second by summing the relative densities of all species belonging to each of the two groups (Table 5–1).

There is little difference between the young and old forest on the basis of the two population structure patterns (Table 5–1). An average of 15 percent of the species belong to the pioneer category in both young and old forest, and the older forest averages only slightly more reproducing species than the younger forest. These results suggest that the forests on BCI are in a transitional stage of succession.

Considering relative density, it is clear that both the young and old stands are in a late stage of secondary succession—the individuals of reproducing species are considerably more abundant than the pioneer species (Table 5–1). Again, however, the young and old stands are not very different.

**Table 5–1.** The abundance of pioneer, or light-gap, species (pattern B, Figure 5–2) and frequently reproducing species (pattern A, Figure 5–2) in the young and old forest on Barro Colorado Island.[a]

| | Younger Forest ($\bar{x}$) | Older Forest ($\bar{x}$) |
|---|---|---|
| Pioneer species | | |
| Percent of species | 15 | 15 |
| Relative density[b] | 9 | 4 |
| Reproducing species | | |
| Percent of species | 37 | 39 |
| Relative density[b] | 59 | 65 |

[a] An average total of 58 species per 1,000 $m^2$ was found in both the young and old forest, regardless of ecological group.
[b] Percent of all individuals greater than 2.5 cm dbh.

The older forest on BCI is frequently referred to as climax, and it is obviously different from the younger vegetation. The differences, however, are primarily physiognomic or in species that tell us nothing (yet) about successional status. An examination of the data shows that many of the pioneer canopy species in the young forest are also canopy species in the older forest, except they are much larger. Reproducing climax species are beginning to occupy the older forest canopy, and of course they are common in the understory, but the younger forests are sufficiently old so that the reproducing species are in the understory there also.

Some of the pioneer species currently in the old forest canopy appear to be in a low vigor, senescent condition with some standing dead trees and decaying trunks. Perhaps these old trees will soon fall, allowing the invasion of other species. On the other hand, it is possible that pioneer species will invade some of these gaps even though the saplings of other species are already established. *Cecropia* spp. is an early pioneer genus that cannot tolerate suppression in the understory, but more individuals of this genus were found in the older forest on BCI, where gaps seem common, than in the younger forest. Richards (1952) made similar observations elsewhere in South America, and Blum (1968), working near BCI, found that 40 percent of a group of soil samples from late secondary forest had viable seeds of *Cecropia* spp. Kramer (1933) and Schulz (1960) suggested that small gaps will most likely be invaded by climax species, but the larger gaps may be filled by the faster growing pioneer, or light-gap, species. If large gaps are sufficiently frequent, it is probable that pioneer species will always occupy a significant portion of the climax, reproducing in pulses triggered by a light gap of sufficient size.

Several investigators have commented on wind as an important cause of gaps in tropical forest (Richards, 1952; Jones, 1955; Webb, 1958; Schultz, 1960), and windfalls are common on BCI. Wind seems to play the same role in the disturbance of the tropical forest as do wind and wilt disease in temperate forest, burrowing mammals in grassland, and frost heaving in tundra.

Considering the probability that some pioneer species are permanent components of the climax vegetation, it is of interest to speculate on whether or not the older forest on BCI is in fact climax, despite the information presented in Table 5–1. Relative densities for the two ecological groups were recalculated for canopy trees only, defined arbitrarily as trees greater than 30 cm dbh. Forty percent of the canopy trees in the older forest are pioneer species and 48 percent are reproducing species. It seems unlikely that gaps are created frequently enough to maintain such a large percentage of pioneer trees in the canopy, suggesting that the older forest on BCI is not climax. The younger forest is definitely seral with only 7 percent of the canopy trees in the reproducing category and 73 percent in the pioneer category.

Although climax forest may not exist on BCI, the probable composition of the climax canopy can be predicted with some confidence. The following are all potentially large trees and have frequent reproduction on BCI: *Alseis blackiana, Calophyllum longifolium, Casearia sylvestris, Guarea glabra, Hirtella triandra, Inga marginata, Lonchocarpus pentaphyllus, Posoqueria latifolia, Poulsenia armata, Prioria copaifera, Protium panamense, Pterocarpus rohrii, Quararibea asterolepis, Tachigalia paniculata, Trichilia cipo, Triplaris cumingiana*, and *Virola sebifera.* At least 20 other potentially large species could be listed, but they are currently less common on the island. The following species are smaller reproducers and would probably be common in the understory: *Casearia guianensis, Cordia lasiocalyx, Coussarea impetiolaris, Desmopsis panamensis, Eugenia nesiotica, Faramea occidentalis, Heisteria concinna, Lacistema aggregatum, Lucuma lucentifolia, Mouriria parviflora, Oenocarpus panamanus, Quassia amara, Randia armata, Rheedia madruno, Sorocea affinis, Stemmadenia grandiflora*, and *Swartzia simplex.*

Canopy emergents exist on BCI, but they all belong to the pioneer category and hence are not at present considered as part of the climax canopy. Others have speculated that the climax canopy of tropical forest may in fact be more uniform than is often depicted in physiognomic descriptions (Eggeling, 1947; Leonard, 1952; Jones, 1955), and this may be true on BCI. The mixed forest association of British Guiana was characterized by emergents, and was referred to as the climatic climax for the area (Richards, 1952). Richards noted, however, that the larger tree species lacked saplings in the understory. Perhaps the mixed forest association is also preclimax. On the other hand, perhaps the exceptionally tall, long-lived emergent growth form represents a successful strategy for survival in an essentially light-limited environment where natural openings occur periodically. This strategy would require efficient seed dispersal and a growth rate that would exceed the growth rate of shade-tolerant species that have already become established. It is known that some early pioneer species have these characteristics, but the emergent species are less studied.

Finally, the method used in this report has been helpful for studying succession in a species-rich tropical forest. Although data on population structure have been gathered by others, they are not common and the data have been analyzed only rarely using the ecological group approach (Eggeling, 1947; Jones, 1955; Schulz, 1960; Vaughan and Wiehe, 1941). The approach offsets some of the problems associated with extracting information from a long list of uncommon species, and the required data on population structure are of value to the forest manager.[1]

## Acknowledgment

The staff of the Smithsonian Tropical Research Institute has been most cooperative in providing facilities on BCI. I gratefully acknowledge their assistance, and the financial support of the National Science Foundation (GB-7778) and the Graduate Research Council of the University of Wyoming. Gerald Lang and Jose Abel Tapia provided valuable field assistance, and the following taxonomists helped with species identification: Tom Croat, Robert Dressler, Jim Duke, John Dwyer, and Tom Elias. Bruce Haines and Mike Evans provided information on soil characteristics; A. Tyrone Harrison made helpful suggestions on an earlier version of the manuscript.

## References

Beard, J. S. 1944. Climax vegetation in tropical America. *Ecology* 25:127–158.

Blum, K. E. 1968. Contributions toward an understanding of vegetational development in the Pacific lowlands of Panama. Ph.D. thesis, Florida State Univ., Tallahassee.

Eggeling, W. J. 1947. Observations on the ecology of Budongo rainforest, Uganda. *J. Ecol.* 34:20–87.

Jones, E. W. 1955. Ecological studies on the rain forest of southern Nigeria. IV. The plateau forest of the Okomu forest reserve. *J. Ecol.* 43:564–594; 44:83–117.

Knight, D. H. 1970. A field guide to the trees of Barro Colorado Island, Panama Canal Zone. *Contr. Dept. Bot. Univ. Wyoming and Smithsonian Trop. Res. Inst.,* 94 pp.

Kramer, F. 1933. De natuurlijke verjonging in het Goenoeng-Gedehcomplex. *Tectona* 26:156–185.

Lang, G. E., D. H. Knight, and D. A. Anderson. 1971. Sampling the density of tree species with quadrats in a species-rich tropical forest. *Forest Sci.* 17:395–400.

Leonard, J. 1952. Les divers types de forets du Congo Belge. *Lejeunia* 16:81–93.

Richards, P. W. 1952. *The Tropical Rainforest.* Cambridge: Cambridge Univ. Press. 450 pp.

Schulz, J. P. 1960. Ecological studies on rainforest in northrn Surinam. *Verhandl. Koninkl. Ned. Akad. Wetenschag.* 53:1–367.

Vaughan, R. E. and P. O. Wiehe. 1941. Studies on the vegetation of Mauritius. III. The structure and development of the upland climax forest. *J. Ecol.* 29:127–160.

[1] Population structure data for specific species on BCI are available from the author.

Webb, L. J. 1958. Cyclones as an ecological factor in tropical lowland rain forest, North Queensland. *Australian J. Bot.* 6:220–228.

Williams, W. T., G. N. Lance, L. J. Webb, J. G. Tracey, and M. B. Dale. 1969. Studies in the numerical analysis of complex rain-forest communities. III. The analysis of successional data. *J. Ecol.* 57:515–535.

# CHAPTER 6

# The Relation Between Rainfall and Mosquito Populations

H. A. M. de Kruijf

## Introduction

Immature stages of mosquitoes are all aquatic and require water in the form of either large permanent pools and streams or temporary rainpools or even small collections of water in fallen leaves or flowers. As the seasons in the tropics are marked by fluctuation in rainfall, it is not surprising to find that populations of many mosquito species follow these seasons very closely. However, while most population studies of adult mosquitoes show a general relationship between density and rainfall, based on monthly figures (see Aitken et al., 1968; de Kruijf, 1972b), they do not explain the detailed fall and rise of populations as related to changes in water supply.

In an earlier study (de Kruijf et al., 1973), catches of adult mosquitoes in Belém, Brazil, were analyzed to determine the day-to-day effect of rainfall on mosquito capture. First, it appeared that maximum catch of rainpool breeding species was related to the total rainfall in the four weeks preceding a catch. Second, either the number of dry days in the same period or the pattern of dry and wet days influenced maximum densities, while swamp breeders did not show such relationships. When the catch of rainpool breeders should have reached a maximum number according to the amount of rainfall, but did not, the number of dry days in the same period appeared to be either too low or too high. In other species there was either not enough or too much alteration between dry and wet days. It appeared that the rainfall total in 14 days preceding the fortnight before the catch and the number of dry days in that period was most critical. These conditions could be related to the duration of the immature stages.

The present paper presents similar data for Surinam.

## Material and Methods

The catches were carried out by handsucking on human bait, usually with three men. All catches took place between 8 AM and 11 AM, mostly three times a week, over a period of 14 months. A description of the area and other particulars of the methods have been published (de Kruijf, 1970, 1972a).

## Results and Discussion

Accumulated rainfall (AR) was measured for 28 days before the catch. In Figures 6–1 to 6–3 the catch of three mosquito species was plotted against accumulated rainfall.

In each species there was great variability in catch associated with any given rainfall amount. While the method of catching was relatively imprecise, the variability probably was due mainly to changes in mosquito densities, since the same procedures were used at all levels of rainfall. In *Aedes serratus* and *Psorophora ferox* maximum variation occurred between 100 and 200 mm accumulated rainfall ( Figures 6–1 and 6–2). In contrast, in *Mansonia titillans* maximum variation occurred at lower rainfall levels (Figure 6–3).

The Surinam observations can also be compared to similar data from Belém, Brazil (de Kruijf et al., 1973). Environmental data for the two areas is presented in Table 6–1. Mean rainfall and the mean number of series of wet days were similar, while the mean number of dry days was slightly greater in Surinam. In

**Table 6–1.** Comparison of rainfall, number of dry days, and the distribution of the length of series of wet days between terra firma forest, Belém, Brazil, and forest Leiding 16A, Surinam.[a]

| | Belém | Surinam |
|---|---|---|
| Mean rainfall per 28 days (mm) | 183.9 | 187.0 |
| Mean number of dry days per 28 days | 11.34 | 13.42 |
| Number of series per 28 days | 9.56 | 10.06 |
| Mean rainfall per rainy day | 11.3 | 12.8 |
| Wet periods: | | |
| 1 or 2 days | 60.1% | 60.2% |
| 3 or 4 days | 19.0% | 19.3% |
| 5 or 6 days | 10.1% | 12.0% |
| Longer | 10.8% | 8.5% |

[a] The Belém figures were calculated over a 28-month period, the Surinam figures over a period of 16 months.

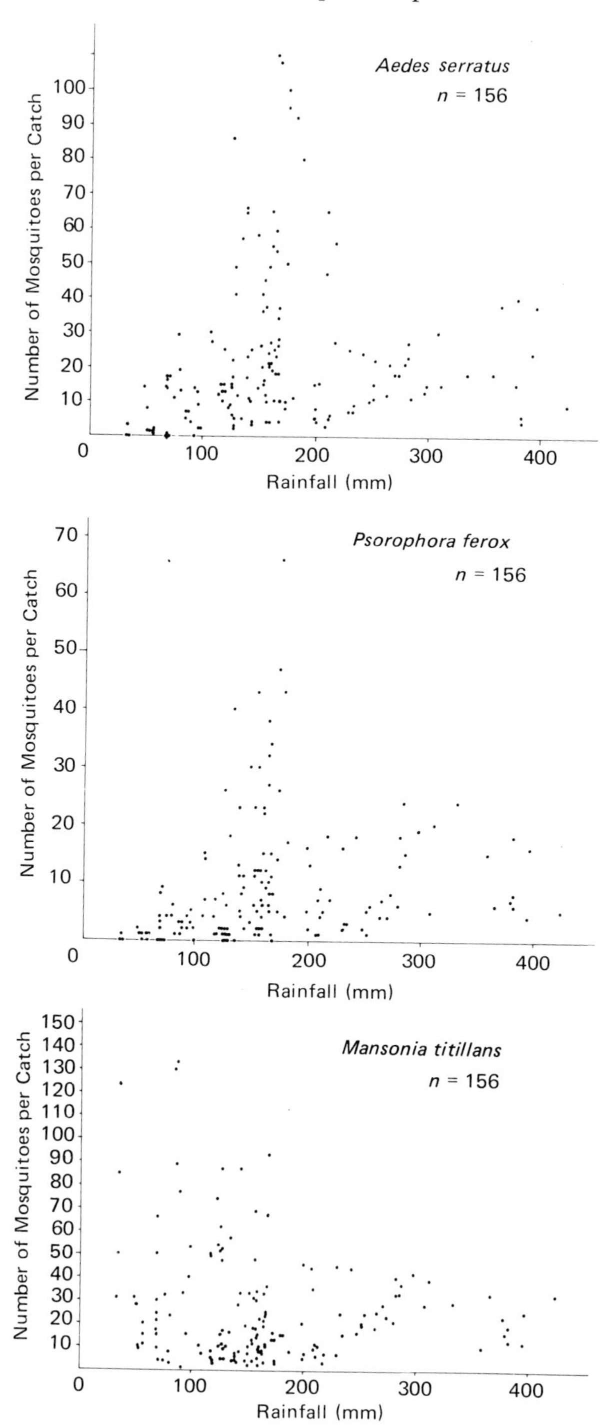


**Fig. 6–1—6–3.** Number of mosquitoes per catch plotted against the accumulated rainfall. Rainfall totals are calculated for 28 days preceding the second day before the day of the catch (see text).

general, the rainfall in the two areas agrees in all major respects and allows comparison of the catch per area based on rainfall figures. Considering the mosquitoes, the mean accumulated rainfall calculated for all catches was significantly lower at Surinam than at Belém, whereas the mean number of dry days is much higher ($p < 0.0005$) (Table 6–2). The mean series of alternating wet and dry days differed only slightly.

In Table 6–2 the means for the three species are also shown. While *Mansonia titillans* did occur in Belém, its numbers were too low to permit comparison. The mean accumulated rainfall for maximum catches was much higher in Surinam for *Aedes serratus* as well as for *Psorophora ferox*. However, the mean number of dry days did not differ significantly for each species individually.

**Table 6–2.** Comparison of the mean value of AR, mean number of dry days, and mean series of alternating wet and dry days related to all catches and maximum catches at Belém and Surinam.[a]

| Total Catches | Belém | Surinam | Probability Level |
|---|---|---|---|
| Mean AR | 210.8 ± 124.1 | 167.0 ± 82.9 | $p < 0.001$ |
| Mean number of dry days | 9.6 ± 5.2 | 13.6 ± 5.5 | $p < 0.0005$ |
| Mean number of series | 9.8 ± 3.1 | 10.4 ± 3.1 | $p < 0.10$ |
| Catches by Species | | | |
| *Aedes serratus* ($n = 22$): | | | |
| Mean AR | 117.5 ± 15.5 | 165.7 ± 25.2 | $p < 0.0005$ |
| Mean number of dry days | 12.4 ± 2.5 | 12.6 ± 3.8 | — |
| *Psorophora ferox* ($n = 13$): | | | |
| Mean AR | 129.2 ± 9.9 | 161.1 ± 17.1 | $p < 0.0005$ |
| Mean number of dry days | 12.3 ± 2.8 | 11.1 ± 3.6 | — |
| *Mansonia titillans* ($n = 22$): | | | |
| Mean AR | — | 112.5 ± 37.2 | — |
| Mean number of dry days | — | 18.9 ± 3.0 | — |

[a] If not stated otherwise, the figures are calculated for the 28-day period (day 30–2) before the catch. If differences between both areas were significant, the probability level is shown in the fourth column.

While the relation between population and rainfall is similar for both species, there is a significant difference in the amount of accumulated rainfall, which causes maximal populations. In Surinam maximal populations are produced by much higher rainfall than in Belém. Since the number of dry days within the 28-day period is similar in both cases, the difference in AR producing maximal populations may be explained by two alternative hypotheses. First, the mosquitoes in Surinam need larger pools than in Brazil, and hence require more rain to fill and maintain such pools. Second, the species in Surinam breed in pools of similar size as those in Belém, but the Surinam forest has a greater outflux of water or the soil is more porous. I am in favor of the second explanation because it seems very unlikely that mosquitoes are concerned about exact measurements of small pools. Of course, they do discriminate between rainpools and large inundated areas mainly because *Aedes serratus* and *Psorophora ferox* deposit their eggs on the edges of the pools and not directly in the water.

The eggs of aedine species (Mattingly, 1969) also require short drying before they can hatch. The optimal number of dry days within a 28-day period is about 12 (±8–16) for Surinam and Belém species. Schober (1966) found that continuous sprinkling of breeding pools causes death of all larvae. If continuous sprinkling also affects the larvae in natural pools, then a number of interruptions by dry days are necessary.

In Belém, catches that were well below the maximum were distributed evenly within the seasons; in Surinam these catches were all observed at the middle or the end of the wet seasons. At this time the study areas were largely inundated; hence even optimal rainfall and optimum number of dry days could not produce maximal populations. This suggests that the catches are related to rainfall even long before the four weeks preceding a catch (de Kruijf, 1970).

Since all maximum catches of *Aedes serratus* and *Psorophora ferox* were obtained during the first half of the wet seasons, I conclude that there is a significant influence of rainfall on population density and that its positive influence in the formation of rainpools can be reduced by too much rain in a short period (accumulated rainfall is too high) or by rain during a prolonged period of few months.

*Mansonia titillans* is a swamp breeder and thus the population of this species should be only slightly influenced by rainfall. However, Figure 6–3 shows that maximum populations are related to low rainfall. This species belongs to the group that glues their eggs in small masses to the underside of the leaves of floating plants. Though not yet determined by observations, it is likely that rainfall will disturb the acrobatic act of ovipositing; therefore high rainfall has a negative influence on *Mansonia titillans* population by preventing oviposition.

## References

Aitken, T. H. G., C. B. Worth, and E. S. Tikasingh. 1968. Arbovirus studies in Bush Bush forest, Trinidad, W. I. September 1959–December 1964. III. Entomologic studies. *Am. J. Trop. Med. Hyg.* 17:253–268.

de Kruijf, H. A. M. 1970. Aspects of the ecology of mosquitoes in relation to the transmission of arboviruses in Surinam. Thesis, Univ. Leiden. 100 pp.

———. 1972a. Aspects of the ecology of mosquitoes in Surinam. I. Biting activity patterns. *Stud. Fauna Surinam Other Guyanas* XIII 51:3–18.

———. 1972b. Aspects of the ecology of mosquitoes in Surinam. III. Seasonal distribution and abundance of some populations. *Stud. Fauna Surinam Other Guyanas* XIII 51:30–48.

———, J. P. Woodall, and A. T. Tang. 1973. The influence of accumulated rainfall and its pattern on mosquito (Diptera) populations in Brazil. *Bulletin Entomological Research* 63:327–333.

Mattingly, P. F. 1969. The biology of mosquito-borne disease. *Sci. Biol. Ser. 1*, 184 pp.

Schober, H. 1966. Agitation of water surfaces by sprinkling to prevent mosquito breeding. *Mosquito News* 26:144–149.

## Interaction Between Species

For many years, population ecologists have been interested in the regulation of population numbers. The study of interactions between populations has not assumed equal importance until quite recently. One step toward this emphasis was the attention paid to food chains and the transfer of energy and materials from population to population (Misra, 1968). Of equal significance has been the study of the relations between plants and animals in dry and wet forests in Central America (Janzen, 1970). These various investigations, while not numerous in the tropics, have resulted in interesting theories that will stimulate tests and further study.

The present data suggest that population interaction is of greater significance in the control of tropical community structure and function than it is in temperate regions. The role of bats, birds, and insects in the pollination of flowers (Baker, 1972) and the role of seed-eating insects in the spatial pattern of tropical trees (Connell, 1971) are examples of these kinds of control phenomena. This rich area of investigation is illustrated in the following two papers. Montgomery and Sunquist describe radio tracking experiments in sloths on Barro Colorado Island, and, with additional data on defecation and feeding, develop hypotheses about the impact of sloths on forest trees. Haines, also working on Barro Colorado Island, discusses the interaction of leaf cutting ants and vegetation. In both cases the writers explore the reciprocal interaction existing between the animal populations and the associated plants.

## References


Baker, H. G. 1972. Evolutionary relationships between flowering plants and animals in American and African tropical forests. In *Tropical Forest Ecosystems in Africa or South America*, B. V. Meggars, E. S. Ayensu, and W. D. Duckworth, eds. pp. 145–159. Washington: Smithsonian Inst. Press.

Connell, J. H. 1971. On the role of natural enemies in preventing competitive exclusion in some marine animals and rain forest trees. In *The Dynamics of Populations*, P. V. Boer and G. R. Gradwell, eds. 611 pp. *Proc. Advan. Study Inst. Dynamics of Numbers on Populations.* Oosterbeek, Netherlands.

Janzen, D. H. 1970. Herbivores and the number of tree species in tropical forests. *Am. Naturalist* 104:501–528.

Misra, R. 1968. Energy transfer along terrestrial food chain. *Trop. Ecol.* 9(2):105–118.

# CHAPTER 7

## Impact of Sloths on Neotropical Forest Energy Flow and Nutrient Cycling[1]

G. G. MONTGOMERY AND M. E. SUNQUIST[2]

### Introduction

We report a field study of the role of two-toed and three-toed sloths (*Choloepus hoffmani* and *Bradypus infuscatus*) in neotropical forest energy flow and nutrient cycling. Sloths are arboreal herbivores with stomachs and digestive processes showing convergence with the ruminants (reviewed by Goffart, 1971). Sloth stomach anatomy and histology, which includes provision for long-term retention of food, keratinous epithelium, and comparative lack of mucosal papillae, indicate that they should be considered as "bulk and roughage eaters" (Hofman and Stewart, 1972), a category reserved in terrestrial ruminants for grazers (grass feeders). Whether sloths are considered grazers on the basis of their stomach characteristics, or browsers because they feed mainly (three-toed) or partially (two-toed) on leaves, many of their effects on neotropical forest processes would be expected to parallel the effects of terrestrial mammalian grazers and browsers of grassland and savanna.

Annual net production of vegetation potentially available to browsers in moist tropical forest ecosystems usually exceeds net production available to browsers and grazers of tropical grassland and savanna (Golley, 1972). However, this potentially large and diverse source of food for herbivorous tropical forest mammals is not readily available because a large part of it grows in upper levels of the forest (cf., Tadaki, 1966) and is out of the reach of terrestrial forms. There are generally fewer mammalian herbivore species with terrestrial, as compared with arboreal, life modes in tropical forest, relating to the lesser amount of food available at the forest floor. Utilization of most of the vegetation available to mammalian browsers in tropical forest requires adaptations both for moving

[1] Dedicated to the memory of I. N. Healey, whose comments during our mutual stay on Barro Colorado Island gave impetus to our work.

[2] Present address: 4660 Heights Drive, Minneapolis, Minnesota 55421.

about in trees (Cloudsley-Thompson, 1972) and for digesting leaves. Requirements relating to these adaptations, such as limitations on maximum body size (Bourliere, 1973), have apparently reduced mammalian radiation as arboreal foliovores, thus there are relatively few mammalian species of arboreal foliovores when compared with the great diversity of terrestrial mammalian herbivores that evolved in tropical savanna and grassland. The great abundance and diversity of potential herbivore food (especially leaves) might have provided for evolution of many more forms.

Among the mammals specialized for arboreal life, relatively few subsist solely or primarily on leaves and buds. For example, 6 of 17 species of Australian arboreal mammals and 4 of 44 species of Malayan arboreal mammals feed primarily on leaves (Harrison, 1972). In general, however, those few arboreal species that are foliovores live at densities that result in rather high species biomass. Total mammalian herbivore biomass for tropical savanna and tropical moist forest may be roughly equivalent (Eisenberg and McKay, 1973), yet the number of species forming the herbivore biomass is much greater in savanna than in moist forest (Bourliere, 1973).

No mammals are so well adapted for living in trees and feeding from them as the sloths. In our attempts to understand the role of arboreal herbivores in forest ecosystems, study of sloths would thus seem to be important. Prior to this study, little was known of how sloths live in the wild and how they relate to the forest systems in which they live. The one extensive field study (Beebe, 1926) did not emphasize the ecological relationships between sloths and plants.

## Methods and Results

The role of sloths in the processes of energy transfer and nutrient cycling involves (1) how they select their food from the array of plants available to them; (2) the amount of food they crop from an area per unit time; and (3) amounts of unused food they return as waste to the forest, where that waste is returned, and rates at which it is recycled. Data on each of these topics will be presented in turn, following the methods applicable to each topic. The data will then be synthesized in the final section of the paper.

Since August 1970, 17 months of fieldwork have been completed on Barro Colorado Island (BCI), Panama Canal Zone. Sloths were studied continuously on BCI from August 1970 through September 1971 and periodically (January–March, August–September) in 1972. BCI is described in the work of Standley (1933), Carpenter (1934), Hladik and Hladik (1969), and others.

The approximately 15 km$^2$ island was formed in 1911–1914 with the flooding of Gatun Lake, which forms a fresh-water portion of the Panama Canal. The ship channel passes to the east and north of BCI, and one or more km of open water separate the mainland and the island there. However, even this expanse of water does not isolate the BCI sloth population from the mainland population. Sloths swim well (Beebe, 1926; Worman, 1946) and a three-toed female swam across the ship channel and set up residence on BCI (Montgomery and Sunquist, unpublished). Furthermore, for about one-fourth of its circumference on the southwest side, the island is separated from the mainland by a much narrower channel which sloths could easily swim and thus move between BCI and the mainland.

The forest on the portion of the island where our study sloths lived (Bennett, 1963) is of young to medium age with a lower stratum reaching to about 40 ft (12 m) and an upper stratum reaching usually to 80–100 ft (24–30 m). Parts of the area have been recently disturbed, which, in combination with treefalls that opened small areas of canopy, resulted in mixing of forest age and vegetation. Perhaps as many as 800 species of plants were available to the sloths as potential food and as support for their movement and other activities.

## Capture, Radio Marking, and Radio Location of Sloths

Radio location of free-living sloths provided much of the data with which we determined home range characteristics, movements, and habitat. Sloths were captured for radio marking by placing a noose around the animal while it was in the tree. The animal was then noosed, pulled back to the climber, placed in a canvas bag, and lowered to the forest floor.

When a sloth could not be captured with the noose, branches were sawed until the sloth fell to the ground with the foliage. After capture, three-toed sloths were easily handled and marked. However, it was necessary to anesthetize two-toed sloths with ether or Halothane (Fluothane; Ayerst Lab. Inc., N.Y.).

Fifty-two three-toed and 12 two-toed sloths were captured. Of these, 31 three-toed and 11 two-toed sloths were radio marked and tracked. Some individuals were handled only once; most were recaptured at least once so that we could change transmitter batteries or examine the animal. One three-toed sloth was captured 11 times. Nine adult three-toed sloths and six adult two-toed sloths were radio tracked for periods of three or more months and provided most of the data for this report.

Sloths were radio marked with transmitters (Model ST-1 and MT-1; AVM Inst. Co., Champaign, Ill.), which were packaged as neck collars. Transmitter packages for adults weighed about 50 g, and those for juveniles 20 g or less. Each transmitter broadcast at two exclusive frequencies (near 150 megahertz and near 450 megahertz), and individual sloths were identifiable by different such pairs of frequencies. We rarely saw the animals, and usually identified them by these frequencies. After a sloth was marked, examined, weighed to the nearest 0.1 lb, and standard body measurements taken, it was released at the base of the tree where captured.

We usually radio located (Montgomery et al., 1973) each animal once per day. Series of hourly locations made for 24-hr periods confirmed that a single daily radio location was adequate to locate a three-toed sloth for that date; they seldom left a tree and then returned to it on the same date. However, two-toed sloths moved from tree to tree at night (Sunquist and Montgomery, 1973) and probably used more trees than the single daily radio locations indicated. When possible, each radio location was visually confirmed.

Each tree used by a sloth was numbered and marked. Later the tree location and elevation at its base were determined. The species of the tree was identified, and its DBH, height, and crown dimensions were measured. Percentage of the crown exposed to sunlight, relative amount of vines in the crown, and relative number of pathways into the crown also were estimated.

## Home Ranges, Movements, and Habitat Use

The home range of a sloth of either species is best characterized as a set of tree crowns; a part of each crown is used for one or more days. A sloth moves among these trees via interconnecting tree crowns and vines. Places on the forest floor to which the animal descends periodically to defecate are included in a home range, as are pathways from the canopy to the forest floor. Vines and the boles of smaller trees are employed by the animal during descent.

An idealized top view of the home range of an adult female three-toed sloth (Figure 7–1) illustrates some features of sloth home ranges. This range is based on 187 radio locations during one year. In this period the animal was located in 63 trees of 35 species. About half of these were in the forest canopy; others

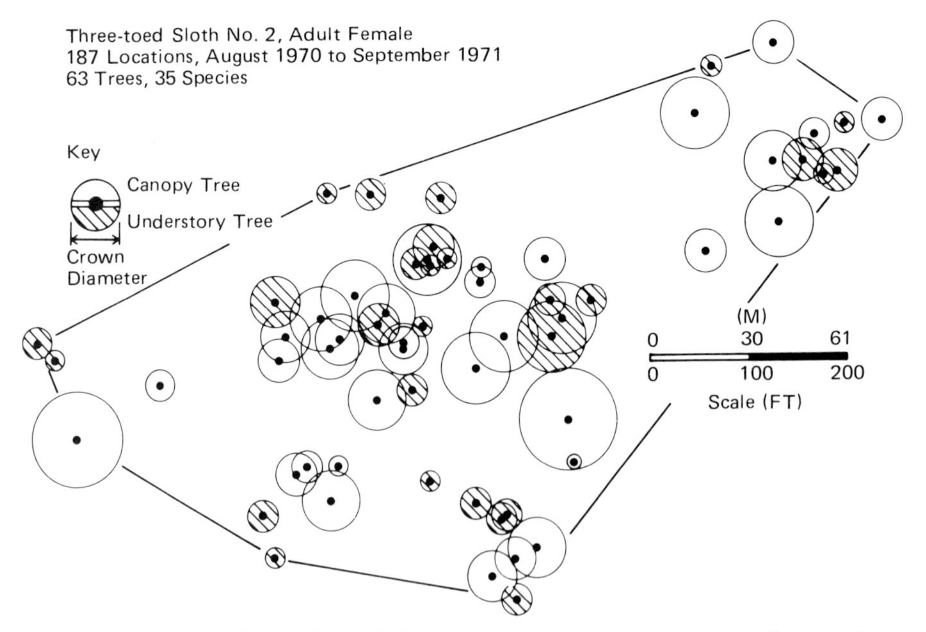


**Fig. 7–1.** Idealized top view of the home range space of an adult female three-toed sloth. Canopy trees were those that had 80 percent of the crown or more exposed to direct sunlight. Tree crowns were drawn on the basis of crown width; not all crowns were circular. Black dots show the location of the bole of each tree.

were in the upper understory. All 63 trees were in an area 250 by 500 ft (76 by 152 m). Thus the area of the home range, based on the minimum-size rectangle that enclosed all trees, was slightly more than 1 ha.

Sloths are not sedentary, but regularly move from tree to tree (Sunquist and Montgomery, 1973). Most two-toed sloths changed trees on successive dates, while three-toed sloths were in the same tree on successive dates about 40 percent of the time. Three-toed sloths moved less than 125 ft (38 m) per day in 89 percent of the cases, whereas two-toed sloths moved more than 125 ft (38 m) per day in 54 percent of applicable measures.

Changes in location by a sloth have vertical as well as horizontal components. For example, an adult female three-toed sloth changed trees 18 times, using 13 trees of 10 species over a period of 33 locations (Figure 7–2). She moved vertically within the range of 25 to 80 ft (7.6 to 24.4 m) above the forest floor (Figure 7–2). Except when defecating, sloths were usually more than 30 ft (9.1 m) above the forest floor and were at times near the tops of the tallest trees. Sloths of both species were located 80 ft (24.4 m) or more above the ground more than 20 percent of the time (Table 7–1).

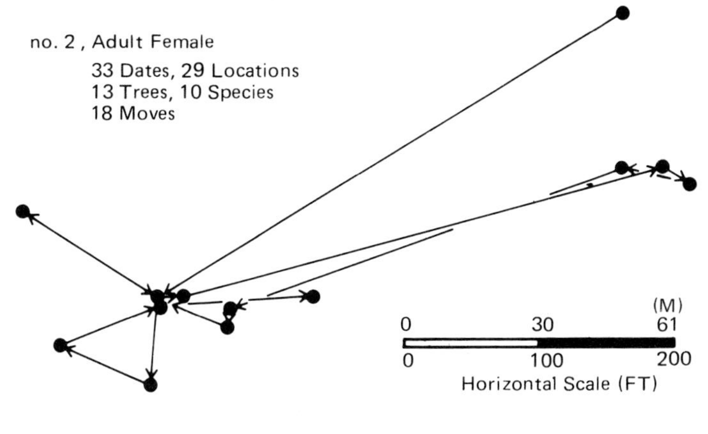


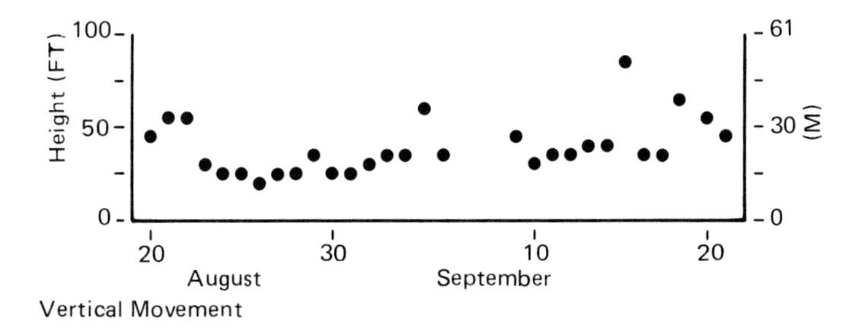


**Fig. 7–2.** Horizontal and vertical movements of an adult female three-toed sloth. Arrows indicate the direction of movement during her moves from tree to tree. Black dots for horizontal movements indicate the locations of the boles of trees she used.

**Table 7–1.** Home range sizes, number of trees used, and number of tree species used by radio-marked sloths.[a]

| Sloth Species | Area of Home Range (ha) | Number of Trees | Number of Tree Species |
|---|---|---|---|
| Two-toed | 1.6 | 28 | 18 |
| | 0.4 | 14 | 12 |
| | 3.9 | 27 | 17 |
| | 3.3 | 34 | 17 |
| | 1.2 | 19 | 16 |
| | 1.4 | 26 | 16 |
| Three-toed | 1.1 | 63 | 32 |
| | 2.8 | 59 | 24 |
| | 1.8 | 33 | 19 |
| | 0.5 | 37 | 28 |
| | 0.8 | 33 | 20 |
| | 1.8 | 41 | 21 |
| | 1.1 | 27 | 21 |
| | 0.7 | 38 | 25 |
| | 3.7 | 42 | 30 |

[a] Data are shown for six adult two-toed sloths for which we had three or more months of data, and for nine adult three-toed sloths for which we had five or more months of data. Size of each home range was estimated by the minimum-size rectangle that enclosed all trees in which the sloth was radio located.

Trees used by a sloth of either species were within a relatively small area. The six two-toed sloths (Table 7–2) had an average home range size of 1.97 ha. The average home range size for nine adult three-toed sloths was 1.60 ha. Total area is probably not an appropriate parameter for characterizing sloth home ranges. Because trees within an area were selectively used, the home range is better typified by the number and species of trees used by an animal. The six two-toed sloths (Table 7–2) used an average of 24.7 trees each and used a consistent number of tree species, averaging 16 species each. The nine three-toed sloths used an average of 41.4 trees each; they averaged 24 species per individual.

Approximately 53 percent of all trees used by sloths of either species were used only once by that sloth (Figure 7–3). Thus a few trees accounted for most of an individual's radio locations. Eight or fewer trees accounted for 50 percent or more of the radio locations of each of the six adult two-toed and each of the nine adult three-toed sloths.

Likewise, a few species of trees accounted for most of an individual's radio locations. Approximately 40 percent of all species used by individuals were used only once (Figure 7–4). For two-toed sloths, four or fewer species of trees accounted for 50 percent or more of the radio locations of each of six animals. Eight or fewer species accounted for 50 percent or more of the locations of each of nine three-toed sloths.

Each sloth had a modal tree in which it was radio located most often (Table 7–2); modal trees were seldom used by other sloths. No two sloths had the same

**Table 7–2.** Modal trees and tree species of six adult two-toed sloths and of 9 adult three-toed sloths that were radio located over periods of three months or more.

| Animal Number | Modal Tree Number | Total Locations | Percent in Modal Tree | Modal Tree Species | Percent in Modal Species |
|---|---|---|---|---|---|
| Two-toed | | | | | |
| 3 | 109 | 47 | 19.1 | *Anacardium excelsum* | 25.5 |
| 4 | 185 | 95 | 57.8 | *Anacardium excelsum* | 70.5 |
| 5 | 288 | 50 | 12.0 | *Anacardium excelsum* | 16.0 |
| 6 | 444 | 69 | 11.6 | *Anacardium excelsum* | 30.4 |
| 7 | 66 | 37 | 16.7 | *Dipteryx panamensis* | 16.2 |
| 8 | 426 | 58 | 13.8 | *Spondias nigrescens* | 17.2 |
| Three-toed | | | | | |
| 202 | 71 | 186 | 9.1 | *Gustavia superba* | 11.5 |
| 205 | 138 | 163 | 13.5 | *Anacardium excelsum* | 15.1 |
| 206 | 18 | 196 | 21.4 | *Tetragastris panamensis* | 28.6 |
| 208 | 103 | 214 | 20.6 | *Ficus trigonata* | 20.6 |
| 219 | 490 | 110 | 14.5 | *Ficus obtusifolia* | 14.5 |
| 221 | 84 | 118 | 10.2 | *Anacardium excelsum* | 14.4 |
| 225 | 379 | 83 | 15.7 | *Inga sp.* | 15.7 |
| 227 | 358 | 131 | 13.0 | *Poulsenia armata* | 17.5 |
| 231 | 205 | 93 | 10.7 | *Dipteryx panamensis* | 10.7 |

modal tree, although the home ranges of many animals overlapped. A two-toed sloth was in its modal tree more than 10 percent of the times it was located, one animal being there about 60 percent of the times. A three-toed sloth used its modal tree an average of 14 percent of the time.

Each sloth also had a modal tree species (Table 7–2) that in some cases was different from the species of the modal tree. Four of six two-toed sloths used *Anacardium excelsum* most often. Three-toed sloths were less consistent. Two of nine were found most often in *Anacardium excelsum.* The remaining seven animals used seven different modal species.

Sloths used a wide variety of trees (Table 7–3) even though the distances they moved were relatively short. In total, 106 tree species were used by sloths. Three-toed sloths were radio located in 98 tree species, and two-toed sloths in 52 tree species. There was a good deal of overlap in tree species, with two-toed sloths using 41.5 percent of the species used by three-toed sloths and three-toed sloths using 84.6 percent of the species used by two-toed sloths.

As an example of the complex sequence of use of various species by sloths, an adult female three-toed sloth (Figure 7–5) regularly left and returned to *Gustavia superba*, staying an average of 1.6 days per visit to that species. *Cecropia eximia* trees were visited twice, while the remaining eight species were visited only once during the 33-day period. Distances the animal moved among these species were small, ranging from 17 to 372 ft (5.2 to 113.4 m), as measured in three dimensions.

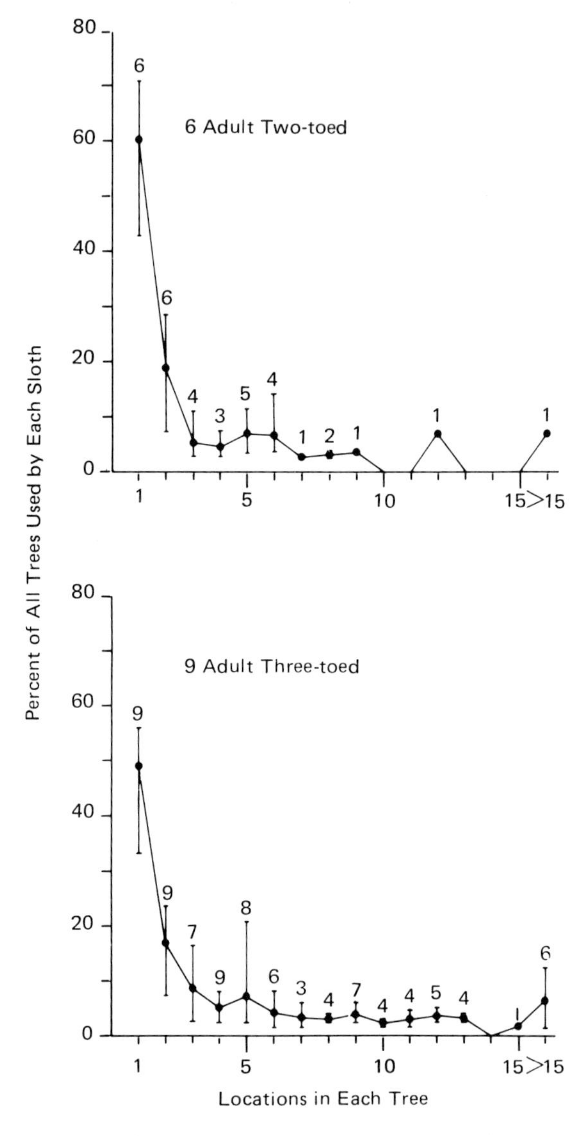


**Fig. 7–3.** Frequency distribution of the percentage of all trees used by each two-toed sloth (*top*) and each three-toed sloth (*bottom*) that were used once, twice, . . . *N* times by the individual. Black dot indicates the mean for the number of animals shown above the line, which indicates the range.

## Three-Toed Sloth Food Habits

We rarely observed three-toed sloths feeding; two-toed sloths are nocturnal (Sunquist and Montgomery, 1973) and were never seen feeding. Samples of stomach contents from three-toed sloths were collected with a tube inserted orally through the esophagus (Montgomery, 1969). We were unable to collect

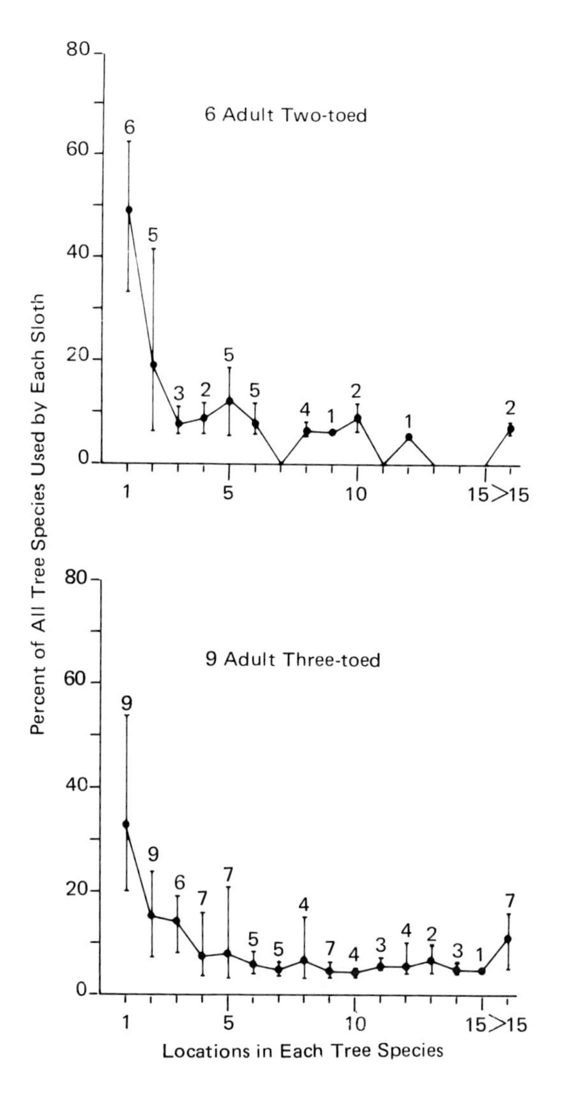


**Fig. 7–4.** Frequency distribution of the percentage of all tree species used by each two-toed sloth (*top*) and each three-toed sloth (*bottom*) that were used once, twice, . . . $N$ times by the individual. Black dot indicates the mean for the number of animals shown above the line, which indicates the range.

such samples from two-toed sloths. Portions of the samples were prepared by methods adapted from Stewart (1967) and compared histologically with leaf material from known species of trees and vines. Identification of unknown leaf material was made on the basis of shape and size of the stomata and other epidermal characteristics (Baumgartner and Martin, 1939). The collection of known leaf material was incomplete; thus many species in samples remained unidentified.

**Table 7–3.** Tree species used by two-toed and three-toed sloths and other arboreal fauna.

| Tree Species | Two-toed N | Two-toed % | Three-toed N | Three-toed % | Three-toed Sloth | Howler Monkey | Spider Monkey | Cebus Monkey |
|---|---|---|---|---|---|---|---|---|
| Acanthaceae | | | | | | | | |
| *Trichanthera gigantea* | 3 | 0.8 | | | | | | |
| Anacardiaceae | | | | | | | | |
| *Anacardium excelsum* | 117 | 30.9 | 92 | 4.8 | O | C | | |
| *Astronium graveolens* | 5 | 1.3 | | | | | | |
| *Mangifera indica* | 1 | 0.3 | 1 | 0.05 | | | | |
| *Spondias mombin* | 1 | 0.3 | 3 | 0.1 | O | C | | |
| *Spondias nigrescens* | 17 | 4.5 | 45 | 2.3 | | | | |
| Annonaceae | | | | | | | | |
| *Guatteria dumetorum* | | | 8 | 0.4 | | | | |
| Apocynaceae | | | | | | | | |
| *Lacmellea panamensis* | 1 | 0.3 | 131 | 6.8 | 0,S | | | |
| Araliaceae | | | | | | | | |
| *Dendropanax arboreus* | | | 1 | 0.05 | | | | |
| *Didymopanax morototoni* | | | 1 | 0.05 | | C | | |
| Bambusa | | | | | | | | |
| *Bamboo* sp. | | | 1 | 0.05 | | | | |
| Bignoniaceae | | | | | | | | |
| *Jacaranda copaia* | | | 19 | 1.0 | O | | | |
| *Tabebuia rosea* | | | 38 | 2.0 | | | | |
| Bombacaceae | | | | | | | | |
| *Bombacopsis quinata* | | | 11 | 0.6 | | C | | |
| *Bombacopsis sessilis* | 15 | 4.0 | 13 | 0.7 | S | | | |
| *Cavanillesia platanifolia* | 2 | 0.5 | | | | | | |
| *Ceiba pentandra* | 5 | 1.3 | 30 | 1.5 | O | H | | |
| *Ochroma pyramidale* | | | 2 | 0.1 | | | | |
| *Pseudobombax septenatum* | | | 30 | 1.5 | O | | | |
| *Quararibea asterolepis* | 1 | 0.3 | 4 | 0.2 | | H | H | |

| | | | | | | |
|---|---|---|---|---|---|---|
| Boraginaceae | | | | | | |
| *Cordia alliodora* | 1 | 0.3 | 9 | 0.5 | | |
| Burseraceae | | | | | | |
| *Protium costaricense* | | | 14 | 0.7 | | |
| *Protium* sp. | 1 | 0.3 | 17 | 0.9 | | |
| *Protium tenuifolium* | 4 | 1.1 | 26 | 1.3 | O | C |
| *Tetragastris panamensis* | 1 | 0.3 | 79 | 4.1 | O | |
| *Trattinickia aspera* | 11 | 2.9 | | | | |
| Combretaceae | | | | | | |
| *Terminalia amazonica* | 5 | 1.3 | 1 | 0.05 | O | |
| Elaeocarpaceae | | | | | | |
| *Slonea terniflora* | 1 | 0.3 | 2 | 0.1 | | |
| Euphorbiaceae | | | | | | |
| *Acalypha diversifolia* | | | 1 | 0.05 | | |
| *Alchornea costaricensis* | 2 | 0.5 | 24 | 1.2 | | |
| *Croton billbergianus* | | | 1 | 0.05 | | |
| *Hura crepitans* | | | 4 | 0.2 | | |
| *Hyeronima laxiflora* | 8 | 2.1 | 38 | 2.0 | | |
| *Sapium caudatum* | 4 | 1.1 | 6 | 0.3 | | |
| Flacourtiaceae | | | | | | |
| *Casearia arborea* | | | 2 | 0.1 | | |
| *Tetrahylacium johansenii* | | | 3 | 0.1 | | |
| *Zuelania guidonia* | | | 11 | 0.6 | | |
| Guttiferae | | | | | | |
| *Calophyllum longifolium* | 1 | 0.3 | 25 | 1.3 | | C |
| *Rheedia madruno* | | | 10 | 0.5 | | |
| *Symphonia globulifera* | | | 8 | 0.4 | | |
| Lauraceae | | | | | | |
| *Beilschmiedia pendula* | 1 | 0.3 | 9 | 0.5 | | |
| *Nectandra* sp. | | | 5 | 0.2 | | |
| Lecythidaceae | | | | | | |
| *Couratari panamensis* | | | 1 | 0.05 | | |
| *Gustavia superba* | 3 | 0.8 | 39 | 2.0 | S | |

**Table 7–3.** *(Continued)*

| Tree Species | Two-toed | | Three-toed | | Three-toed | Howler | Spider | Cebus |
|---|---|---|---|---|---|---|---|---|
| | N | % | N | % | Sloth | Monkey | Monkey | Monkey |
| Leguminosae | | | | | | | | |
| *Andira inermis* | | | 2 | 0.1 | O | | | |
| *Dipteryx panamensis* | 15 | 4.0 | 73 | 3.8 | S | | | |
| *Inga goldmanii* | 2 | 0.5 | 47 | 2.4 | | | | |
| *Inga marginata* | | | 1 | 0.05 | | | | |
| *Inga guaternata* | | | 1 | 0.05 | | | | |
| *Inga* sp. | | | 37 | 1.9 | O | | | |
| *Lonchocarpus* sp. | | | 4 | 0.2 | | | | |
| *Platymiscium polystachyum* | 3 | 0.8 | 7 | 0.4 | | C | | |
| *Platypodium elegans* | 1 | 0.3 | 10 | 0.5 | | | | |
| *Prioria copaifera* | 10 | 2.6 | 2 | 0.1 | | | | |
| *Pterocarpus hayesii* | 1 | 0.3 | | | | | | |
| Meliaceae | | | | | | | | |
| *Guarea guidonia* | | | 2 | 0.1 | O | | | |
| *Trichilia cipo* | 1 | 0.3 | 79 | 4.1 | O | | | |
| Moraceae | | | | | | | | |
| *Artocarpus communis* | | | 1 | 0.05 | | | | |
| *Brosium bernadetteae* | 3 | 0.8 | | | | | | |
| *Cecropia eximia* | | | 146 | 7.6 | O,S | H | H | |
| *Cecropia obtusifolia* | | | 14 | 0.7 | O,S | | | |
| *Coussapoa panamensis* | | | 2 | 0.1 | | | | |
| *Ficus costaricana* | | | 1 | 0.05 | | C | | |
| *Ficus insipida* | 4 | 1.1 | 11 | 0.6 | | H | | |
| *Ficus obtusifolia* | 1 | 0.3 | 21 | 1.1 | | | | |
| *Ficus popenoei* | | | 3 | 0.1 | | | | |
| *Ficus* sp. | | | 20 | 1.0 | O | | | |
| *Ficus tonduzii* | | | 3 | 0.1 | | C | | |
| *Ficus trigonata* | | | 62 | 3.2 | O | | | |
| *Ficus yoponensis* | 1 | 0.3 | 20 | 1.0 | O | | | |

| | | | | | | | |
|---|---|---|---|---|---|---|---|
| *Maguira costaricana* | | | 10 | 0.5 | | | |
| *Poulsenia armata* | 10 | 2.6 | 114 | 5.9 | O | H | H |
| *Pourouma aspera* | | | 2 | 0.1 | | | |
| *Trophis racemosa* | | | 6 | 0.3 | | | |
| Myristicaceae | | | | | | | |
| *Virola nobilis* | 13 | 3.4 | 49 | 2.5 | O | | |
| *Virola sebifera* | 4 | 1.1 | 28 | 1.4 | O | | |
| Myrtaceae | | | | | | | |
| *Eugenia nesiotica* | 1 | 0.3 | 1 | 0.05 | | | |
| *Eugenia* sp. | | | 2 | 0.1 | | | |
| Nyctaginaceae | | | | | | | |
| *Guapira standleyanum* | | | 2 | 0.1 | | | |
| Olacaceae | | | | | | | |
| *Heisteria concinna* | 9 | 2.4 | 3 | 0.1 | | | |
| Palmae | | | | | | | |
| *Astrocaryum standleyanum* | | | 10 | 0.5 | | | H |
| *Scheelia zonesis* | | | 1 | 0.05 | | | |
| Rhizophoraceae | | | | | | | |
| *Cassipourea elliptica* | | | 2 | 0.1 | | | |
| Rosaceae | | | | | | | |
| *Hirtella americana* | | | 2 | 0.1 | | | |
| *Hirtella* sp. | | | 3 | 0.1 | O | | |
| *Hirtella triandra* | | | 3 | 0.1 | | | |
| *Licania hypoleuca* | | | 5 | 0.2 | | | |
| *Licania platypus* | 7 | 1.8 | 26 | 1.3 | O | | |
| Rubiaceae | | | | | | | |
| *Alseis blackiana* | 7 | 1.8 | 27 | 1.4 | | | |
| *Genipa americana* | 10 | 2.6 | 6 | 0.3 | | | |
| *Guettarda foliacea* | | | 1 | 0.05 | | | |
| *Macrocnemum glabrescens* | 5 | 1.3 | 30 | 1.5 | | | |
| *Posogueria latifolia* | | | 2 | 0.1 | | | |
| Rutaceae | | | | | | | |
| *Zanthoxylum panamense* | 1 | 0.3 | 6 | 0.3 | | | |
| *Zanthoxylum procerum* | 1 | 0.3 | | | | | |

**Table 7–3.** (*Continued*)

| Tree Species | Two-toed N | Two-toed % | Three-toed N | Three-toed % | Three-toed Sloth | Howler Monkey | Spider Monkey | Cebus Monkey |
|---|---|---|---|---|---|---|---|---|
| Sapindaceae | | | | | | | | |
| *Cupania sylvatica* | | | 1 | 0.05 | | | | |
| Sapotaceae | | | | | | | | |
| *Chrysophyllum cainito* | | | 1 | 0.05 | | C | | |
| *Chrysophyllum panamense* | 7 | 1.8 | 33 | 1.7 | | | | |
| *Pouteria stipitata* | 4 | 1.1 | 20 | 1.0 | | | | |
| Solanaceae | | | | | | | | |
| *Solanum hayesii* | 3 | 0.8 | | | | | | |
| Sterculiaceae | | | | | | | | |
| *Guazuma ulmifolia* | | | 1 | 0.05 | | | | |
| *Sterculia apetala* | 1 | 0.3 | 4 | 0.2 | | | | |
| Tiliaceae | | | | | | | | |
| *Apeiba membranacea* | 15 | 4.0 | 74 | 3.8 | O | | | |
| *Apeiba tibourbou* | | | 14 | 0.7 | | | | |
| *Luehea seemannii* | 19 | 5.0 | 40 | 2.1 | O | | | |
| Ulmaceae | | | | | | | | |
| *Trema micrantha* | | | 1 | 0.05 | | | | |

Evidence for use of leaves as food:
S = stomach contents, this study
O = observed; this study
H = observed; Hladik and Hladik (1969)
C = observed; Carpenter (1939)

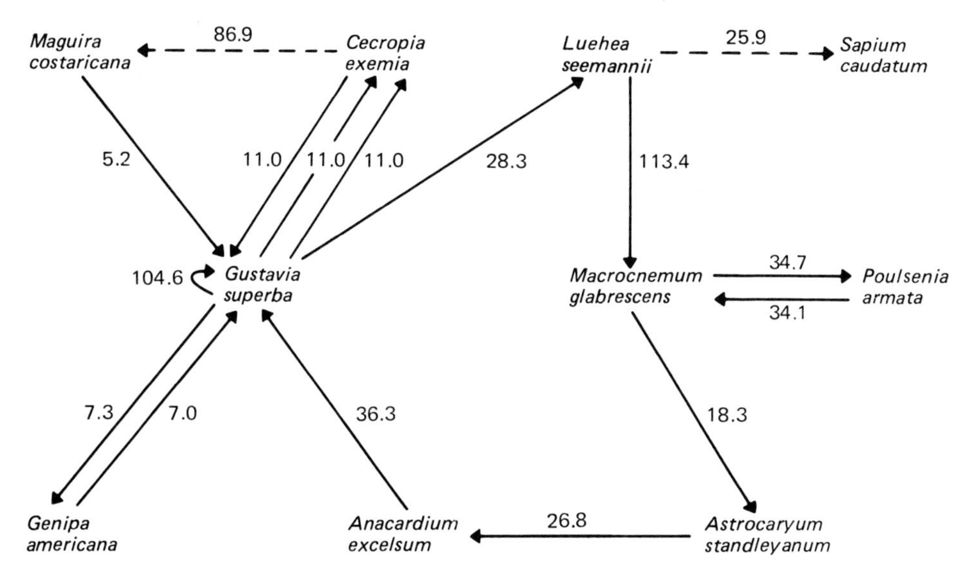


**Fig. 7–5.** Sequential use of various tree species by an adult female three-toed sloth during a 33-day period. Distance in meters that the sloth moved between trees of the various species were measured in three dimensions. Sequence of use of the tree species is indicated by arrows; dashed lines indicate missing data.

Twenty-eight tree species and three vine species were eaten by three-toed sloths (Table 7–3). We observed three-toed sloths feeding on the leaves of 25 different tree species and one vine species. Three additional tree species and two vine species were identified from three-toed sloth stomach contents.

Our finding that sloths fed on at least 31 plant species and probably many additional ones is in contrast to the conclusion (Beebe, 1926) that three-toed sloths readily accept only leaves from *Cecropia* trees, but will eat *Spondias* leaves when *Cecropia* is not available. Lundy (1952) also felt that three-toed sloths lived almost exclusively on *Cecropia* leaves. Others have observed three-toed sloths feeding on several plant species (reviewed by Goffart, 1971), including leaves from the genera *Ficus, Spondias, Eriobotyra, Luehea, Bombax, Chorisia*, and *Clusia*. Not included in Goffart's review was the observation of sloths feeding on additional species of the genera *Hevea, Elizabetha*, and *Ceiba* by Carvalho (1960), who noted that *Cecropia* trees were abundant but that three-toed sloths were not often found in them.

In spite of these reports, the idea that three-toed sloths prefer *Cecropia*, an idea not originally based on systematic field study, is still strongly entrenched (cf. Goffart, 1971). Although some study animals were never radio located in *Cecropia* trees, our data are not complete enough to completely refute the idea that sloths are dependent on *Cecropia*. They do, however, certainly suggest major roles for many other tree species in three-toed sloth food habits.

Sloths fed on and used some of the same species of trees that were used by primates on BCI (Hladik and Hladik, 1969; Carpenter, 1934). Of the 28 tree species known to have been used by three-toed sloths as food, six were used as food by primates. Of the 98 tree species in which three-toed sloths were radio located, 16 were fed on by primates.

Hladik and Hladik (1969) estimated that 50 percent of the diet of howler monkeys (*Alouatta*), 30 percent of that of night monkeys (*Aotus*), 20 percent of that of spider monkeys (*Ateles*), and maximums of 15 and 10 percent of those of *Cebus* and *Saguinus*, respectively, were leaves; thus there is potential competition for food between primates and sloths. On numerous occasions we radio located sloths in trees that were occupied also by howler or spider monkeys, and saw no behavioral manifestations of this competition. A more important point for the forest, however, is that the effects of cropping by sloths, primates, and perhaps iguanas (*Iguana iguana*), as well as by invertebrate herbivores, are additive.

A home range was not exclusively occupied by one sloth, but was shared with several other sloths of both species. As an example of the overlap of home ranges, 22 of the 63 trees used by an adult female three-toed sloth were also used by at least one other radio-marked sloth, none of them at the same time as the female. A total of nine different radio-marked three-toed sloths and four different radio-marked two-toed sloths used portions of the female's home range. Possibly other sloths shared her home range as well.

Adult sloths of both species tend to be solitary, and we very rarely saw two or more of the same species in the same tree. Beebe (1926) noted that three-toed sloths were solitary, and described antagonism that occurred when two or more were placed in the same cage. Two-toed sloths show more tolerance and are kept in groups of two or more per cage (Veselovsky, 1966; McCrane, 1966; Van Doorn, 1971). There were several instances when we observed a two-toed and a three-toed sloth in the same tree at the same time, but because their activity patterns are different (Sunquist and Montgomery, 1973), they were probably in the tree for different reasons. The two-toed sloths rested during the daytime when radio locations were made, while the three-toed sloths fed, groomed, rested, and carried out other activities.

Because of the overlap in home ranges, several sloths often visited the same tree, but as a rule only one animal was a frequent visitor to the tree, other individuals visiting it for one to two days at most during the study. Some trees were much visited during the study, such as a small *Lacmellea panamensis*, which was known to have been used by four sloths for a total of 87 days.

## Leaf Production

Estimates of the proportion of the total leaf production used by sloths were based on two areas (0.336 and 0.434 ha). Three-toed sloths were located on the sample areas 438 times, while two-toed sloths were located there 49 times. In these areas all trees with more than 0.5 ft (15.2 g) DBH were identified and measured. In this area there were an average of 358 trees per hectare. Of these trees, 76 were used by one or more sloths (Table 7–4).

**Table 7–4.** Summary of biomass and cropping rate estimates for population classes of two-toed and three-toed sloths.[a]

| | Two-toed Sloths | | Three-toed Sloths | | | |
|---|---|---|---|---|---|---|
| | Adult | Juvenile | Ad. Male | Ad. Female | Juvenile | Young |
| Total Sloths Captured | 10 | 2 | 17 | 18 | 6 | 11 |
| Percent of Species Population | 94.3 | 5.7 | 32.7 | 34.6 | 11.5 | 21.1 |
| Average weight (Kg)[b] | 3.50 ± .79 | 1.05 ± .10 | 2.78 ± .42 | 3.18 ± .74 | 1.06 ± .50 | 0.73[c] |
| Individuals/ha | 0.99 | 0.06 | 2.77 | 2.93 | 0.97 | 1.79 |
| Biomass/ha (Kg)[b] | 3.46 | 0.06 | | | | |
| Percent of Species Biomass | 98.2 | 1.8 | 7.70 | 9.32 | 1.03 | 1.31 |
| Percent of Total Sloth Biomass | 15.1 | 0.3 | 39.8 | 48.1 | 5.3 | 6.8 |
| Annual Cropping Rate | | | 33.6 | 40.7 | 4.5 | 5.7 |
| (Kg of leaf per ha) | 4.29 | 0.04[d] | 14.24 | 17.24 | 0.95[d] | 1.21[d] |

[a] Sloth weights were averaged (95 percent confidence intervals are shown) from weights of all animals captured during the study, except that weights of young three-toed sloths were estimated on the basis of average weight when they became independent of their mothers. No young two-toed sloths were captured. See text for explanation of methods for calculating the other entries in the table.

[b] Weights are corrected for gut contents by gross weight —30 percent.

[c] All young assumed to weigh 1050 g, the average weight at social weaning (Montomery and Sunquist in ms.).

[d] Assummed cropping rate is ½ of adult cropping rate.

Leaf production for each tree was assumed to be proportional to the cross-sectional area of the tree crown, computed by multiplying the measured width and depth of the crown. The sum of the cross-sectional areas of the crowns was 379,597 sq ft (35.255 $m^2$) per hectare. The sum of the crown area used by sloths was 40 percent of the total for all trees.

Our best estimate of the amount of leaf material produced annually per hectare on BCI comes from the leaves that fall to the forest floor. This estimate is lower than the amount of leaf available to sloths, because cropping occurs before the leaves fall. Foster (unpublished) estimated annual leaf production at about 6,000 kg/ha on the basis of leaf-fall traps samples. Ovington and Olson (1970) estimated an average leaf biomass in Puerto Rican montane forest at 7,880 kg/ha. Odum et al. (1970) gave an estimate of 8,120 kg/ha for lower montane forest in Puerto Rico. Odum (1970) also used an estimated average of 9,390 kg/ha in modeling the Puerto Rican forest.

Sloths of both species descend to the forest floor to defecate. Feces are accumulated by an animal for a period of time and passed periodically on the forest floor as small, hard pellets. Defecation rates were used as a basis for censusing sloths, while weights of fecal masses were used for estimates of amounts of leaf processed by sloths per unit time. The defecation rate of three-toed sloths was estimated in a field experiment involving five adult animals. All animals were radio located daily during the experiment.

Each animal was recaptured at the beginning of the experiment and again within two days after it defecated. A radio transmitter, measuring about 5 by 15 mm and weighing about 3 g, was inserted into the rectum through the anus. Transmitters were about twice as large as fecal pellets produced by the sloths. The transmitter was passed with the feces when the sloth defecated, and allowed radio location of the fecal mass. Estimates of defecation rate were based on intervals, to the nearest day, between successive defecations. Feces were cleaned in tap water and dried in an oven at 80° C for two weeks before the mass was weighed to the nearest 0.01 g.

In this study all feces were deposited within 10 ft of a tree used by the sloth. Three-toed sloths buried their feces by punching a depression in the leaf litter or soil with their tail. Two-toed sloths, which lack a tail, usually left their fecal masses on the soil or litter surface. Feces of two-toed sloths were distinguishable from that of three-toed sloths on the basis of size and shape of the fecal pellets.

Thirteen intervals between successive defecations were determined for five adult three-toed sloths. Based on these intervals, the sloths defecated on the average once every eight days ($\pm 2$ days; $p < 0.05$). Dry weights of fecal masses ranged from 45 to 68 g. The mean dry weight was 56 g ($\pm 14$ g; $p < 0.05$). The mean amount of feces accumulated per sloth per day was 7.4 g ($\pm 1.6$ g; $p < 0.05$).

Assuming a "digestive efficiency" (Kleiber, 1961, p. 253) of 50 percent, each sloth cropped an average of 14.7 g of leaves (dry weight). The assumed efficiency is similar to that for tropical ruminants fed on leaves (Hussain et al., 1960a, 1960b). We avoided many of the sources of error that can occur when food intake rates are estimated on the basis of feces weight (Golley, 1967, pp.

111–115) by using free-living sloths in the experiment. The normal feeding pattern of the animals was probably not changed significantly, either in choice of diet or amounts cropped.

Sloths used in the experiment ranged in weight from 3.63 to 4.99 kg (mean 4.16 kg $\pm$ 0.96 kg; $p < 0.05$). Goffart (1971) suggested a downward correction of 30 percent of the body weight to account for the weight of the gut contents. Biomass estimates and sloth body weights used throughout this paper were calculated on the basis of gross body weight minus 30 percent. After correction, sloths in the experiment had a mean body weight of 2.91 kg.

The three-toed sloths accumulated and passed feces at an average rate of 2.5 g/kg of sloth per day. This represents an estimated cropping rate of 5.1 g of leaf per kg of sloth per day, using an assumed digestive efficiency of 50 percent.

With the best available published information, we estimate that howler monkeys may crop leaves at a rate (38 g/kg per day) more than seven times the three-toed sloth cropping rate. This rough estimate is based on visually estimated quantities of fresh leaves consumed during daily periods in two examples by an adult monkey (Hladik and Hladik, 1969) converted to dry weight by using an estimated leaf water content of 60 percent (Lieth, 1970), and assuming an adult howler monkey body weight of 5.5 kg (Eisenberg and Thorington, 1973).

Three-toed sloths cropped leaves at a rate one-fourth or less the intake rates of terrestrial mammalian grazers in Africa (reviewed by Petersen and Casebeer, 1971). As examples, Grant's and Thompson's gazelles (*Gazella granti* and *G. thomsonii*) crop at rates of 22.2 and 26.5 g/kg body weight per day, while hartebeests (*Alcelaphus buselaphus cokii*) consume 19.9 to 31.8 g/kg body weight per day.

The lower intake rate we found for three-toed sloths is probably related to (1) their lower basal metabolic rate, which is 51 percent lower than those of mammals of comparative size (Goffart, 1971), and (2) the daily lowering of their body temperature with reduction in ambient temperature and available sunlight (Montgomery and Sunquist, unpublished). The combination lowers their total metabolic demand, although their total daily activity is comparable with that of other mammals; they are active for about 10 hr of each diel (Sunquist and Montgomery, 1973). The lower individual metabolic demand and related reduced intake of food, in comparison with those of more metabolically active animals such as the howler monkey, result in a reduced individual effect on the forest and would allow maintenance of higher sloth population densities with the same total effect on the food supply. Retention or reevolution of the primative trait of poikilothermy might thus be a part of a life history strategy that allows three-toed sloths to remain cryptic, relatively sedentary forms while maximizing opportunities for contact and mating.

## Census of Sloths

The census was based on counts of the fecal masses deposited by sloths in an area of known size during a known time period. The average defecation rate was then used to estimate the number of individuals in the area during the census period. This method involved many of the sources of error inherent in pellet-group count censuses of ungulates (Overton and Davis, 1969). These sources are considered below.

Sloth feces decompose very slowly over several months, and differential decomposition during the period should have been negligible. Dung beetle or other animal activity was not important. The influence of rain also was not important since the census was conducted during the dry season. Some fecal masses may have been missed, but bias in successive searches should have been equal. Disturbance of normal defecation patterns of the animals was probably minimal. The sloths were many meters above our heads while we searched the area and were not obviously disturbed by our presence. Radio-marked animals continued to use the census area during and after the searches.

An area of 0.324 ha was selected for census. We twice searched the entire area after moving leaves and other material aside with rakes; each search required more than a week of effort. The average interval between searches was 35 days (27 January to 2 March 1972). Leaves were replaced to minimize disturbance of the forest floor. When fecal masses were found, they were marked and covered with nylon netting. Thus we could distinguish masses found during the first search from those placed in the census area between searches, and to examine fecal masses six months and more after their initial discovery.

A total of 14 fecal masses (twelve three-toed, two two-toed) were found between the first and second search. In no case was there evidence that two or more sloths had defecated in exactly the same place, as noted for semiwild three-toed sloths by Krieg (1939). Using the average defecation rate, each would have defecated about four times during the 35-day interval. Our estimate of the number of three-toed sloths on the area was 12 fecal masses per 4.4 defecations = 2.4 three-toed sloths. Conversion to unit area gives 2.4 sloths per 0.324 ha = 8.5 three-toed sloths per hectare.

If two-toed sloths defecate in the wild at about the same rate as they do in captivity (Goffart, 1971; Enders, 1940), each should have defecated once every six days, or 5.8 times during the census period. Our estimate of the number of two-toed sloths on the area was 2 fecal masses per 5.8 defecations = 0.3 two-toed sloths. Conversion to unit area gives 0.3 sloths per 0.324 ha = 1.1 two-toed sloths per hectare.

The size of the area used for the census was small relative to the home range, and the number of fecal masses found was smaller than would be desired for an adequate census. Because we censused only one area by this method, there is no satisfactory way of deriving an estimate of the variances associated with our population estimates. Until better information is available, we use the above results as best estimates of sloth population densities. However, the numbers of radio-marked animals present in areas of known size on particular dates support the census results. For example, a contiguous area of forest measuring 1.3 ha contained three two-toed and ten three-toed sloths on 5 April 1971. These animals represented known population densities of 7.7 three-toed and 2.7 two-toed sloths per hectare.

No previous attempt to census sloths has been reported. Beebe (1926) felt that there were 20 three-toed sloths in a square mile (0.013/ha), although he saw only three. Walsh and Gannon (1967) reported rescue of 2,104 three-toed and 840 two-toed sloths during flooding of moist tropical forest in Surinam. An analysis of numbers of all animals rescued indicates that three-toed sloths contributed about 15 percent of total mammalian biomass, being the commonest species rescued (Eisenberg and Thorington, 1973); two-toed sloths contributed about half this amount of biomass and were the fourth commonest animal rescued. Goodwin (1946) states that two-toed sloths are the more abundant of the two species in Costa Rica. Enders (1935) was able to purchase twice as many three-toed as two-toed sloths when he offered to buy live animals for the same price each. Lundy (1952) felt that three-toed were five times as abundant as two-toed sloths in the Canal Zone.

## Discussion

Discussion will center on some features of the role of sloths in forest energy processes. Because this synthesis is based on order of magnitude estimates and limited data, the reader should keep in mind that sources of variance in the estimates were ignored for the purposes of discussion.

When each three-toed sloth crops leaves at a rate of 5.1 g (dry weight) per kg of sloth per day, the annual cropping rate is 1.9 kg of leaves of leaves per kg of sloth. If we assume that two-toed sloths have the same rate of food intake as three-toed sloths, but that only two-third of their diet is leaves and the remainder fruit, then they annually crop 1.2 kg of leaves and 0.6 kg of fruit per kg of sloth. Juveniles and young of both species were assumed to crop at half the rate of adults.

### Sloth Biomass Ratios and Total Biomass

Biomass ratios among the individuals of each sloth population were used for estimating sloth cropping rates. Three-toed sloths were captured (Table 7–5) in proportions of 33 adult males, 35 adult females, 12 juvenile, and 21 young per 100 three-toed sloths. Two-toed sloths, which could not be sexed, were captured in proportions of 94 adults and 6 juveniles per 100 two-toed sloths; no young were captured. Total sloth biomass per unit area was estimated on the basis of the proportion of sloths in each population class, the weight of the average animal in each class, and the estimated numbers of sloths per hectare.

Biomass estimates per hectare of the three-toed sloth population classes were adult female, 9.3 kg; juvenile, 1.0 kg; young, 1.3 kg. Three-toed sloths had a total estimated biomass of 19.4 kg/ha. For two-toed sloths, biomass estimates per hectare were 3.5 kg for adults and 0.1 kg for juveniles. The total for the species combined was 22.9 kg/ha.

**Table 7–5.** Estimated proportion of the total annual leaf production cropped by three-toed sloths from their modal trees.[a]

| Animal Number | Modal Tree Number | Estimated Cropping by Sloth (kg) | Estimated Annual Leaf Production (kg) | Percent Cropped |
|---|---|---|---|---|
| 202 | 71 | 0.498 | 25.8 | 1.9 |
| 205 | 138 | 0.903 | 51.4 | 1.8 |
| 206 | 18 | 1.172 | 17.1 | 6.8 |
| 208 | 103 | 1.128 | 56.9 | 2.0 |
| 219 | 490 | 0.794 | 64.8 | 1.2 |
| 221 | 84 | 0.558 | 58.6 | 0.9 |
| 225 | 379 | 0.859 | 11.1 | 7.7 |
| 227 | 358 | 0.712 | 112.2 | 0.6 |
| 231 | 205 | 0.586 | 78.0 | 0.7 |

[a] Annual cropping by each sloth was estimated on the basis of the frequency with which the sloth was radio located in its modal tree. Leaf production by modal trees was estimated on the basis of the cross-sectional area of the crown, and the proprtion of total cross-sectional area per hectare which it represented.

The nonvolant mammalian biomass of BCI has been estimated as 53 kg/ha (Eisenberg and Thorington, 1973), with 60 percent contributed by arboreal species. Sloths thus contribute about 73 percent of the arboreal mammalian biomass, while Howler monkeys (*Alouatta palliata*) contribute only about 12 percent (Eisenberg and Thorington, 1973). By comparison with arboreal primate herbivores elsewhere, *Colobus* forms 79 percent of arboreal mammalian biomass in Ghanian forest, while *Presbytis* forms 92 percent of that on Ceylon (Eisenberg et al., 1972).

Sloth densities are roughly equivalent to those of single species of terrestrial grazing mammals, in terms of biomass, or are higher. Mentis (1970) calculated species biomass for data on the numbers of several species of East African large mammals removed during a control removal operation. The commonest animal killed, the zebra (*Equus*), was apparently present in densities equivalent to 23 kg/ha. White rhino (*Ceratotherium simum*) were present at about 25 kg/ha, while all other species biomasses of grazers and browsers were less than 10 kg/ha.

Total terrestrial herbivore biomass on two Ceylonese national parks was 8.6 kg/ha and 7.5/ha (McKay and Eisenberg, 1973). Elephants (*Elaphas maximus*) contributed the greatest species biomass (4.0 kg/ha) in one area (McKay, 1973), while axis deer (*Axis axis*) contributed the greatest (2.6 kg/ha) in the other (Eisenberg and Lockhart, 1972).

The total animal biomass in Puerto Rican forest has been estimated at 68 kg/ha (Odum et al., 1970) and the arboreal herbivores as 3.21 kg/ha. Sloths on BCI represent a biomass about six times as great as the arboreal animal biomass (including rats) in Puerto Rican forest. Finally, the BCI sloth biomass is nearly equal to the total herbivore biomass of 30 kg/ha given for a central Amazonian forest ecosystem (Fittkau and Klinge, 1973).

## Total Cropping Rate by Sloths

Cropping rate by sloths per hectare of forest was estimated by multiplying the biomass for each population class by the annual cropping rate and summing the results. Three-toed sloth adults cropped 31.5 kg and juveniles and young 2.2 kg, giving a total cropping rate of 33.6 kg/ha/yr. Adult two-toed sloths cropped 4.3 kg and juveniles 0.04 kg, for a total rate of 4.3 kg/ha/yr. Sloths in aggregate cropped leaves at an annual rate of 38 kg/ha.

Howler monkeys, despite lower density, lower biomass, and lower proportion of leaves in the diet, may annually crop about one-third more leaves than sloths. Based on the figures given above for a leaf intake rate of 38 g/kg of howler monkey per day, or 13.9 kg of leaf per kg of howler monkey per year, and on the estimated biomass of 3.8 kg of howler monkey per hectare, they may crop 52.8 kg of leaves per hectare of forest per year.

This estimated leaf cropping rate is based on combining several field estimates which were themselves subject to large sources of error, and which need validation. The estimated biomass and particularly the estimated leaf intake rates may be higher than they in fact are for the BCI howler monkey population. If this is not the case, then a higher total leaf cropping rate by howler monkeys may be related to (1) higher metabolic demands imposed by being homiotherms, and (2) greater amount of leaf required by an animal with a gut less specialized for ruminant-like digestion (Hladik and Hladik, 1969) to derive the same nutritional benefits as sloths, which are relatively specialized for such digestion (Denis et al., 1967; Goffart, 1971).

## Proportion of Forest Leaf Production Cropped by Sloths

Sloths cropped an estimated 0.63 percent of the total annual leaf production of the forest (Figure 7–6). By comparison, Odum and Ruiz-Reyes 1970) estimated total cropping rates by folivorous insects at about 7 percent. For temperate forest Bray (1964) reported mean annual utilization of portions of attached leaves by insects ranging from 5.9 to 10.6 percent of the total leaf surface. Hladik and Hladik (1972, Table 3) estimated amounts of leaf and other food eaten by *Presbytis senex* on Ceylon and the total production of leaves by various food species. Based on those data, the monkeys cropped from 5.5 to 29.5 percent of the total leaf production on their territories. Furthermore, the data would indicate that from 15 to 47 percent of the annual leaf production of all *Adina cordifolia* (Rubiaceae) on each territory was ingested by the *Presbytis senex* that lived on that territory, as well as high percentages of the annual leaf production of other tree species. These cropping rates seem high at least by order

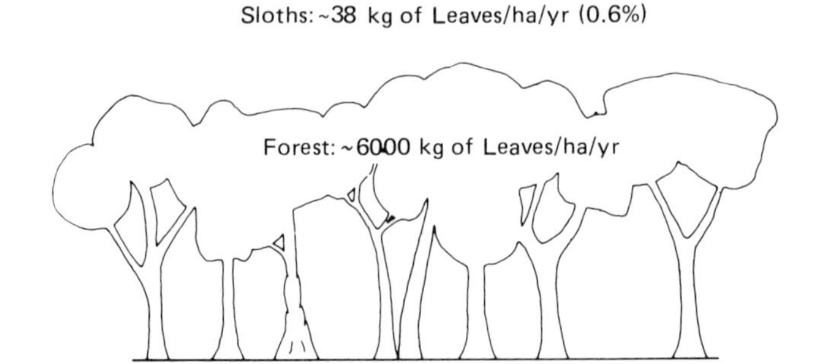


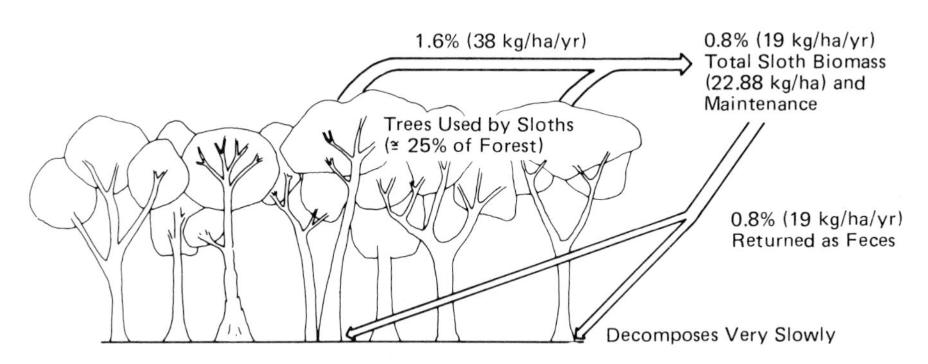


**Fig. 7–6.** Schematic summary of the proportion of the total forest leaf production that sloths annually crop (*top*) and the proportion of the leaf production of those trees used by sloths that is cropped annually by both sloth species in aggregate (*bottom*). Both the proportions of leaf production cropped (shown as percentages) and the amount per unit area (kilogram per hectare) are shown. Note that half of the leaf cropped by sloths is returned to the forest as slowly decomposing feces.

of magnitude, particularly when it is considered that other animals, vertebrate and invertebrate, probably cropped leaves from those trees. Validation of these results by more exact methods is necessary before it can be concluded that a single species of mammalian arboreal foliovore crops a percentage of its food supply comparable to percentages cropped by terrestrial foliovores.

Terrestrial grazers in aggregate crop as much as 45 percent of the annual net production on managed rangeland (reviewed by Wiegert and Evans, 1967) and 30 percent or more from wild grassland. A single wild species of large mammalian terrestrial grazer, such as the kob (*Adenota kob*), may ingest 10 percent of the primary production (Buechner and Golley, 1967). However, two to five terrestrial species usually total at least 70 percent of the herbivore biomass, even though there may be as many as 18 herbivorous species present (McKay and Eisenberg, 1973). High total cropping values for terrestrial mammalian grazers in the tropics are primarily due to several species cropping a particular area, rather than to high cropping rates by a single species.

**Cropping of Trees Used by Sloths**

Many quantitative and qualitative factors affect the suitability of plants as food for a sloth or other herbivore (McCullough, 1970). Depending on nutritional requirements and digestive capabilities, each species will select more or less different foods, or will select them in different proportions, from a given area in which several species of herbivore coexist (cf. Hladik et al., 1971). Both the kind and amount of food the herbivore selects may be an important facet of the life history strategy (cf. McKay, 1973, p. 95).

This selection of food makes it somewhat unrealistic to compute cropping rates on the basis of total forest leaf production. More meaningful figures for the impact of the sloths on the forest can perhaps be derived by relating the cropping rate to the portion of the forest used by sloths. Sloths used 25 percent of the trees on our sample areas; these trees produce 40 percent of the total annual leaf production, or 2,400 kg of leaves per hectare. In cropping these trees, sloths would remove an estimated 1.6 percent of the annual leaf production (Figure 7–6).

Even more intensive cropping of a tree results when it is the modal tree of a sloth. We estimated the total annual production of each of the modal trees and the proportion of the year the sloth spent in the tree on the basis of radio locations Table (7–5). Some modal trees lost as much as 7.7 percent of their estimated annual leaf production to a single sloth (Fig. 7–7). In addition, some modal trees were used by other sloths, increasing the total cropping of the tree. The maximum estimated total cropping for a single tree was 20 percent annually for a small *Lacmellea panamensis*, which received 87 sloth-days of use by a total of four animals.

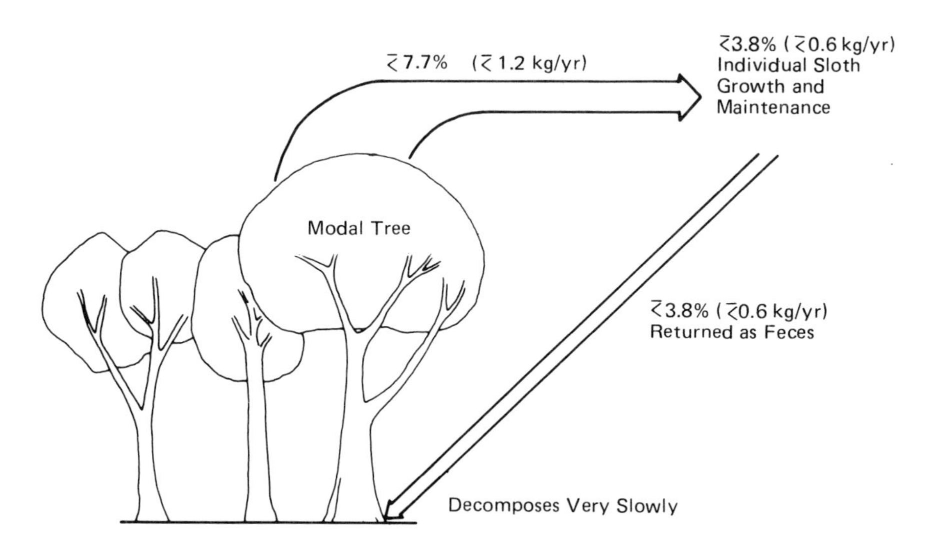


**Fig. 7–7.** Schematic summary of the proportion of the annual leaf production of individual trees that are used most often by a sloth (modal tree) that is cropped by the sloth. Note that half the leaf material is returned to the modal tree as slowly decomposing feces.

**Role of Sloths in Forest Processes**

Intensive cropping of a tree by sloths is not entirely disadvantageous because the sloth selectively returns waste material to it. We estimate that half the leaf material eaten by sloths is returned to the forest system as feces. Fecal masses are selectively placed near the boles of trees most often used by sloths. Each fecal mass is a package of nutrients and minerals that, as it leaches and decomposes, becomes available to the plants near it. The uric acid concentration of sloth urine is relatively high (Goffart, 1971), and much of the nitrogenous end products of digestion are passed in urine. Presumably these products are recycled relatively quickly from the fecal mass. Other digestive by-products and undigested leaf passed as feces would become available more slowly.

Roots of tropical trees may selectively grow toward sources of nutrients (Odum, 1970b) and form an absorptive mat at the nutrient source (Odum, 1970c), possibly as an adaptation where there are mineral deficiencies. Trees that have sloth feces near their boles may have an advantage in competing for nutrients and minerals from sloth feces because their roots have shorter distances to grow than the roots of neighboring trees.

Of themselves, the amounts of nutrients and minerals that reach a tree from sloth feces may not appear to be of sufficient quantity to affect the tree. However, sloth feces represent one of the few stable, long-term sources of nutrients and minerals in the forest system. Partly because three-toed sloths bury their fecal masses, sloth feces remain where they are placed by the animal for long periods of time (six or more months). The feces decomposes very slowly in place; we could readly recognize more than 90 percent of fecal masses more than six months after they were originally found, and much of the original weight of the feces remained. By comparison, leaves decompose rapidly in moist tropical forest (Jenny et al., 1949). For example, Luse (1970) noted that leaves of *Sloanea* and *Dacryodes* were completely decomposed 51 days after being placed on the forest floor, and noted that minerals (P) moved laterally (downstream) from where the leaves decomposed.

In a system where most of the nutrients and minerals are bound up in plant and animal tissue and the remainder recycled rapidly, sloth feces stands out as a long-term stable source that is available only to certain trees. These trees pay the price of being cropped by the sloths but receive a limited resource in return. The effect of sloth feces may be related to stabilizing some components of the forest system. Jordan et al. (1972) indicate that low rates of flow of minerals relative to input rates will result in more stable systems; sloths slow the normally high recycling rates for certain trees. Sloths may thus represent not a chemical but a behavioral specialist that (Odum, 1970, p. I-242): ". . . provide the forest special services in holding or recycling particular elements."

In the coevolution of sloths and their food trees, the rate at which trees evolve defense mechanisms against sloths may have been slowed by two factors. In reducing the cropping pressure on single species by utilizing a broad spectrum (each three-toed sloth tends to use a different species as a modal tree), predator pressure against a single tree species is reduced. By selectively returning feces to the cropped trees sloths may promote growth and survival of the trees, even while they act as predators on them. This situation differs from the influence of mammalian seed predators (Smith, 1970) where predation may relatively rapidly influence the pattern of reproduction of the trees.

## Acknowledgment

The study was supported by Smithsonian Research Foundation Grant No. 435090 to J. F. Eisenberg, and by the Environmental Sciences Program, Smithsonian Institution. Smithsonian Tropical Research Institute provided logistic support, and personnel of the Institute gave advice and field help. R. A. Foster identified trees used by study sloths and gave other invaluable field help and advice. Judith A. Montgomery prepared the histological material for identification of stomach contents and assisted with fieldwork. Many visitors to Barro Colorado Island contributed by their comments while work was in progress, and in some cases helped to capture sloths. E. Leigh offered welcome comments on the manuscripts.

## References

Baumgartner, L. L., and A. C. Martin. 1939. Plant histology as an aid in squirrel food-habits studies. *J. Wildlife Mgmt.* 3:266–268.

Beebe, W. 1926. The three-toed sloth, *Bradypus cuculliger cuculliger* Wagler. *Zoologica* 7:1–67.

Bennett, C. F., Jr. 1963. A phytophysiognomic reconnaissance of Barro Colorado Island, Canal Zone. *Smith. Misc. Coll.* 145, 7:1–8.

Bourliere, F. 1973. The comparative ecology of rain forest mammals in Africa and tropical America: Some introductory remarks. In *Tropical Forest Ecosystems in Africa and South America: A Comparative Review*, B. J. Meggers, E. S. Ayensu, and W. D. Duckworth, eds. pp. 279–292. Washington, D.C.: Smithsonian Inst. Press.

Bray, J. R. 1964. Primary consumption in three forest canopies. *Ecology* 45:165–167.

Buechner, H. K., and F. B. Golley. 1967. Preliminary estimation of energy flow in Uganda kob (*Adenota kob thomasi* Newman). In *Secondary Productivity of Terrestrial Ecosystems*, K. Petrusewicz, ed. pp. 243–260. Vol. 1, pp. 1–379. Warsaw: Inst. Ecology Polish Acad. Sci.

Carpenter, C. R. 1934 (1964). A field study of the behavior and social relations of howling monkeys. (*Alouatta palliata*). In *Naturalistic Behavior of Nonhuman Primates*, C. R. Carpenter, ed. pp. 3–92. University Park, Pa.: Pennsylvania State Univ. Press.

Carvalho, C. T. de. 1960. Notes on the three-toed sloth, *Bradypus tridactylus*. *Mammalia* 24:155–156.

Cloudsley-Thompson, J. L. 1972. The habitat and its influence on the evolutionary development. In *Biology of Nutrition*, R. N. T.-W. Fiennes, ed. pp. 351–373. New York: Pergamon Press.

Denis, C., C. Jeuniaux, M. A. Gerebtzoff, and M. Goffart. 1967. La digestion stomacale chez un paresseux: l'unau *Choloepus hoffmanni* Peters. *Ann. Soc. Zool. Belg.* 97:9–29.

Eisenberg, J. F., and M. Lockhart. 1972. An ecological reconnaissance of Wilpattu National Park, Ceylon. *Smithsonian Contr. Zool. 101.*

———, and G. M. McKay. 1974. Comparison of ungulate adaptations in the New World and Old World tropical forests with special reference to Ceylon and the rainforests of Central America. In *The Behavior of Ungulates and Its Relation to Management*, V. Geist and F. Walther, eds. Int. Union Conservation Nature (in press).

———, N. A. Muckenhirn, and R. Rudran. 1972. The relation between ecology and social structure in primates. *Science* 176:863–874.

———, and R. W. Thorington. 1973. A preliminary analysis of a neotropical mammal fauna. *Biotropica* 5:150–161.

Enders, R. K. 1935. Mammalian life histories from Barro Colorado Island, Panama. *Bull. Mus. Comp. Zool.* 78:385–502.

———. 1940. Observations on sloths in captivity at higher altitudes in the tropics and in Pennsylvania. *J. Mammal.* 17:165–166.

Fittkau, E. J., and H. Klinge. 1973. On biomass and trophic structure of the central Amazonian rain forest ecosystem. *Biotropica* 5:1–14.

Goffart, M. 1971. *Function and Form in the Sloth.* 225 pp. New York: Pergamon Press.

Golley, F. B. 1967. Methods of measuring secondary productivity in terrestrial vertebrate populations. In *Secondary Productivity of Terrestrial Ecosystems*, K. Petrusewicz, ed. pp. 99–124. Inst. Ecol. Polish Acad. Sci.

———. 1972. Summary. In *Tropical Ecology with an Emphasis on Organic Production*, P. M. Golley and F. B. Golley, compilers. pp. 407–413. Athens: Inst. Ecol., Univ. Georgia.

Goodwin, G. G. 1946. Mammals of Costa Rica. *Bull. Am. Mus. Nat. Hist.* 87:275–478.

Harrison, J. L. 1972. Feeding habits of animals in arboreal habitats. In *Biology of Nutrition*, R. N. T.-W. Fiennes, ed. pp. 505–512. New York: Pergamon Press.

Hladik, A., and C. M. Hladik. 1969. Rapports trophiques entre vegetation et primates dan la foret de Barro Colorado (Panama). *Terre Vie* 1:29–117.

Hladik, C. M., J. Bousset, P. Valdebouze, G. Viroben, and J. Delort-Laval. 1971. Le regime alimentaire des primates de lile de Barro-Colorado (Panama). *Folia Primat.* 16:85–122.

———, and A. Hladik. 1972. Disponibilites alimentaires et domaines vitaux des primates a Ceylan. *Terre Vie* 2:149–215.

Hofman, R. R., and D. R. M. Stewart. 1972. Grazer or browser; a classification based on the stomach-structure and feeding habits of East African ruminants. *Mammalia* 36:226–240.

Hussain, M., B. N. Majumdar, B. Sahai, and N. D. Kehar. 1960a. Studies on tree leaves as cattle fodder. II. The nutritive value of Bargad leaves (*Ficus bengalensis*). *Ind. J. Dairy Sci.* 13:1–8.

———. 1960b. Studies on tree leaves as cattle fodder. IV. The nutritive value of Pipal leaves (*Ficus religiosa*). *Ind. J. Dairy Sci.* 13:9–15.

Jenny, H., S. P. Gessel, and F. T. Bingham. 1949. Comparative study of decomposition rates of organic matter in temperate and tropical regions. *Soil Sci.* 68:419–432.

Jordan, C. F., J. R. Kline, and D. S. Sasscer. 1972. Relative stability of mineral cycles in forest ecosystems. *Amer. Natur.* 106:237–253.

Kleiber, M. 1961. *The Fire of Life.* 454 pp. New York: John Wiley and Sons.

Krieg, H. 1939. Begegnungen mit Ameisenbären und Faultieren in freier Wildbahn. *Z. Tierpsychol.* 2:282–292.

———. 1961. Das Verhalten der Faultiere (Bradypodidae). *Hanbuch Zool.* 8/27. 10(12b):20–23.

Lieth, H. 1970. The water content of some leaves at El Verde. Appendix A. In *A Tropical Rain Forest*, H. T. Odum, ed. p. I-281. Washington, D.C.: Div. Tech. Information, U.S. Atomic Energy Comm.

Lundy, W. E. 1952. The upside-down animal. *Nat. Hist.* 61:114–119.

Luse, R. A. 1970. The phosphorus cycle in a tropical rain forest. In *A Tropical Rain Forest*, H. T. Odum, ed. pp. H-161 to H-166. Washington, D.C.: Div. Tech. Information, U.S. Atomic Energy Comm.

McCrane, M. P. 1966. Birth, behaviour and development of a hand-reared two-toed sloth, *Choloepus didactylus. Intern. Zoo Yearbook* 6:153–163.

McCullough, D. R. 1970. Secondary production of birds and mammals. In *Analysis of Temperate Forest Ecosystems*, D. E. Richle, ed. pp. 107–130. New York: Springer-Verlag.

McKay, G. M. 1973. Behavior and ecology of the Asiatic elephant in southeastern Ceylon. *Smithsonian Contr. Zool. 125.*

———, and J. F. Eisenberg. 1974. Movement patterns and habitat utilization of ungulates in Ceylon. In *The Behavior of Ungulates and Its Relation to Management*, V. Geist and F. Walther, eds. I.U.C.N. (Publ.) (in press).

Mentis, M. T. 1970. Estimates of natural biomasses of large herbivores in the Umfolozi Game Reserve area. *Mammalia* 34:363–393.

Montgomery, G. G. 1969. Weaning of captive raccoons. *J Wildlife Mgmt.* 33:154–169.

———, W. W. Cochran, and M. E. Sunquist. 1973. Radio-locating arboreal vertebrates in tropical forest. *J. Wildlife Mgmt.* 37:426–428.

Odum, H. T. 1970a. Summary: An emerging view of the ecological system at El Verde. In *A Tropical Rain Forest*, H. T. Odum, ed. pp. I-191 to I-289. Washington, D.C.: Div. Tech. Information, U.S. Atomic Energy Comm.

———. 1970b. Introduction to section F. In *A Tropical Rain Forest*, H. T. Odum, ed. pp. F-3 to F-7. Washington, D.C.: Div. Tech. Information, U.S. Atomic Energy Comm.

———. 1970c. Rain forest structure and mineral-cycling homeostatis. In *A Tropical Rain Forest*, H. T. Odum, ed. pp. H-3 to H-52. Washington, D.C.: Div. Tech. Information, U.S. Atomic Energy Comm.

———, W. Abbott, R. K. Selander, F. B. Golley, and R. F. Wilson. 1970. Estimates of chlorophyll and biomass of the Tabonuco forest of Puerto Rico. In *A Tropical Rain Forest*, H. T. Odum, ed. pp. I-3 to I-19. Washington, D.C.: Div. Information, U.S. Atomic Energy Comm.

———, and J. Ruiz-Reyes. 1970. Holes in leaves and the grazing control mechanism. In *A Tropical Rain Forest*, H. T. Odum, ed. pp. I-69 to I-80. Washington, D.C.: Div. Tech. Information, U.S. Atomic Energy Comm.

Overton, W. S., and D. E. Davis. 1969. Estimating the numbers of animals in wildlife populations. In *Wildlife Management Techniques*, R. H. Giles, Jr., ed. pp. 403–456. Washington, D.C.: The Wildlife Society.

Ovington, J. D., and J. S. Olson. 1970. Biomass and chemical content of El Verde lower montane rain forest plants. In *A Tropical Rain Forest*, H. T. Odum, ed. pp. H-53 to H-61. Washington, D.C.: Div. Tech. Information, U.S. Atomic Energy Comm.

Ozorio de Almeida, A., and Branco de A. Fialho. 1924. Metabolisme, temperature et quelques autres determinations physiologiques faites sur le paresseux (*Bradypus tridactylus*). *Comp. Rend. Soc. Biol.* 91:1124–1125.

Petersen, J. C. B., and R. L. Casebeer. 1971. A bibliography relating to the ecology and energetics of East African large mammals. *E. Afr. Wildlife J.* 9:1–23.

Smith, C. C. 1970. The coevolution of pine squirrels (*Tamiasciurus*) and conifers. *Ecol. Monographs* 40:349–371.

Standley, P. C. 1933. The flora of Barro Colorado Island, Panama. *Contr. Arnold Arboretum* 5:1–178.

Stewart, D. R. M. 1967. Analysis of plant epidermis in faeces: A technique for studying the food preferences of grazing herbivores. *J. Appl. Ecol.* 4:83–111.

Sunquist, M. E., and G. G. Montgomery. 1973. Activity patterns and rates of movement of two-toed and three-toed sloths (*Choloepus hoffmanni* and *Bradypus infuscatus*). *J. Mammal.* 54:946–954.

Tadaki, Y. 1966. Some discussions on the leaf biomass of forest stands and trees. *Bull. Gov. For. Exp. Sta. (Tokyo)* 184: 135–161.

Van Doorn, C. 1971. Verzogerter Zyklus der Fortpflanzung bei Faultieren. *Z. Kölner Zoo* 14:15–22.

Veselovsky, Z. 1966. A contribution to the knowledge of the reproduction and growth of the two-toed sloth, *Choloepus didactylus*, at the Prague Zoo. *Intern. Zoo Yearbook* 6:147–153.

Walsh, J., and R. Gannon. 1967. *The Time Is Short and the Water Rises*. 224 pp. London: Thomas Nelson and Sons.

Wiegert, R. G., and F. C. Evans. 1967. Investigations of secondary productivity in grasslands. In *Secondary Productivity of Terrestrial Ecosystems*, K. Petrusewicz, ed. pp. 499–518. *Inst. Ecol. Polish Acad. Sci.* V 2:383–879.

Worman, R. G. 1946. *Swimming Sloth*. Nat. Hist. N.Y. 55:49.

# CHAPTER 8

## Impact of Leaf-Cutting Ants on Vegetation Development at Barro Colorado Island

BRUCE HAINES

### Introduction

Invertebrates can influence vegetation development through modification of primary productivity, gene flow, and nutrient cycling. Invertebrate consumption of leaves, twigs, bark, and wood can reduce primary productivity. Selective consumption of these organs, selective seed predation, and selective pollination can alter the species composition of the vegetation. Invertebrate activities in detritus decomposition influence the pattern and rates of nutrient cycling.

This report assesses the role of the leaf-cutting ant, *Atta colombica tonsipes* Santschi, in vegetation development. Compared with other invertebrates, the trail-following behavior of *Atta* makes the study of its impact on vegetation relatively simple. Leaf, flower, and fruit materials are cut within the forest canopy and transported over distances as long as 180 m to the colony (Figure 8–1), which may reach 10 m diameter at the soil surface. Harvested plant material is carried to the underground fungus gardens within the colony nest. Here, in a complex of symbiotic interactions between fungus and ant (Martin, 1970), the harvested plant material is digested by the fungus. The fungus is, in turn, eaten by the ant. The degraded remains of the harvested plant material, along with the dead and dying ants, are carried to the nest surface and then down the slope to some structure such as a vine, rock, or tree trunk. From there the refuse is dropped to the ground where it accumulates in what become smelly organic refuse dumps (Figure 8–1). Alternatively, when water is near, the refuse is dropped into water (Figure 8–1).

The present study was directed toward two questions: (1) Do harvesting, transporting, and refuse dumping activities result in localized mineral nutrient accumulations; and (2) on nutrient-impoverished Panamanian latosols (Brown and Wolfschoon, 1960; Martini, 1966), do localized nutrient accumulations result in localized modification of vegetation?

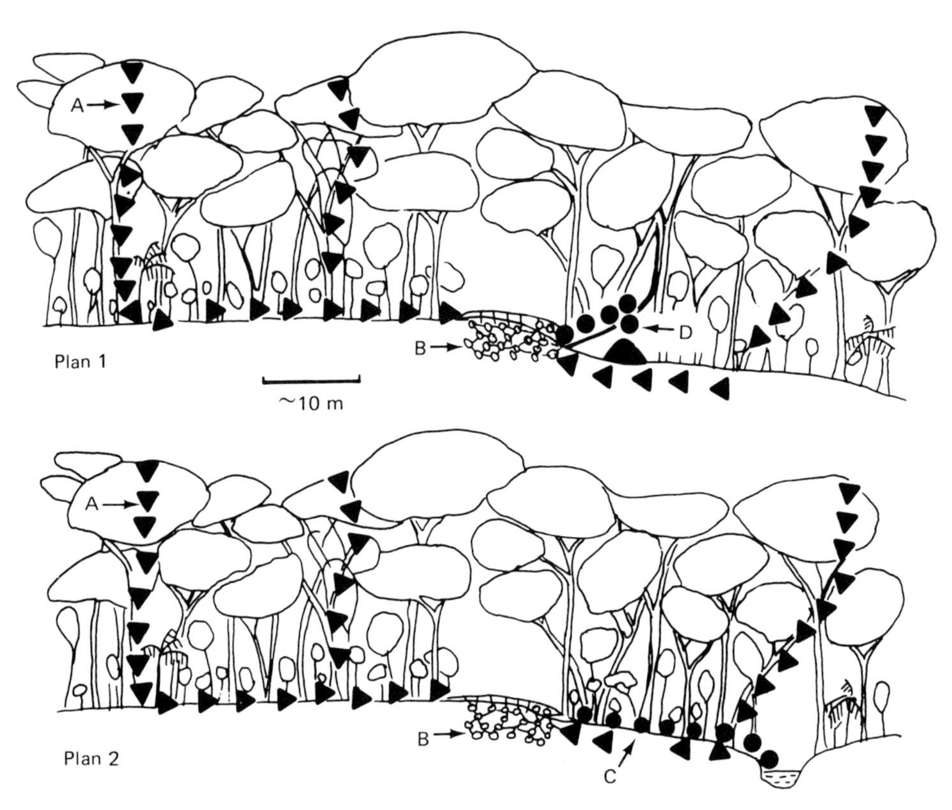


**Fig. 8–1.** Diagramatic representation of energy and mineral nutrient flows to and through an *Atta colombica tonsipes* Santschi colony. Leaf, flower, and fruit materials (▼) are carried from canopy to ground, then from 0 to 200 m along distinct trails to fungus gardens (*B*) within the nest. Organic refuse from fungus gardens (*C*) is carried (●) to a stone, vine, or tree trunk from which it is dropped (*D*) onto ground where it accumulates in dumps (plan 1) or into running water (plan 2). Canopy height and nest diameter are drawn to scale.

## The Nutrient Accumulation Hypothesis

The mineral nutrient accumulation hypothesis was tested by removing pairs of soil samples from the 0 to 20 cm and 50 to 70 cm depths of forest, nest, and dump areas associated with each of five nest sites. Samples were dried over light bulbs, ground, and shipped to the U. S. Department of Agriculture, Miami Plant Quarantine Inspection Station, for steam sterilization and then to Duke University for chemical analysis. Sterilization was assumed to affect all samples equally.

Total potassium, calcium, and magnesium concentrations were determined by atomic absorption spectrophotometry on nitric perchloric acid digests of soils (Jackson, 1965, p. 331). Total phosphorus was determined colorimetrically on the same digests. Available phosphorus was extracted by the Bray and Kurtz method (Jackson, 1965, p. 159) and determined colorimetrically. Total nitrogen was determined by the Kjeldahl method on paired samples from one nest. The results (Figure 8–2) show that total nitrogen, phosphorus, potassium, magnesium, and calcium, and available phosphorus are accumulated in the *Atta* organic refuse dumps above their respective background levels in the forest floor.

## The Vegetation Modication Hypothesis

Pot studies, field measurements, and field experiments were made to investigate the impact of the nutrient accumulations on plant establishment and growth.

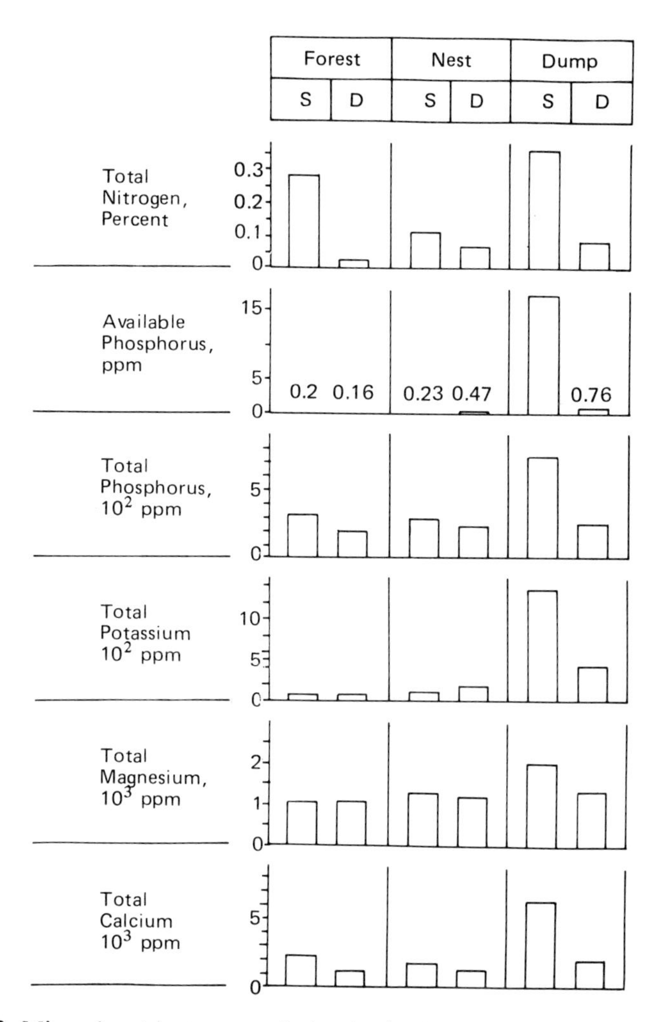


**Fig. 8–2.** Mineral nutrient accumulation in the forest, ant nest, and ant dumps in surface ($S$) and deep ($D$) soils.

## Pot Studies

Pot studies were performed to determine whether biologically significant differences existed between surface (0 to 20 cm) and the deep (50 to 70 cm) soils of the forest, nest, and organic refuse dump. Soils were potted and placed in a random block design. They were first planted with fast-growing garden herbs to determine whether longer range studies utilizing slower-growing native woody species might be instructive. Herb growth showed definite differentiation with respect to soils (Figure 8–3). Plants on dump soils exhibited the greatest growth. Repetition with native woody species showed similar trends (Figure 8–4).

Another test was carried out to determine on which soils nutrients are limiting under full sunlight conditions. One set of pots received supplementary 5–10–5 N–P–K fertilizer at a rate calculated to raise the P content of the forest soils by 100 ppm. A second set received no fertilizer. Nutrients were limiting on all

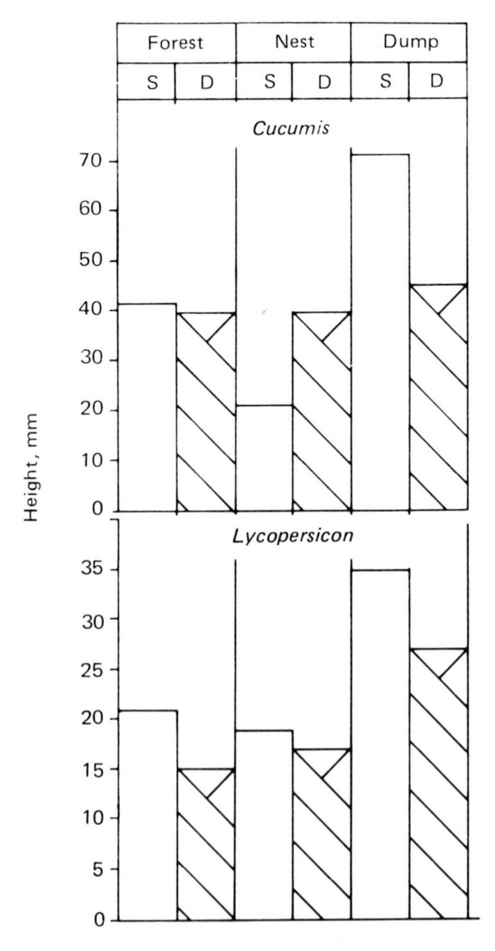


**Fig. 8–3.** Height of *Cucumis* and *Lycopersicon* grown on surface (*S*) and deep (*D*) soils from forest, nest, and dump for 12 days.

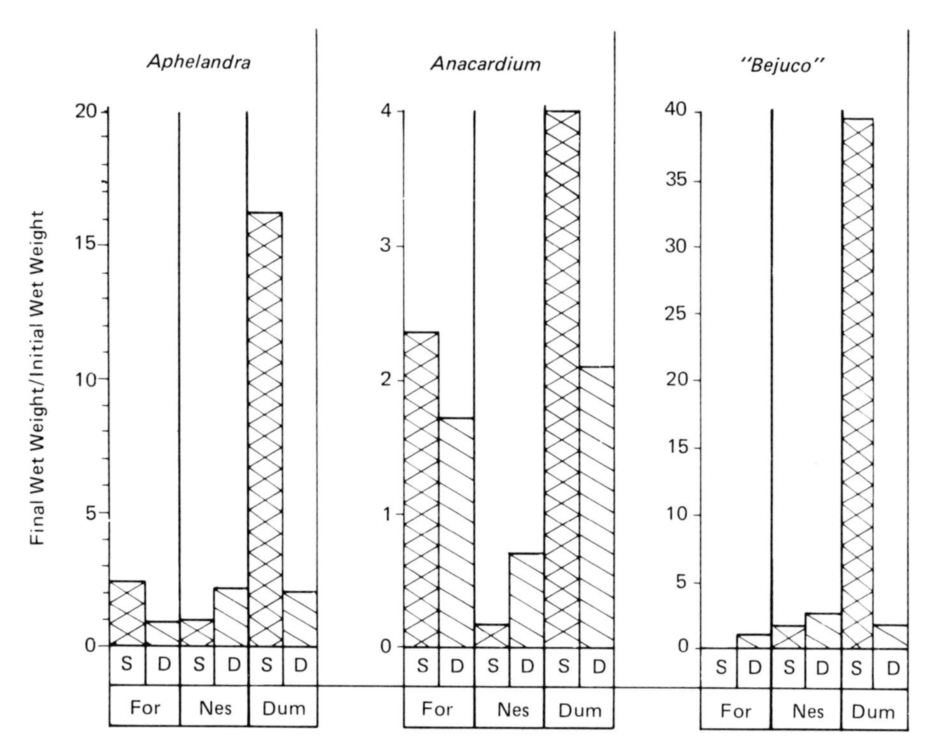


**Fig. 8–4.** Growth of *Aphelandra*, *Anacardium*, and "Bejuco" on surface (*S*) and deep (*D*) soils of forest (*For*), the nest (*Nes*), and the dump (*Dum*).

soils except the dump surface soils (Figure 8–5). Repetition with woody plants produced poor growth on dump soils, probably due to poor water retention.

## Field Measurements

Partly eroded *Atta* refuse dumps exhibited larger concentrations of small roots than eroded portions of forest soil. Was root growth stimulated by the refuse dumps? The distribution of root biomass was measured by washing roots from 5-liter volumes of soil taken from the surface and deep regions of forest, nest, and dump areas associated with ten *Atta* nests. The smallest (0 to 2 mm diameter) and probably most physiologically active roots attained their greatest biomass in the dump surface soils (Figure 8–6). While these values may be underestimated because the samples were taken during the dry season, clearly the growth of roots was stimulated by the dump.

Relative plant growth was measured as numbers of seedlings per square meter on forest, nest, and dump surface areas associated with 11 *Atta* colonies. Seedling frequency was plotted against seedling height in Figure 8–7. The plot shows that (1) on the dump there were many small seedlings but few large ones, (2) on the nests there were especially few large seedlings, and (3) on the

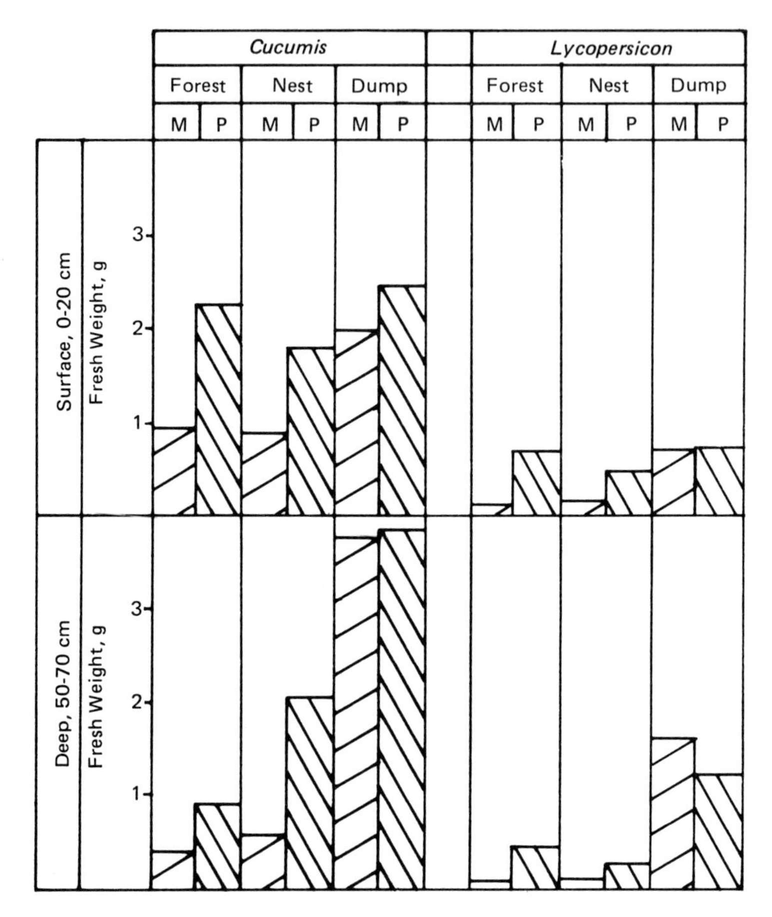


**Fig. 8–5.** Growth of *Cucumis* and *Lycopersicon* on soils indicated, plus (*P*) and minus (*M*) supplementary fertilizer.

forest floor all heights of seedlings were represented. Only the forest floor appears to have the potential to contribute trees to the canopy.

The biomass of small diameter roots and the numbers of seedlings on forest, nest, and dump sites reveal a negative relationship. Where the biomass of small roots is greatest, the size of seedlings is least. This important pattern appeared again in a field experiment.

## Field Experiment

Simulated *Atta* dumps were constructed for comparison of plant response to *Atta* organic refuse and to commercial fertilizer. Also, the effect of roots was evaluated by trenching. The plots, 8 by 8 m, were split into two 3 by 8 m subplots oriented upslope and separated by a 2-m wide strip. The trench was dug around one of the two subplots to a depth of 0.5 m. The trench, before refilling,

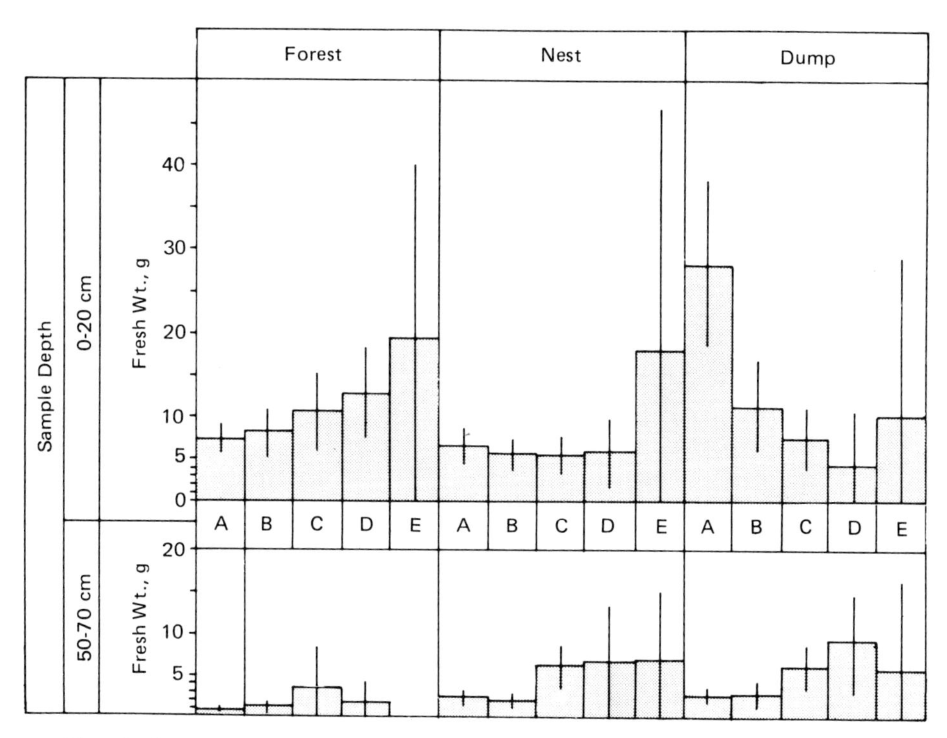


**Fig. 8–6.** Mean root fresh weight in forest, nest, and dump soils. Roots were washed from 5-liter soil volumes and sorted into the following diameter classes: A, 0–2 mm; B, 2.1–5 mm; C, 5.1–10 mm; D, 10.1–20 mm; and E, greater than 20.1 mm. Vertical bars are 95% confidence limits to mean weights obtained from ten active *A. colombica* nest areas.

was lined with 4 mil black polyethylene plastic sheeting to retard root reinvasion. The trenched and untrenched subplots were further subdivided into three 3 by 2 m plots.

The treatments were randomly allocated as follows: (1) *Atta* organic refuse, 100 kilos estimated dry weight; (2) fertilizer at a rate equaling the P applications in the *Atta* refuse treatment; and (3) a control. Five individuals of each of five different native wood species were planted within a 1 by 2 m area centered within each sub-subplot. Also, light intensity plots were set up to determine the role of light in seedling growth. Four plots were established; two in full shade on Barro Colorado Island, one in full shade, and one under a hole through the canopy on the adjacent mainland.

The growth responses of the volunteer plants (Figure 8–8) are in several ways more instructive and clear-cut than the responses of the planted seedlings (Haines, 1971). The responses showed the dependence of the volunteer flora on light, as well as a negative relation between root mass and seedling growth. Where root biomass is great, seedling mass is small. This negative relation between root growth and seedling growth is a repetition of the pattern observed at *Atta* colony sites. Soil moisture data largely explain this negative relationship.

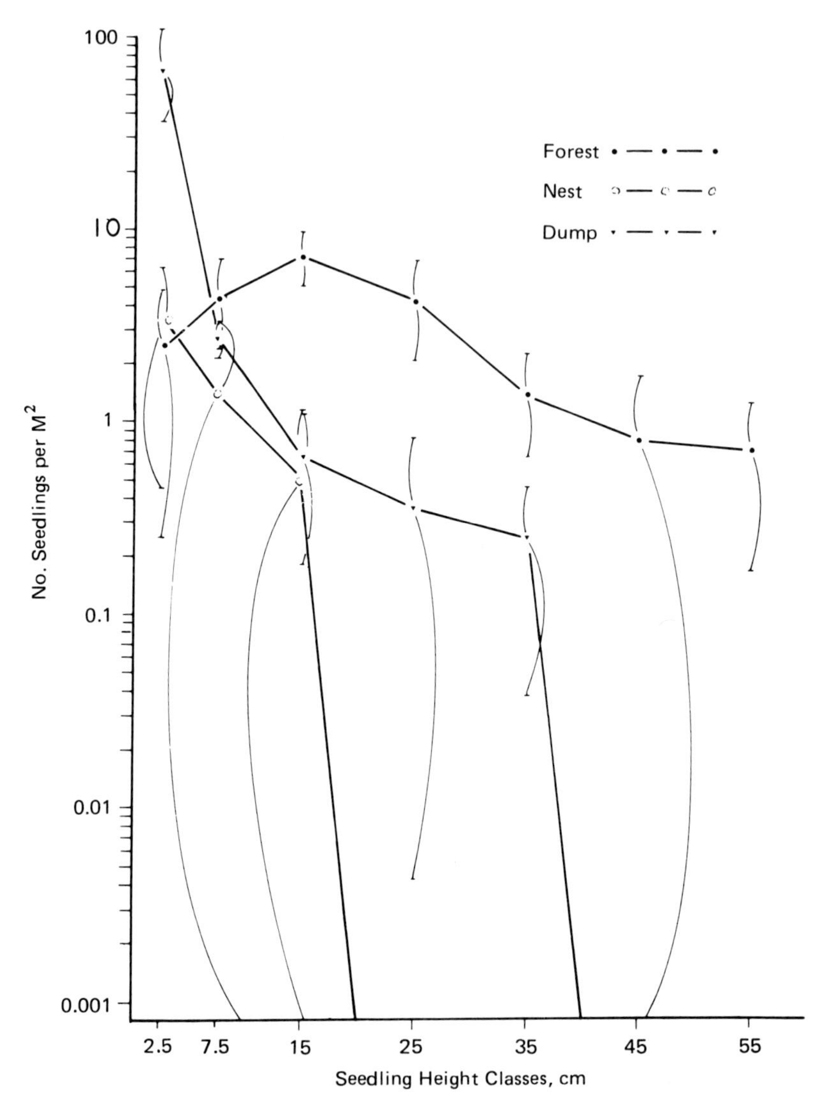


**Fig. 8–7.** Seedling survival on forest, nest, and dump soil. Seedling height is plotted against seedling density. Vertical lines are 95% confidence limits to means from 11 nest areas.

*Atta* refuse at a matric potential of 1/1,000 bar holds about 1 cc of water per cc dry soil, while forest soil at the same matric potential holds about 0.7 cc of water per cc dry soil. Thus, at high potentials, *Atta* refuse can deliver relatively a large volume of water. However, as the matric potential drops, the deliverable water volume (of the *Atta* refuse) drops very rapidly. At very low matric potentials the *Atta* refuse can deliver very little water. The forest soil, on the other hand, can deliver moderate volumes of water over a wide range of matric water potentials.

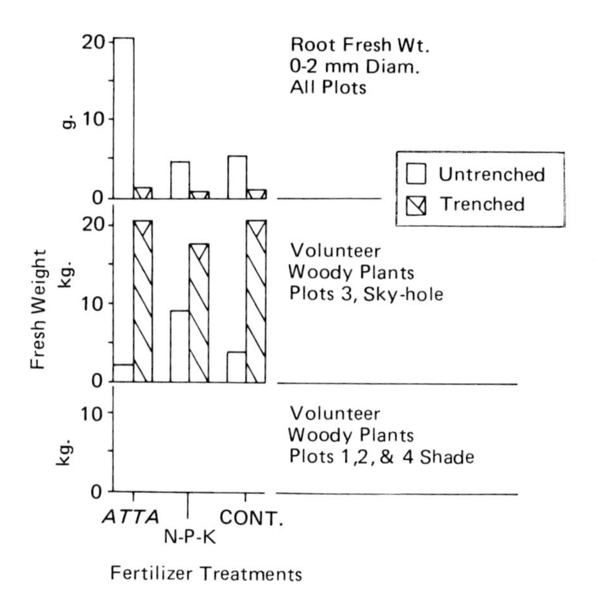


**Fig. 8–8.** Growth of volunteer woody plants in relation to light intensity, root biomass, and type of fertilizer treatment. Results are shown for fertilizer treatments: *Atta* refuse (Atta); chemical fertilizer (N–P–K); and control (Cont) at high light intensity (Sky-Hole) and in shade. Note negative relation of volunteer woody plant growth to root biomass.

Soil moisture data (Haines, 1971) reveal that tree roots remove water from the refuse dumps to levels detrimental to seedlings. On forest soils the matric potentials in trenched and in rooted plots never dropped below −15 bars, the so-called wilting point for many plants. In contrast, in the rooted dump soils, matric potentials well below −15 bars were commonly observed. In trenched or root-free dump soils matric potentials were 1 bar or higher. Thus, on dump soils, the severing of the roots from their trees by trenching made a great difference in water availability. The *Atta* refuse dump soils stimulated the production of a large mass of small diameter tree roots. The effect of the water removal by this root mass, amplified by the unusual water-retention properties of the dump soil and by the four-month dry season, made the dump, in effect, a desert for the seedlings during the dry season.

## Discussion

The various observations made in this study suggest that there is a series of interactions between *Atta* and successional development. What is the impact of succession on *Atta* and of *Atta* on succession? After several years of work on the second question, the first still appears more easily answered.

## The Impact of Succession on *Atta*

My observations in Panama suggest that the abundance of a particular *Atta* species in an area is related to the successional status of the vegetation. With increase in agriculture, the abundance of *Atta* colonies seems to increase. Non-agricultural disturbance can bring similar results. For example, while searching for *Atta* nests in forest on the Atlantic side of the isthmus, I encountered 7.7 nests per kilometer of road and 1.6 per kilometer of forest footpath in similar terrain. Colony abundance seemed to increase where there was a belt of early secondary vegetation in the road right-of-way between the road proper and the forest. A similar pattern was also observed in mid-isthmus.

As succession proceeds, the abundance of colonies seems to decrease with one *Atta* species gradually replacing another. The second species in turn became less abundant until, in very old forests, *Atta* colonies are very hard to find. Decreasing abundance during succession seems to be important in the shifting agriculture practiced by the Boni Bush Negroes in Guiana. Depending on leaf-cutting ant abundance, fallow times ranging from 2 to 20 years were required before land could be reclaimed for farming (Hurault, 1965).

## The Impact of *Atta* on Succession

Ideally, the assessment of the impact of *Atta* on successional development would entail quantification of their influence on energy flow, plant gene flow, and upon nutrient cycling. The number of flowers destroyed by *Atta* is relatively easily determined, but the fraction of the total flower production destroyed in each of the many forest tree species is not. Similarly, assessment of energy and nutrient flows through the colonies is relatively easy, but without data on forest productivity and nutrient cycling with which to interpret the *Atta* data, detailed flow measurements at *Atta* colonies are not very useful. This is because the significance of an organism to an ecosystem is not always directly proportional to either the nutrients or the energy it processes (Kuenzler, 1961). As a selective harvester and point refuse dumper, *Atta* is very likely to have a disproportionately large impact on the ecosystem.

### *Selective Harvest*

Evidence that *Atta colombica* is selective was obtained by placing piles of young and old leaves from seven species on 19 trails of four different nests. The percentage of each pile removed to the fungus garden was recorded after 24 hr. The ants had to cut through the leaf piles to maintain their trails; thus no pile was ignored. Among accepted species, young leaves were consistently preferred (Table 8–1).

Selective harvest by insects is ecologically important (Franklin, 1970). Through the preferential removal of certain leaf age classes, the ants may have

**Table 8–1.** Leaf preference expressed as summed percentages of materials removed from 19 sets of 14 leaf types to four *A. colombica* nests in 24 hr.

| Leaf Source Species[a] | Relative Leaf Age Young | Old |
|---|---|---|
| *Ochroma limonensis* Rowlee | 340 | 10 |
| *Didymopanax morototoni* (Aubl.) Dcne. & Planch. | 225 | 95 |
| *Cecropia* sp. | 0 | 0 |
| *Zanthoxylum* sp. | 260 | 10 |
| *Gustavia superba* (H.B.K.) Berg. | 360 | 260 |
| Palm | 0 | 0 |
| *Mangifera indica* L. | 1,600 | 1,470 |
| Total | 2,785 | 1,845 |

[a] Nomenclature follows Standley (1933).

a greater impact on tree hormone balance, and hence upon tree productivity, than if they removed an equal mass of randomly selected leaves. For example, experiments on *Pinus* have implicated (O'Neil, 1962) and have shown (Onaka, 1950) changes in hormonal activity and in the amount and pattern of growth following removal of younger needles.

### *Refuse Disposal*

As a point refuse dumper, *Atta* has a greater modifying impact on the pattern of nutrient cycling than other herbivores, which tend to be scatter refuse dumpers. The increased mineral nutrient and small root concentrations in refuse dumps strongly suggest that *Atta* activities result in nutrient fluxes through the dumps in excess of background rates for the forest.

The small concentrations of nutrients and roots beneath the *A. colombica* refuse dumps suggest that little if any of the nutrients are lost through deep leaching. The concentrations of roots in the refuse have a negative impact on the seedling population. If the seedlings do not benefit from the increased nutrient fluxes, the nutrients must be taken up by the roots that have proliferated in the dump. However, further study is needed to determine if the leaves harvested by *Atta* are from trees partly rooted in the dump and whether increased nutrient cycling might lead to modified forest structure.

### *Shifting Agriculture*

*Atta*, through its activities as a herbivore and refuse dumper, may also have an impact upon agriculture. The shifting agriculturalist often finds *Atta* among the reasons for moving his fields. For example, Weber (1947) thought that the nomadic habits of Indian farmers on the Orinoco Delta in Venezuela were partly the result of the crop-damaging activities of *A. cephalotes*. Fauterae (1952) advances the same argument with respect to *A. sexdens* and nomadism among

Amazonian natives. Goncalves (1967) endorses this idea because, he says, superstition-free farmers with techniques for combating *A. sexdens* are still pushed from their cultivated lands by the ants. If the fallow time is shortened and ant abundance has not yet subsided from a previous increase, the newly planted crop may be attacked sooner by the ants. This interaction might explain phenomena like the Peruvian Ant Irruption (Smithsonian Center for Short Lived Phenomena, Events Nos. 449 and 498, 1960) in which the *Atta* population was said to have reached plague proportions in agricultural areas. A quantitative investigation of the population dynamics of the ants and shifting agriculturalists as they interact with successional vegetation would be very useful in developing long-term land use management strategies in regions where *Atta* occurs.

## Acknowledgment

This work was carried out while the author was associated with Duke University, Department of Botany and the Smithsonian Tropical Research Institute. The support of a Smithsonian Pre-doctoral Internship and the assistance of NSF GB3698 (Billings and Oosting) and GB12558 (Billings) to Duke University are gratefully acknowledged.

## References

Brown, J. W., and T. A. Wolfschoon G. 1960. Some chemical and physical properties of representative soils of the Republic of Panama. *7th Intern. Congr. Soil Sci.* Madison, Wisc. 4:271–277.

Fauterae, E. 1952. Etudes d'ecologia humaine dans l'aire amazonienne. See Goncalves, 1967.

Franklin, R. T. 1970. Insect influences on the forest canopy. In *Analysis of Temperate Forest Ecosystems*, D. E. Reichle, ed. pp. 86–99. New York: Springer-Verlag.

Goncalves, C. R. 1967. As formigas cortadeiras da Amazonia, dos generos *"Atta"* Fabr. e *"Acromyrmex"* Mayr (Hym., Formicidae). *Atas Simposio Biota Amazonica (Zoologia)* 5:181–202.

Haines, B. L. 1971. Plant responses to mineral nutrient accumulations in refuse dumps of a leaf-cutting ant in Panama. Ph.D. thesis, Duke Univ. (Univ. Microfilms, Ann Arbor, No. 72-317).

Hurault, J. 1965. *La vie materielle des Noirs Réfugiés Boni et des Indiens Wayana du Haut-Maroni (Guyane Francois). Agriculture, economie et habitat.* 142 pp. Paris: ORSTOM.

Jackson, M. L. 1965. *Soil Chemical Analysis.* Englewood Cliffs, N.J.: Prentice-Hall.

Kuenzler, E. J. 1961. Phosphorus budget of a mussel population. *Limnol. Oceanogr.* 6(4):400–415.

Martin, M. M. 1970. The biochemical basis of the fungus-attine ant symbiosis. *Science* 169:16–20.

Martini, J. A. 1966. Chemical, mineralogical, and physical properties of seven surface soils from Panama with special reference to cation exchange capacity and potassium status. Ph.D. dissertation. Cornell Univ. University Microfilms. Ann Arbor, Mich. No. 66-5603.

Onaka, F. 1950. The effects of defoliation, disbudding, girdling and other treatments upon growth, especially radial growth in evergreen conifers (in Japanese, English summary). *Bull. Kyoto Univ. Forests* 18:55–95.

O'Neil, L. C. 1962. Some effects of artificial defoliation on the growth of Jack Pine, *Pinus banksiana* Lamb. *Can. J. Bot.* 40(2):273–280.

Standley, P. C. 1933. The flora of Barro Colorado Island, Panama. *Contr. Arnold Arboretum No. 5.*

Weber, N. A. 1947. Lower Orinoco River fungus-growing ants (Hymenoptera: Formicidae, Attini) *Bol. Entomol. Venez.* 6:143–161.

# Tropical Forest Analysis

Forests are the steady state communities over relatively large portions of the tropical terrestrial landscape. They vary greatly in structure and function, depending upon the periodicity and amount of rainfall and other environmental factors. Ecologists have established the range of biomass, diversity of higher plants, and primary production in these forests, although much more descriptive data are needed in order to predict these characteristics for any local habitat. Fewer data are available on other functional processes such as mineral cycling. Also, even fewer data exist on the role of species populations in control of ecosystem structure and processes. Indeed, this gap between species ecology and ecosystem ecology is a subject worthy of much greater attention by ecologits everywhere.

Data from tropical forests have been used to develop a series of hypotheses on ecosystem structure and function. These include the concepts that diversity of species populations and stability of structure and function are related; that rate processes (such as productivity) are higher in tropical communities; and that mineral conservation mechanisms are better developed in tropical environments.

The following papers touch on these and other concepts. Klinge and his associates provide a summary of their extensive studies of forest in Amazonia, with information on the vegetation biomass, litter fall, and other structural features of the forest. Huttel presents data on root mass in forests in Ivory Coast. These data are especially useful, since usually only the more easily harvested aboveground plant mass is measured. Garg and Vyas give data on litter fall in an Indian deciduous forest, and Malaisse and co-workers provide similar data, with an extensive analysis of litter decomposition in the African Miombo ecosystem. Bernhard-Reversat presents a comparison of input and output of phosphorus, potassium, and calcium in a moist evergreen forest in Ivory Coast. Finally, Karr uses productivity data to account for bird species diversity in tropical forests.

# CHAPTER 9

## Biomass and Structure in a Central Amazonian Rain Forest

H. KLINGE, W. A. RODRIGUES, E. BRUNIG, AND E. J. FITTKAU

### Introduction

Klinge and Rodrigues (1971) analyzed the structure and floristic composition of a central Amazonian rain forest stand northeast of Manaus, state of Amazonas, Brazil. The 0.2-ha study site was on level solid ground, adjacent to the Walter Egler Forest Reserve, which forms the western boundary of a 137,000-ha forest area inventoried by Rodrigues (1967). Structural studies of forest stands on level solid ground of the Manaus area have been conducted by Lechthaler (1956), Takeuchi (1960, 1961), and Aubreville (1961) on plots of 1 ha or less.

Litter production of a rain forest stand in the Walter Egler Forest Reserve was studied by Klinge and Rodrigues (1968) in the vicinity of the biomass estimation plot; litter decomposition was also studied at this location (Klinge, 1974). Stark (1971) and Coutinho and Lamberti (1971) reported on litter decomposition in forests of the lower Rio Negro region. There are no other studies on litter of Amazonian forests or on their biomass. This report summarizes the available data into a synoptic picture of the central Amazonian rain forest.

### Biomass of the Forest

As the technique of destructive sampling and the results will be described in detail elsewhere (Klinge and Rodrigues, 1974), here we will summarize the data on fresh phytomass and dead organic matter (see Tables 9–1 and 9–2), and show in Table 9–3 the vertical organization of dicotyledonous tree and palm phytomass.

It is easy to see that the relatively few trees of the tallest strata A and B represent most of the living aerial tree and palm phytomass (85.6 percent) and

**Table 9–1.** Phytomass (fresh matter) of a central Amazonian rain forest (metric tons per hectare).

| | | | |
|---|---|---|---|
| Leaf matter: | | | |
| Dicotyledonous trees above 1.5 m height | 14.1 | | 1.9% |
| Palms above 1.5 m height | .34 | | 0.5% |
| Total above 1.5 m height | 17.5 | 17.5 | 2.4% |
| Plants below 1.5 m height | 0.6 | 0.6 | 0.1% |
| Total leaf matter of dicotyledonous trees and palms | 18.1 | 18.1 | 2.5% |
| Branches and twigs: | | | |
| Dicotyledonous trees above 1.5 m height | 202.2 | | 27.5% |
| Plants below 1.5 m height | 0.2 | | — |
| Total branches and twigs | 202.4 | 202.4 | 27.5% |
| Stems: | | | |
| Dicotyledonous trees above 1.5 m height | 465.5 | | 63.3% |
| Palms above 1.5 m height | 2.1 | | 0.3% |
| Total above 1.5 m height | 467.6 | 467.6 | 63.6% |
| Plants below 1.5 m height | 0.6 | 0.6 | 0.1% |
| Total stems | 468.2 | 468.2 | 63.7% |
| Other plants (total phytomass): | | | |
| Lianas | 46.0 | | 6.3% |
| Epiphytes | 0.1 | | — |
| Parasites | 0.1 | | — |
| Total other plants | 46.2 | 46.2 | 6.3% |
| Total aerial phytomass | | 743.9 | 100% |
| Root mass: | | | |
| Fine roots | 49.0 | | |
| Other roots | 206.0 | | |
| Total root mass | 255.0 | 255.0 | |
| Total phytomass | | 989.9 | |

that the trees of lower strata, such as palms, lianas, vascular epiphytes, and parasites, make up only very low percentages.

Tree leaf matter is 14 tons/ha. This low figure indicates, as discussed by Fittkau and Klinge (1973), that there is no broad base for feeding phytophagous animals. According to our estimate, this group of animals represents only 30 kg/ha (14 percent) compared with 15 kg/ha (7 percent) of carnivorous animals, or 165 kg/ha (79 percent) of soil fauna zoomass. When evaluating the amount of leaf matter in comparison with phytophagous zoomass, it should be taken into consideration that the phytophagous zoomass estimate includes a number of animals that feed on fruits and plant liquids, while others are omnivorous. That portion of the phytophagous zoomass that feeds on green leaf matter exclusively is consequently extremely low.

Wood phytomass represents 97.4 percent of aboveground tree and palm phytomass, and root phytomass (Klinge 1973a) accounts for 25.8 percent of total phytomass of the stand. The wood fraction of dead organic matter (Table

**Table 9–2.** Dead organic matter (dry matter) of a central Amazonian rain forest.

| Fraction | Metric Tons per Hectare | Percentage | |
|---|---|---|---|
| | Mean Range | Total | Total |
| Standing dead wood: | | | |
| (25 ind./0.2 hr) | | 7.6 | 2.7 |
| Dead wood of litter layer: | | | |
| Stems (10 ind./0.2 ha) | 12.5 | | |
| Branches | 5.7 ( 0.4 –18.2) | | |
| | 18.2 (12.9 –30.7) | 18.2 | 6.5 |
| Fine litter: | | | |
| Leaf matter | 4.0 ( 2.4 – 5.9) | | |
| Wood matter | 3.0 ( 1.5 – 4.9) | | |
| Fruits, flowers | 0.2 ( 0.01– 0.6) | | |
| | 7.2 ( 3.9 –11.4) | 7.2 | 2.5 |
| Soil organic matter: | | 250 | 88.3 |
| Total dead matter | | 283 | 100 |

**Table 9–3.** Vertical organization of aerial phytomass of dicotyledonous trees and palms of a central Amazonian rain forest.

| Stratum | Individuals/Ha Dicotyledonous Trees | | Aerial phytomass (tons/ha) Palms | Leaves | Branches + Twigs | Stems | Total | Total % |
|---|---|---|---|---|---|---|---|---|
| A | 50 | | 0 | 2.3 | 48.7 | 139.2 | 190.2 | 27.6 |
| B | 315 | | 0 | 7.1 | 123.1 | 269.3 | 399.5 | 58.0 |
| $C_1$ | 760 | | 15 | 3.9 | 26.1 | 47.3 | 77.3 | 11.2 |
| $C_2$ | 2,765 | | 155 | 2.0 | 3.6 | 10.0 | 15.6 | 2.3 |
| D | 5,265 | | 805 | 2.2 | 0.7 | 1.8 | 4.7 | 0.7 |
| E | | 83,650 | | 0.6 | 0.2 | 0.6 | 1.4 | 0.2 |
| Total | | 93,780 | | 18.1 | 202.4 | 468.2 | 688.7 | 100 |
| % | | | | 2.6 | 29.4 | 68 | 100 | |

9–2) (21.2 tons dry matter/ha) is also large, but is much less than the soil organic matter. The overwhelming dominance of wood matter in the phytomass is also reflected by a relatively large proportion of zoomass representing woodeaters. This fraction of zoomass is about equal to the phytophagous zoomass, but is composed by only two types of life forms: termites and coleopterous larvae.

A comparison between tree leaf litter fall [4.8 to 6.4 tons dry matter/ha/year (Klinge and Rodrigues, 1968)] and the tree leaf mass of the stand shows that approximately 80 percent of leaf matter returns annually to the soil. Because the forest is evergreen this amount of litter is replaced by new leaves. The high rate of leaf matter turnover suggests the intensity of matter and nutrient cycling in this type of lowland tropical rain forest as well as the importance of dynamic processes.

Conversion of the total phytomass (Table 9–1) into dry matter gives a total of about 450 to 500 tons/ha. Adding the dead organic matter results in a total forest dry matter of about 750 tons/ha, which is within the norms for tropical forests calculated by Whittaker and Woodwell (1971). This similarity of the central Amazonian rain forest in the Manaus area with other tropical rain forests is unexpected, because the Manaus area of central Amazonia is, in terms of limnological and geochemical conditions (Fittkau, 1971a, 1971b), extremely poor.

The abundance of both living and dead phytomass show that a forest growing on a poor soil, such as the pale yellow latosol of the Manaus area (Anon., 1969), may build up a large amount of living phytomass and produce a large quantity of dead matter and detritus. If there are few or no nutrient reserves in the acidic leached soil (Klinge, 1973a, 1974) there must be a strong recycling of nutrients via the detritus pathway and via canopy leaching, in order to guarantee the mineral nutrition of the vegetation.

One prerequisite for recycling is a high production of detritus. Detritus production should be the larger as the detritus is poorer in nutrients. Another prerequisite is a rapid decomposition of the detritus. Compared with other tropical forests the central Amazonian litter is poor in nutrients (Klinge and Rodrigues, 1968), and its decomposition is rapid (Klinge, 1974; Stark, 1971; Coutinho and Lamberti, 1971). A third prerequisite for establishing an effective nutrient recycling is the immediate absorption of nutrients liberated from detritus decomposition or added by crown leaching and stem flow in order to minimize nutrient loss by leaching. This prerequisite appears to hold true in our area, as judged from the analyses of its natural waters, which have been classified as "slightly contaminated distilled water" (Fittkau et al., 1974). Furthermore, it seems that fungi, which are well adapted to the acidic humid environment, play an important role in both the rapid decomposition of nutrient-poor forest detritus and in the absorption of nutrients via mycorrhiza, as pointed out by Stark (1969, 1971) and Went and Stark (1968).

## Stem Wood Volume of the Forest

The total stem wood volume of trees 15 or more cm in diameter is 385 $m^3$/ha (Table 9–4); the mean wood volume per tree is about 1 $m^3$. A similar figure for trees of 18 or more cm dbh was given by Lechthaler( 1956), whose corresponding figure for trees of 28 or more cm dbh was 292 $m^3$/ha. For this dbh class we found a volume of 304 $m^3$/ha. Aubreville (1961), referring to unpublished data for A. Ducke Forest Reserve by R. Onety Soares, mentioned an average stem wood volume of 142.4 $m^3$/ha for trees of 15 or more cm dbh, and for trees 25 cm or more in diameter, an average of 108 $m^3$/ha. These latter data agree well with those given by Rodrigues (1967).

The over 400 trees/ha (dbh $>$ 15 cm) represents 90.7 percent of the total

**Table 9–4.** Stem wood volume and phytomass per plant families represented by individuals with 15 or more cm dbh in the central Amazonian rain forest.

| Plant Family | Stem Wood Volume | | Phytomass | |
|---|---|---|---|---|
| | $m^3$/ha | % | tons/ha | % |
| Leguminosae | 87.1 | 22.7 | 153.1 | 24.5 |
| Euphorbiaceae | 57.3 | 14.9 | 89.0 | 14.3 |
| Sapotaceae | 42.6 | 11.1 | 76.6 | 12.3 |
| Vochysiaceae | 40.0 | 10.4 | 63.9 | 10.2 |
| Apocynaceae | 24.4 | 6.3 | 25.8 | 4.1 |
| Lecythidaceae | 20.7 | 5.4 | 41.1 | 6.6 |
| Indeterminata | 39.9 | 10.1 | 61.6 | 9.9 |
| Total of 14 other families (Myristicaceae, Violaceae, Myrtaceae, Palmae, Olacaceae, Icacinaceae, Burseraceae, Humiriaceae, Lauraceae, Nyctaginaceae, Annonaceae, Celastraceae, Moraceae, Sterculiaceae) | 73.3 | 19.1 | 113.4 | 18.1 |
| Total | 385.3 | 100 | 624.5 | 100 |

aboveground phytomass. Leguminosae occur most frequently with Euphorbiaceae, Sapotaceae, and Vochysiaceae also occurring often. Together these families represent about 60 percent of the total stem wood volume.

The differences described above between the results of a few authors working in the same region, each only some kilometers apart from the other, show how far we are from a well-founded knowledge of the central Amazonian forest (Klinge, 1973b).

## Structure of the Forest

The high number of trees and palms of strata C, D, and E, which have diameters mostly below 10 cm, contribute so little to the total phytomass that it seems preferable to study only the economically important tree fractions of strata A and B, but on a much larger surface than we did. By such a restriction, however, we would neglect a very large portion of plants belonging to a high number of species, and the study would then not be directed toward the ecosystem, its functions, and its populations. In other words, the structure of the forest would not be considered.

In strata A and B Leguminosae account for 20 percent of the individuals; Sapotaceae and Euphorbiaceae each, 18 percent; Lecythidaceae, 9 percent; Vochysiaceae, 6 percent; Burseraceae, Violaceae, Nyctaginaceae, and Humiriaceae, each 3 percent; and representatives of 11 other families each, 1.5 percent. Some 50 species of 20 families are thus involved in the formation of the high canopy, but the total plant population comprises easily over 600 species/ha of which about 500 species were identified in the 0.2-ha plot. These lower strata plants belong to over 30 plant families. Therefore 90 percent of the forest species are in the lower strata C and D.

Black et al. (1950), Cain et al. (1956), Pires et al. (1953), and others working in the Amazon region found a similar richness in plant families and species. Considering the Manaus area, Rodrigues (1967) found an average of 65 species/ha and a total of 470 species belonging to 47 plant families (dbh: 25 or more cm). Lechthaler (1956) observed more than 30 families with 75 species per hectare (dbh: 8 cm or more). Takeuchi (1960, 1961) sampled 32 genera and 15 families per hectare among the trees over 10 cm dbh (10 genera not being identified).

As stated elsewhere (Klinge, 1974; Fittkau and Klinge, 1973), 14 families, each with nine or more species, form the most conspicuous group, as judged by the number of individuals and species. This group comprises 74 percent of species, or 86 percent of individuals, counted on the plot. Richest in species are the Leguminosae, which occur in all strata. Next are Sapotaceae, Lauraceae, Chrysobalanaceae, Rubiaceae, and Myristicaceae, which are typical components of strata $C_2$ and D; they also occur in upper strata. Myristicaceae, Burseraceae, and Annonaceae are found in all strata except A, like Lecythidaceae, Moraceae, and Violaceae. Low individuals of Euphorbiaceae are very frequent, but there are also very tall individuals. Palms are a very important component of strata C and D. Representatives of 13 other plant families occur with a few individuals in these lower strata, and a group of 13 other families is typical for stratum D exclusively. In terms of biomass, Leguminosae, Euphorbiaceae, Sapotaceae, and Vochysiaceae, which are families rich in species and individuals, are the most important. They are also important families from an economical point of view because of the high proportions of trees 15 or more cm in diameter. Similar observations on the dominance of families in Amazon forests, regarding their numbers of species and individuals and biomass, were made by all authors working in the region. The high proportion of Euphorbiaceae in our plot, however, is a conspicuous feature of the stand.

Taking into account the high number of species and individuals, it can hardly be expected that rules of species distribution within the boundaries of the small plot will be found. It will be the task of future studies to clarify these distribution patterns of the species involved and the interrelationships between the patterns of different species.

## References

Anon. 1969. *Os solos da area Manaus-Itacoatiara.* Ensaios 1. Estado do Amazonas, Manaus.

Aubreville, A. 1961. *Etude ecologique des principales formations vegetales du Bresil et contribution a la connaissance des forets de l'Amazonie bresilienne.* 268 pp. Nogent-sur-Marne: C.T.F.T.

Black, G. A., Th. Dobzhansky, and C. Pavan. 1950. Some attempts to estimate species diversity and population density of trees in Amazonian forests. *Bot. Gaz.* 111: 413–425.

Cain, St A., G. G. de Oliveira Castro, J. Murca Pires, and N. T. da Silva. 1956. Application of some phytosociological techniques to Brazilian rain forest. *Am. J. Bot.* 43:911–941.

Coutinho, L. M., and A. Lamberti. 1971. Respiracao edafica e produtividade primaria numa comunidade amazonica de mata de terra-firme. *Ciencia Cultura* 23(3): 411–419.

Fittkau, E. J. 1971a. Esboco de uma divisao ecologico-paisagistica da regiao amozonica. In *II Simposio y Foro de Biologia Tropical Amazonica, Florencia (Caqueta) y Leticia (Amazonas) 1969*, J. M. Idrobo, ed. Bogota.

———. 1971b. Ökologische Gliederung des Amazonasbeckens auf geochemischer Grundlage. *Münst. Forsch. Geol. Paläontol.* 20/21:35–50.

———, and H. Klinge. 1973. On biomass and trophic structure of the Central Amazonian rain forest ecosystem. *Biotropica* 5(1):2–15.

———, U. Irmler, W. Junk, F. Reiss, and G. W. Schmidt. 1974. Productivity, biomass and population dynamics in Amazonian water bodies. In *Tropical Ecological Systems*, Chapter 20, F. Golley and E. Medina, eds. New York: Springer-Verlag.

Klinge, H. 1974. Biomasa y materia organica del suelo en el ecosistema de la pluviselva centro-amazonica. Paper presented to IV Congr. Latino Americano y II Reunion Nacional de la Ciencia del Suelo, Maracay 1972. *Acta Cientif. Venezolano* (in press).

———. 1973a. Root mass estimation in lowland tropical rain forests of Central Amazonia, Brazil. I. Fine root masses of a pale yellow latosol and a giant humus podzol. *Trop. Ecol.* 14:29–38.

———. 1973b. Über den Terra firme-Wald von Manaus. Paper presented to Symp. Geogr. Inst. Max Planck Inst. Limno. Saarbrücken.

———, and W. A. Rodrigues. 1968. Litter production in an area of Amazonian terra firme forest. I, II. *Amazoniana, Kiel* 1(4):287–310.

———. 1971. Materia organica e nutrientes na mata de terra firme perto de Manaus. *Acta Amazonica, Manaus* 1(1):69–72.

———. 1974. Phytomass estimation in a central Amazonian rain forest. Paper presented to IUFRO Working Group on Forest Biomass. Vancouver 1973. IURO Biomass Studies, Orono 339–350.

Lechthaer, E. 1956. Inventario das arvores de una hectare de terra firme da zona Reserva Florestal Ducke, Municipio de Manaus. *Amazonia, Rio de Janeiro Bot.* 3, 10 pp.

Pires, J. Murca, Th. Dobzhansky, and G. A. Black. 1953. An estimate of the number of species of trees in an Amazonian forest community. *Bot. Gaz.* 114:467–477.

Rodrigues, W. A. 1967. Inventario florestal piloto ao longo da estrada Manaus—Itacoatiara, Estado do Amazonas: Dados preliminares. In *Atas Simposio sobre a biota amazonica, Belem 1966*, H. Lent, ed. Vol. 7, pp. 257–267. Rio de Janeiro: Conselho Nacional de Pesquisas.

Stark, N. M. 1969. Mycorrhiza and nutrient cycling in the tropics. Proc. First North American Conference on Mycorrhizae, April 1969. *Misc. Publ. 1189*, U.S. Dept. Agr., Forest Serv. Separate No. FS-303:228–229.

———. 1971. Nutrient cycling. II. Nutrient distribution in Amazonian vegetation. *Trop. Ecol.* 12(2):177–201.

Takeuchi, M. 1960. A estrutura da vegetacao na Amazonia. I. A mata pluvial tropical. *Bol. Mus. 'E.Goeldi.' Bot.* 6:1–17.

Takeuchi, M. 1961. The structure of the Amazonian vegetation. II. Tropical rain forest. *J. Fac. Sci. Univ. Tokyo Sect. III Bot.* 8(1–3):1–26.

Went, F. W., and N. Stark. 1968. The biological and mechanical role of soil fungi. *Proc. Natl. Acad. Sci.* 60:497–504.

Whittaker, E. H., and G. M. Woodwell. 1971. Measurement of net primary productivity of forest ecosystems. *Proc. Brussels Symp.* pp. 159–175. UNESCO-IBP 1969. Paris.

# CHAPTER 10

## Root Distribution and Biomass in Three Ivory Coast Rain Forest Plots

CHARLES HUTTEL

Little information on root distribution and biomass in tropical rain forests is available (Bartholomew et al., 1953; Greenland and Kowal, 1960). Rodin and Bazilevich (1967) gave an average estimated biomass value of roots in tropical forests; more recently, Jenik (1971) published data on root systems observed in Ghana forests.

For our ecological program, which includes studies on vegetation structure and water balance, it was necessary to have an estimate of the biomass and the distribution of the underground part of the forest.

### Site Description

Roots were sampled at three stations in the evergreen forest of lower Ivory Coast. Two samples were taken in the Banco forest located near Abidjan. This forest is representative of the Turraeantho-Heisterietum type developed on sandy soils formed of tertiary continental sediments (Mangenot, 1955). Climatic, floristic, and some edaphic factors of this site are described elsewhere (Huttel, 1972; Anon., 1969). Two plots were sampled, one on the plateau and the other in the thalweg. The textural analysis of the soils of these two plots is described in Figure 10–1. Clay and silt do not exceed 25 percent, and the coarse sand fraction represents more than 50 percent over the whole profile. The thalweg soil is richer in clay than the plateau soil. Elements over 2 mm in diameter are absent or negligible in this type of soil.

Another root sample was taken in the Yapo forest situated 50 km north of Abidjan, which is another rain forest with soils richer in fine elements. While many species are common to the two forest types, the Yapo type has more species. It can be characterized by forest floor Cyperacae, *Mapania*, and some species of *Diospyros* (Mangenot, 1955; Guillaumet and Adjanohoun, 1971). Clay content of the soil is high (30 to 50 percent), and the two silt classes are

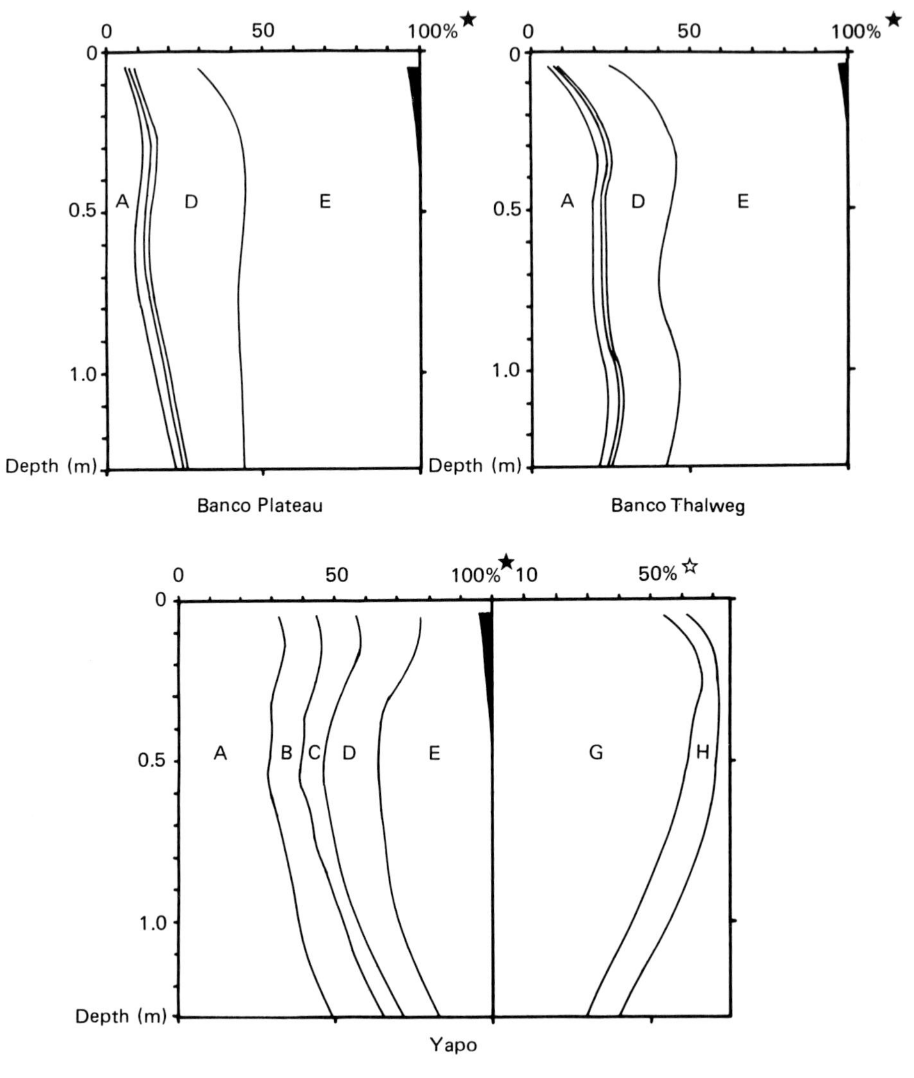


*A*—Clay, particle diameter under 2 *r.*
*B*—Fine silt, particle diameter between 2 and 20 *r.*
*C*—Coarse silt, particle diameter between 20 and 50*r.*
*D*—Fine sand, particle diameter between 50 *r* and 0.2 mm.
*E*—Coarse sand, particle diameter between 0.2 and 2 mm.
*F*—Organic matter (in black).
*G*—Fine gravel, particle diameter between 2 and 20 mm.
*H*—Coarse gravel, particle diameter over 20 mm.
★—Sifted soil weight percentage. ☆—Total soil weight percentage.

**Fig. 10–1.** Texture of the soils of the three plots.

important. The sand proportion reaches a maximum of 50 percent and the coarse sand fraction is under 40 percent. Another difference in the two soils is the high proportion of sandstones or quartzitic gravels present in the parent rock or fragments of an old ferrallitic crust. The gravel fraction, expressed in volume percentage, varies from 20 percent in the top horizon to a maximum of 44 percent in the 30- to 50-cm deep layer and decreases to 23 percent in the lowest investigated layer (130-cm depth). Similarly, the fine soil fraction (all the particles under 2 mm in diameter) varies from 20 percent to a minimum of 10 percent and increases with depth to 23 percent in the last horizon. The porosity is the same in the three plots, varying from 55 to 70 percent in the first layer to 45 percent in deeper layers.

## Methods

The methods used in the study included boring and unearthing roots. The boring technique was developed by Bonzon and Picard (1969). An engine takes core samples without borer rotation by cutting roots with the sharp edge of a percussion-driven auger. The apparatus is strong enough to pass through stones and large roots. Below a depth of 130 cm, which is the limit of this borer, we used shorter horizontal borings starting from the sides of excavations. Under 40 cm depth there is no difference in the results between vertical and horizontal borings (Picard, 1969).

Samples were taken every 10 cm to a depth of 50 cm, and then every 20 cm to 130-cm depth. Horizontal samples were taken every 20 cm to 250-cm depth. The vertical borer takes 237 $cm^3$ soil in a 10-cm progression.

The core samples were washed under a low pressure water jet on a sieve, and the residue handpicked to separate roots from coarse sand and gravels and dead plant detritus. The core samples from clay-rich soils were dried and treated with a sodium chloride solution before washing. The roots were dried and weighed. Root weights are expressed in grams or milligrams in 100 $cm^3$ of soil.

The method of unearthing the roots was carried out at Banco plateau by two students, César and Menaut. Starting from tree trunks they followed with pick and shovel the roots in the first 30-cm depth. All the roots over 1 cm in diameter in this layer were drawn on a precise map. By planimetry it was possible to know the root volume, and with specific weight values the root weight in the plot could then be calculated.

## Root Vertical Distribution

In all three cases roots decrease quickly with depth (Figures 10–2 and 10–3). There is a concentration of large roots in the upper soil layers; the 0- to 10-cm horizon contains 21 percent of the roots of the whole profile at Banco plateau,

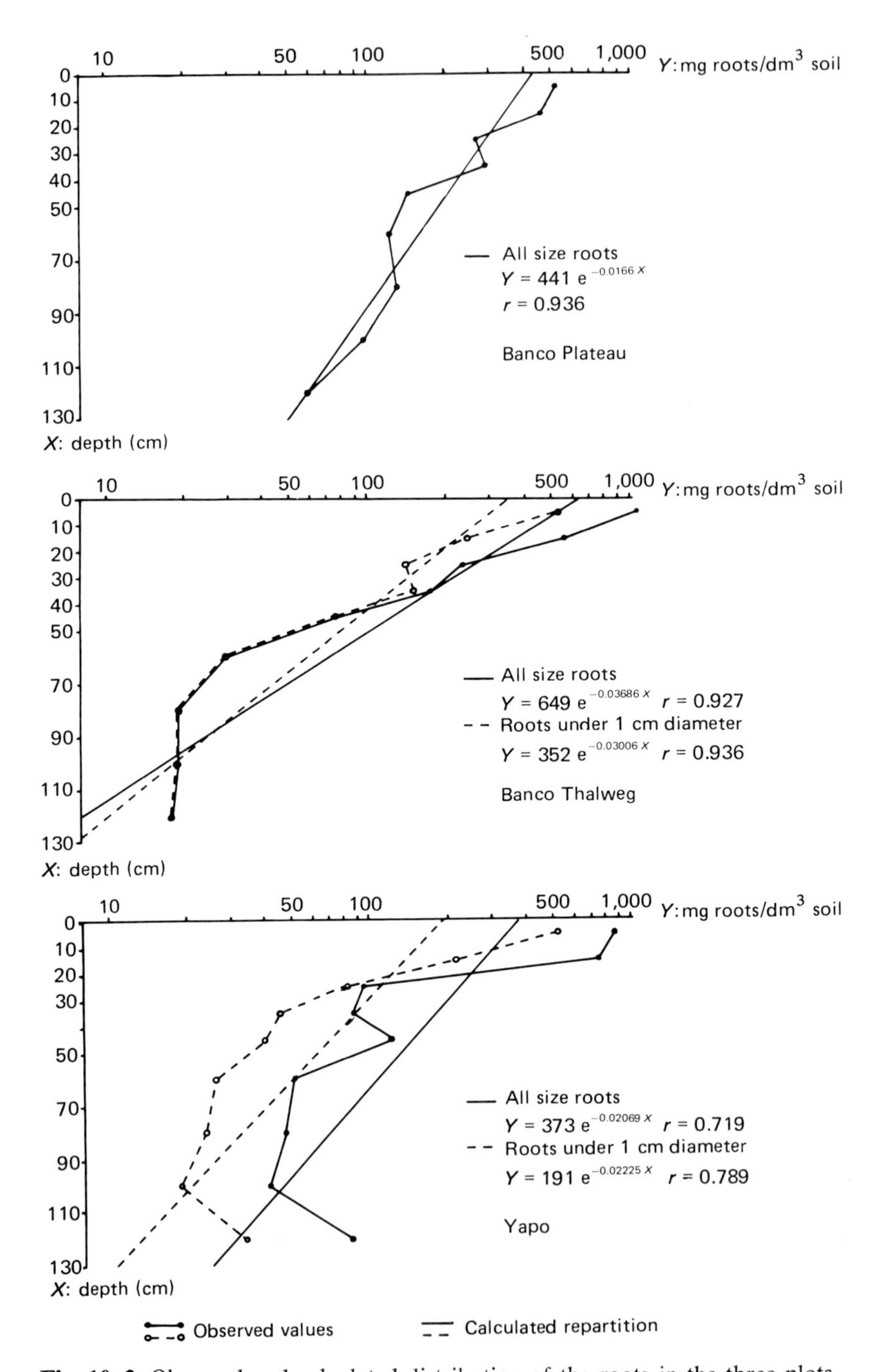


**Fig. 10–2.** Observed and calculated distribution of the roots in the three plots.

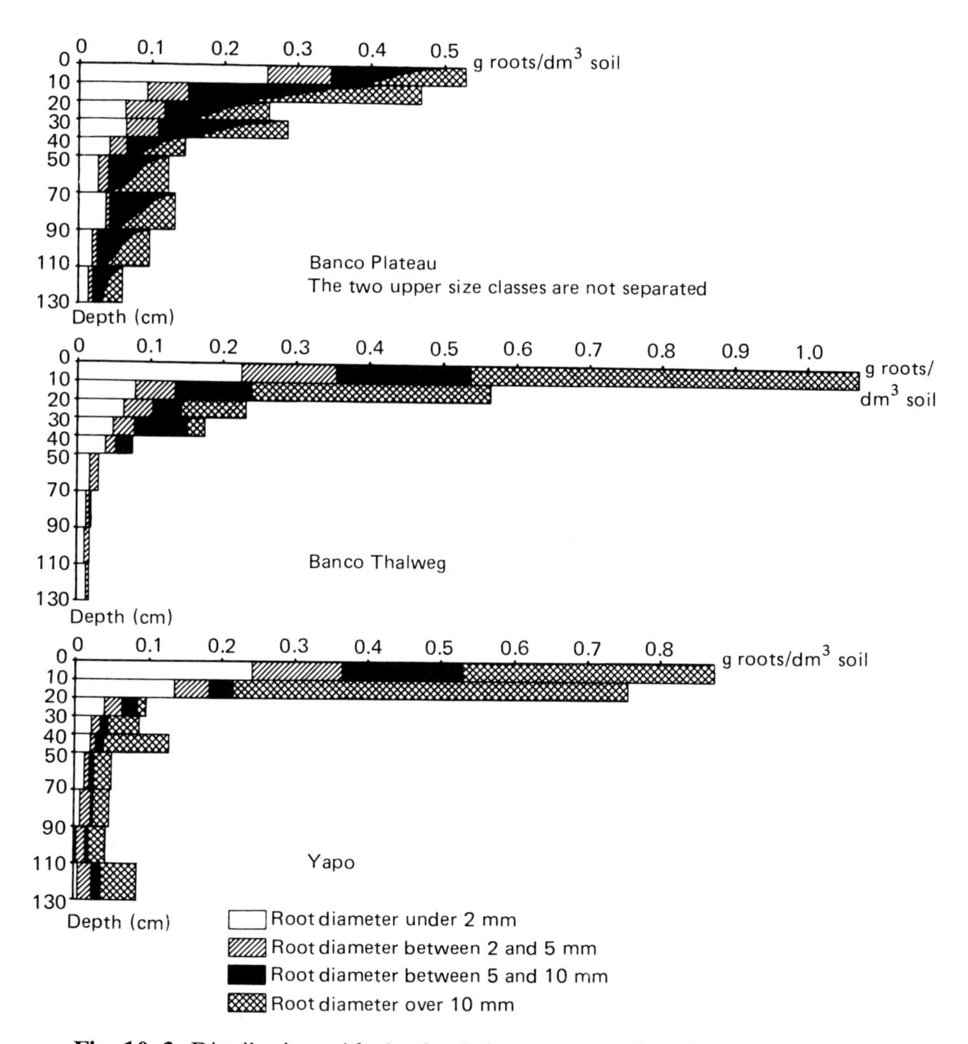


**Fig. 10–3.** Distribution with depth of the roots sorted in size classes.

37 percent at Banco thalweg, and 47 percent at Yapo. Half of the roots are located in the first 30 cm at Banco plateau, in the first 15 cm at Banco thalweg, and in the first 10 cm at Yapo. At Banco plateau and Banco thalweg the root quantity found in the 130- to 250-cm layer represents 5 to 6 percent of the root weight of the whole profile.

The observed root distribution to 130 cm was used to calculate the relation between depth and root abundance. The form of the curve suggests an adjustment to an exponential function and the calculation confirms this hypothesis. Figure 10–2 shows the observed and calculated curves. The correlation coefficient is good, and $\alpha$ is in all cases under 0.05. If the root weights under 130 cm are used, we also obtain good values for $r$: 0.968 at Banco plateau and 0.719 at Banco thalweg ($\alpha$ less than 0.01).

However, the regression curves do not represent suitably the observed distribution (linearity test). There are two zones with an excess of roots, one in the first 20 cm and another in the deepest layers. These two zones delimit a zone of root depression. The upper zone of abundant roots may be related to the relative richness of exchangeable cations (Table 10–1) in the surface soil. The lower zone seems to be related to an increase in the available water at Banco plateau and Yapo. The deficit zone corresponds to a soil layer poor in mineral elements and with unfavorable hydric characteristics.

Roots were also sorted into size classes, which were distributed as shown in Figure 10–3. The quantity of smaller absorptive roots (less than 2 mm in diameter) decreases regularly with depth, and their distributions can be fitted to exponential curves with correlation coefficients between 0.913 and 0.921. Over the whole profile the proportion of roots of this size is about 25 percent.

The most remarkable observation was the absence of roots over 5 mm in diameter below 50 cm at Banco thalweg. It seems that growth in root thickness is inhibited below this level. However, considering the whole profile, the proportion of roots over 5 mm diameter (61 percent) is comparable to that observed at the two other sites (59 percent at Banco plateau and 65 percent at Yapo).

**Table 10–1.** Some analytical characteristics of the soils.

| Site | Depth (cm) | | | | | | | | |
|---|---|---|---|---|---|---|---|---|---|
| | 0<br>10 | 10<br>20 | 20<br>30 | 30<br>40 | 40<br>50 | 50<br>70 | 70<br>90 | 90<br>110 | 110<br>130 |
| Banco plateau | | | | | | | | | |
| E. C. | 1.05 | 0.40 | 0.25 | 0.42 | 0.15 | 0.23 | 0.24 | 0.05 | 0.10 |
| B. S. | 8.0 | 5.7 | 3.7 | 6.8 | 2.7 | 4.2 | 4.8 | 1.2 | 2.3 |
| A. W. | 4.0 | 2.9 | 3.0 | 4.0 | 5.3 | 4.1 | 4.1 | 5.5 | 6.7 |
| Banco thalweg | | | | | | | | | |
| E. C. | 0.51 | 0.15 | 0.19 | 0.45 | 0.11 | 0.68 | 0.24 | 0.55 | 0.39 |
| B. S. | 10.8 | 3.6 | 3.9 | 10.3 | 2.8 | 17.1 | 6.4 | 19.9 | 10.0 |
| A. W. | 5.5 | 6.0 | 7.4 | 6.2 | 8.9 | 7.5 | 5.0 | 4.7 | 4.4 |
| Yapo | | | | | | | | | |
| E. C. | 2.74 | 0.87 | 0.61 | 0.42 | 0.62 | 0.41 | 0.58 | 0.19 | 0.35 |
| B. S. | 19.6 | 12.2 | 10.7 | 7.6 | 11.0 | 9.5 | 16.7 | 4.3 | 4.9 |
| A. W. | 9.8 | 11.6 | 14.9 | 15.5 | 11.4 | 10.8 | 11.3 | 11.4 | 13.1 |

E. C. = Exchangeable cations (me/100 g soil).
B. S. = Degree of base saturation (%).
A. W. = Available water (% of dry soil).

Leaving Banco thalweg aside, the proportion of roots above 5 mm in diameter in each layer is not related to the depth. The average is about 62 percent, and at the very most we can note a significant difference between this average and the values observed at Banco plateau in the first layer and at Yapo in the third one. The proportion of the different root size classes is stable in spite of the differences of soils and vegetation between the sites.

## Root Biomass

The root weights in soil volumes can be converted to root biomass. We found 24.8 tons/ha of roots at Banco plateau, 22.9 tons/ha at Banco thalweg, and 24.0 tons/ha at Yapo. The confidence intervals calculated from 40 profiles at the 0.05 level are, respectively, 27, 21, and 31 percent of the total biomass. These values illustrate the great variability between the profiles; the richest profile can hold a root weight 40 times as large as the poorest one. However, the three values for root biomass do not differ significantly, even though substantial differences in soil and vegetation were observed between sites.

The method of unearthing roots gives yet another result. Figure 10–4 is a simplified map of root distribution drown from the data of César and Menaut. By planimetry and data obtained on root volumes, the amount of roots present in the top 30 cm was calculated to be 1.93 $m^3$ in 175 $m^2$, or about 110 $m^3$/ha. Converting volume to weight, using the specific weights of roots measured on the plots, we obtain a root biomass of 49 tons/ha, or about twice the values obtained by boring.

While these biomass values are less than the estimate given by Jenik (1971), they are nearer the data of Rodin and Bazilevich (1967). Even so, the two

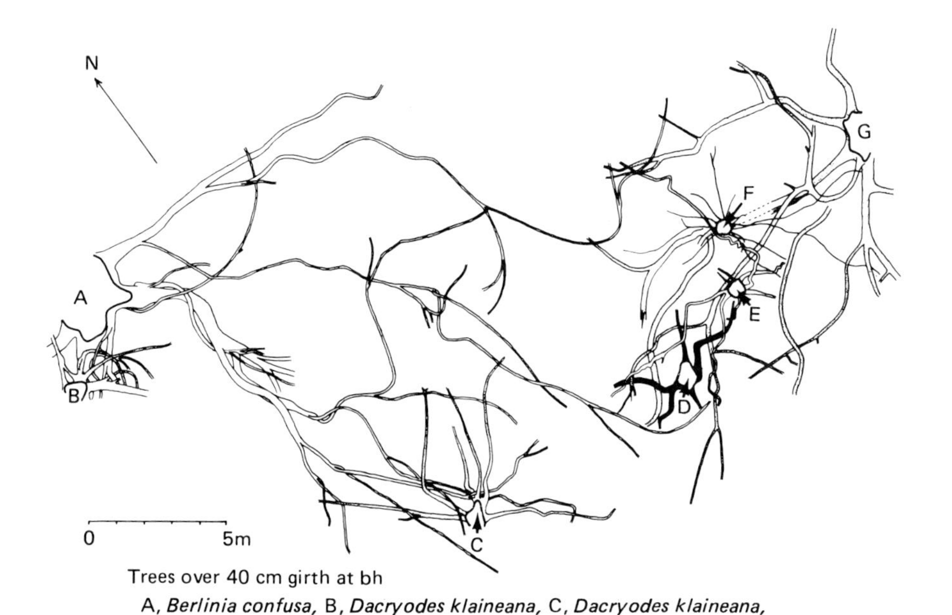


**Fig. 10–4.** Map of the roots in the top 30 cm at Banco plateau. (Simplified after César and Menaut.)

methods underestimate root biomass since neither takes into account the taproots. The method of unearthing also leaves out all the root under 1 cm in diameter.

## References

Anon. 1969. Nouvelles du Programme Biologique International (Participation francaise). Etudes sur la productivité primaire de la forêt dense humide sempervirente de basse Côte d'Ivoire. *Oecol. Plant.* 4:321–324.

Bartholomew, W. V., J. Meyer, and H. Laudelout. 1953. Mineral nutrient immobilization under forest and grass fallow in the Yamgambi region, with some preliminary results on the decomposition of plant material on the forest floor. *Publ. INEAC Ser. Sci. No. 57.* 27 pp.

Bonzon, B., and D. Picard. 1969. Matériel et méthodes pour l'étude de la croissance et du développement en pleine terre des systèmes racinaires. *Cahiers Orstom Sér. Biol. No. 9* 3–18.

Greenland, D. J., and J. I. L. Kowal. 1960. Nutrient content of the moist tropical forest of Ghana. *Plant Soil* 12:154–174.

Guillaumet, J.-L., and E. Adjanohoun. 1971. La végétation de la Côte d'Ivoire. In: "Le milieu naturel de la Côte d'Ivoire." *Mém. Orstom No. 50* 157–263.

Huttel, C. 1972. Estimation du bilan hydrique dans une forêt sempervirente de basse Côte d'Ivoire. Isotopes and radiation in soil-plant relationships including forestry. *IAEA Proc. Symp. Vienna 1971*, 439–452.

Jenik, J. 1971. Root. Root structure and underground biomass in equatorial forests. Productivity of forest ecosystems. *Proc. Brussels Symp. 1969*, 323–331.

Mangenot, G. 1955. Etudes sur les forêts des plaines et plateaux de la Côte d'Ivoire. *IFAN, Etud. éburnéenes* 4, 5–61.

Picard, D. 1969. Comparaison de deux techniques de prelèvement d'echantillons de racines. *Cahiers Orstom Sér. Biol. No. 9* 19–31.

Rodin, L. E., and N. I. Bazilevich. 1967. *Production and Mineral Cycling in Terrestrial Vegetation.* 288 pp. Edinburgh and London: Oliver and Boyd.

# CHAPTER 11

# Litter Production in Deciduous Forest Near Udaipur (South Rajasthan), India

R. K. GARG AND L. N. VYAS

## Introduction

Litter on the forest floor affects the moisture status, runoff pattern, and nutritional character of the land. Yet there are few data on the annual litter production in tropical forests (Bray and Gorham, 1964). In India annual litter production in deciduous and evergreen forests has been studied by Puri (1953), Upadhyaya (1955), Singh (1968), Seth et al. (1963), and Subba Rao et al. (1972). No data are available from the semiarid region of Rajasthan.

The present study was conducted in Kewara-Nal forest situated 16 km southeast of Udaipur (24° 35′N latitude and 75° 49′E longitude) at an altitude of 587 m above sea level. The climate of the area is typical monsoonal, with an average annual rainfall of 660 mm distributed over 38 to 41 rainy days. Mean maximum temperatures (38.6°C) are generally highest in April, May, and June. The lowest mean minimum temperature (7.8°C) is recorded in the month of January.

The soils of the area form a part of the Great-Indo-Gangetic plain, and are black or red; they are sandy-clay-loam in texture, with a pH varying from 6.6 to 8.6

Leaf litter fall was estimated by randomly placing 20 traps (50 by 50 cm) 1.35 m above the ground throughout the forest from September 1971 to March 1972. The traps were provided with muslin cloth funnels and a heavy stone was tied to the base of the cloth to minimize wind disturbance. The traps were harvested twice a month. The litter was separated by species and oven dried at 85°C.

## Observations

Two groups of species could be recognized on the basis of their leaf fall (Table 11–1).

**Table 11–1.** Contribution of leaf litter by different species in Kewara-Nal forest community ($g/m^2$), October 1971–March 1972.

| Plant species | Oct. 71 | Nov. 71 | Dec. 71 | Jan. 72 | Feb. 72 | Mar. 72 | Total |
|---|---|---|---|---|---|---|---|
| *Aegle marmelos* | — | 1.440 | 8.000 | 2.244 | — | 0.044 | 11.728 |
| *Albizzia lebbek* | — | — | — | 0.444 | 0.444 | 0.888 | 1.776 |
| *Anogeissus latifolia* | 0.222 | 13.200 | 45.199 | 13.932 | 4.444 | 0.901 | 77.898 |
| *Bauhinia purpurea* | — | 0.400 | 0.977 | 0.533 | 1.644 | 0.009 | 3.563 |
| *Boswellia serrata* | 2.222 | 72.000 | 40.177 | 2.577 | 0.444 | 0.122 | 117.542 |
| *Butea monosperma* | 0.888 | — | — | 1.333 | 3.555 | — | 5.776 |
| *Celastrus paniculata* | — | — | — | 0.889 | — | — | 0.889 |
| *Dalbergia sissoo* | 0.200 | 11.200 | 0.644 | 4.089 | — | 1.111 | 18.244 |
| *Grewia tiliaefolia* | 0.044 | 2.700 | 2.844 | 0.711 | — | — | 6.299 |
| *Lannea coromandelica* | 16.500 | 28.600 | 3.333 | — | — | — | 48.433 |
| *Nyctanthes arbor-tristis* | 1.333 | 2.600 | 2.577 | 0.489 | — | — | 6.999 |
| *Soymida febrifuga* | 0.088 | 2.000 | 1.955 | 1.511 | 4.444 | 6.399 | 16.397 |
| *Sterculia urens* | 12.000 | 32.000 | 3.777 | — | — | — | 47.777 |
| *Wrightia tinctoria* | 9.650 | 12.800 | 3.360 | 0.400 | — | — | 26.210 |
| *Zizyphus jujuba* | 2.222 | 4.040 | 8.240 | 0.911 | — | — | 15.413 |
| Total | 45.369 | 182.980 | 121.283 | 30.063 | 14.975 | 9.474 | 404.144 |

1. Winter leaf shedding type: The plants of this group shed their leaves during winter season only (October–January). Examples are *Lannea coromandelica, Sterculia urens*, and *Wrightia tinctoria.*

2. Winter and spring leaf shedding type: The plants of this group continue shedding their leaves from October to March. Common plants of this group are *Boswellia serrata, Anogeissus latifolia*, and *Soymida febrifuga.*

Maximum leaf litter was contributed by *Boswellia serrata* (28 percent), while *Boswellia serrata, Anogeissus latifolia, Lannea coromandelica*, and *Sterculia urens* jointly made up 72 percent of the total.

The annual trend in litter production is shown in Figure 11–1. Litter fall is minimal in March and reaches a peak (182.9 $g/m^2$) in November. Litter fall during November contributes 45 percent of the total, while November and December jointly contribute 75 percent of the total amount of litter.

## Discussion

Bhatnagar (1971) has recognized two phenological groups in mixed teak forests in India—winter and summer leaf shedding types. In the drier environment of the present study, all tree species complete their litter fall by the end of spring season. Consequently no litter can be collected in the summer season.

Average annual leaf litter production in different climatic zones of the world is presented in Table 11–2. These data show that leaf litter production in tropi-

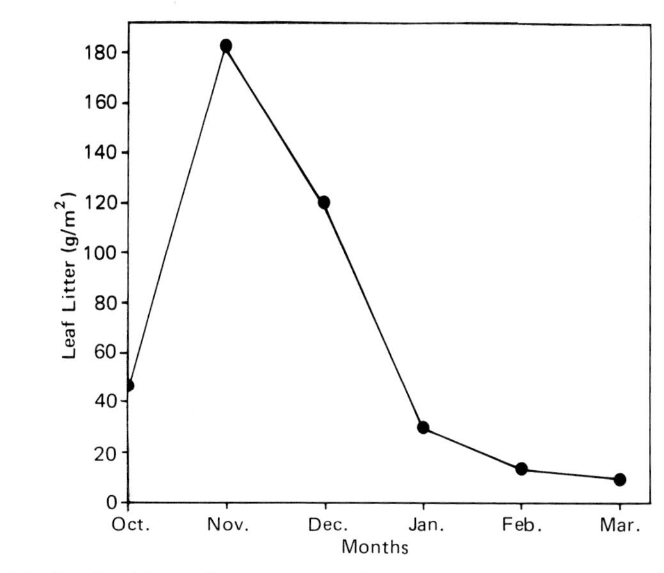


**Fig. 11–1.** Monthly leaf litter production in Kewara-Nal forest, Udaipur.

**Table 11–2.** Annual leaf litter production in different parts of India and major climatic zones of the world.

| Vegetation | Location | Leaf Litter (M ton/ ha/yr) | Source |
|---|---|---|---|
| Climatic zones of the world: | | | |
| Alpine | 67° N Lat. | 0.7 | Bray and Gorham (1964) |
| Cool temperate | 37°–62° N Lat. | 2.5 | Bray and Gorham (1964) |
| Warm temperate | 30°–40° N & S Lat. | 3.6 | Bray and Gorham (1964) |
| Equatorial | Within 10° N & S of equator | 6.8 | Bray and Gorham |
| Indian subcontinent: | | | |
| *Shorea robusta* (Plantation) | 30°19′ N Lat. | 5.9 | Puri (1953) |
| *Tectona grandis* (Plantation) | 30°19′ N Lat. | 5.3 | Seth et al. (1963) |
| *Shorea robusta* (Plantation) | 30°19′ N Lat. | 5.0 | Seth et al. (1963) |
| Deciduous forest (Sagar) | 23°50′ N Lat. | 2.6–9.3 | Upadhyaya (1955) |
| *Tectona grandis* (Plantation) | 26°45′ N Lat. | 6.06–17.3 | Singh (1962) |
| Deciduous forest (Varanasi) | 24°42′–25.5 N Lat. | 1.01–6.21 | Singh (1968) |
| Deciduous forest (Udaipur) | 24°32′ N Lat. | 4.04 | Present authors |

cal deciduous forests is approximately one and a half to two times that of cool temperate and warm temperate forests, respectively. Higher production in tropical forest is probably due to higher temperature, longer growing season, and greater amount of insolation (Singh, 1968).

Bray and Gorham (1964) have shown an inverse relationship between amount of total annual litter production and the latitude of the locality, and have suggested that leaf litter constitutes roughly 70 percent of the total litter. Locating the position of Udaipur in the figure presented by Bray and Gorham (Figure 11–2), the annual litter production for Udaipur is calculated to be 7.75 tons/ha, representing 5.42 tons/ha of leaf litter. The observed amount of litter (4 tons/ha) is slightly less than this expected value. The lower value of observed litter production may reflect unfavorable climatic conditions during the study period.

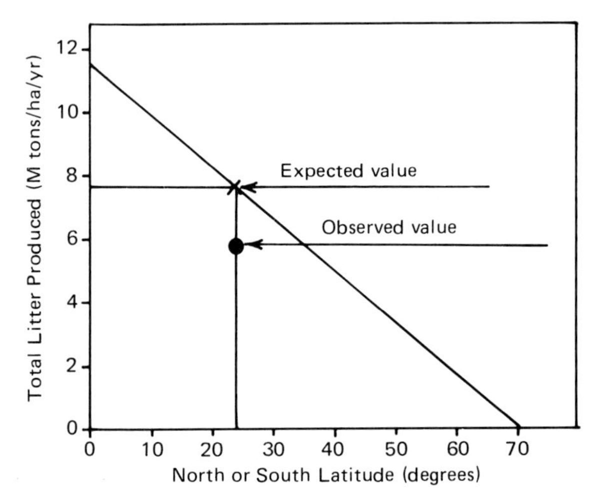


**Fig. 11–2.** Comparison of expected and observed values of leaf litter production for Udaipur.

## Acknowledgment

We are thankful to Shri A. L. Sankhla, divisional forest officer, Udaipur, for permitting us to work in the Kewara forest block, and to Professor H. D. Kumar for laboratory facilities.

## References

Bhatnagar, S. 1971. The ecological changes in the litter layer of some mixed teak forests. In *Proc. School Plant Ecol.*, R. Misra and R. R. Das, eds. Oxford IBH.

Bray, J. R., and E. Gorham. 1964. Litter production in forests of the world. *Advan. Ecol. Res.* 2:101–157.

Puri, G. S. 1953. Leaf fall in Dehradun forests. *Bull. Bot. Soc. Uni. Saugar* 9:28–34.

Seth, S. K., O. N. Kaul, and A. C. Gupta. 1963. Some observations on nutrition cycle and return of nutrients in plantations at New Forest. *Ind. For.* 89:90–98.

Singh, J. S. 1962. Preliminary studies on the humus status of some forest communities of Bashahar Himalayas. Proc. nat. Acad. Sci. India., Sec. B 32: 403–407.

Singh, K. P. 1968. Litter production and nutrient turnover in deciduous forests of Varanasi. *Proc. Symp. Recent Advan. Trop. Ecol.* 655–665.

Subba Rao, B. K., B. G. Dabral, and S. K. Pande. 1972. Litter production in forest plantation of chir (*Pinus roxburghii*), teak (*Tectona grandis*) and sal (*Shorea robusta*) at New Forest, Dehra Dun. *Symp. Trop. Ecol. With an Emphasis on Organic Productivity*. pp. 235–243. Athens, Ga.

Upadhyaya, S. D. 1955. Soil formation in relation to plant cover. Ph.D. thesis, Univ. Saugar.

# CHAPTER 12

## Litter Fall and Litter Breakdown in Miombo

F. Malaisse, R. Freson, G. Goffinet, and M. Malaisse-Mousset

### Introduction

The area of Africa where woodland is dominant covers 12.1 percent, or 3.8 $\times$ $10^6$ $km^2$, of the land area (Malaisse et al., 1972). The main part of this area corresponds closely to the Zambezian Domain (White, 1965). In the vicinity of Lubumbashi miombo woodland comprises over 90 percent of the country; *Brachystegia* spp. and *Julbernardia paniculata* (Benth.) Troupin are the most dominant trees.

Miombo is generally regarded as a comparatively recent vegetation type, largely depending on destruction of dry evergreen forest (Fanshawe, 1971), "muhulu." Cumulative effects of repeated annual fires and cutting produce savanna woodland and finally savanna, while total absence of fire results in muhulu (Malaisse, 1973). The miombo ecosystem may or may not be in equilibrium. Its large distribution suggests that miombo is a pyroclimax. However, some types of miombo belong to the forest associations of the Berlinio-Marquesion alliance, on the one hand, and others to the Mesobrachystegion alliance, on the other hand, and suggest that progressive changes may occur between them.

The aim of the present report is to analyze and discuss litter breakdown in miombo. Research was carried out on the Luiswishi miombo and the Kasapa miombo, the first belonging to the Berlinio-Marquesion alliance, while the second belongs to the Mesobrachystegion (Schmitz, 1963). General description and floristic composition of these communities have been published (Malaisse and Malaisse-Mousset, 1970; Malaisse, 1973; Malaisse et al., 1972). Unless otherwise noted, results are for the Luiswishi miombo.

The macroclimate of Lubumbashi (formerly Elisabethville) is characterized by one rainy season (November to March), one dry season (May to September), and two transition months (October and April). July and August are always without rainfall. Total rainfall is about 1,270 mm with a range of 895 to 1,551 mm for the period 1918 to 1970.

The average yearly temperature is about 20°C. Temperature is lowest at the beginning of the dry season. September, October, and sometimes November are the warmest months. There also is a distinct cold dry season (May to July) and a warm dry season (August to October). During the last period, heating of the soil induces the austral spring characterized by the flowering and flushing of trees before the beginning of rain (Lebrun and Gilbert, 1954). Temperature amplitude is low in rainy seasons and high in dry cold seasons. The relative humidity of the air follows the rainfall pattern, with a minimum in October and a maximum in February.

The mean yearly solar radiation is $16.8 \times 10^9$ kcal/ha/yr, of which $7.75 \times 10^9$ kcal/ha/yr is between 400 and 700 m$\mu$ (Freson, 1971). The minimum mean daily radiation occurs during the rainy season due to the high cloudiness (394 cal/cm$^2$/day in December) and increases progressively until May. The maximum occurs during the warm dry season (533 cal/cm$^2$/day in September), the sky's clarity overruling the additional effects of atmospheric opacity due to bushfires and the decrease in incoming solar radiation (Bernard, 1950).

## Litter Fall

At Kasapa miombo litter fall was studied over a period of five years starting in January 1968. Receptacles for catching litter (Newbould, 1970) were not used, because they were attractive to the local people. Litter was collected monthly on the ground on 10 m$^2$ distributed at random. As the experimental plots were cleaned once a month, litter may have lost weight between collection periods as the result of leaching, microbial decomposition, and feeding by animals (Kirita and Hozumi, 1969). Thus the measured amount is an underestimate of litter fall. Data are in metric units (grams, tons) on an oven-dry basis.

### Tree and Shrub Strata

#### *Leaves*

As most of the trees and shrubs of miombo (92 percent) are deciduous, yearly leaf fall is strongly related to yearly leaf production. The amount taken by phytophages before leaf fall has not been estimated; however locally it may reach high values during caterpillar pullulations (Malaisse-Mousset et al., 1970). Annual tree and shrub leaf fall varies from 2.5 to 3.4 tons/ha, and averages 2.9 tons/ha. Leaf fall represents 74 percent of the tree-shrub litter fall, except for 1972, when it fell to 47 percent because of heavy fruiting. Leaf fall is lowest from December to February (Figure 12–1), then increases progressively until August or September, and afterward decreases (Malaisse and Malaisse-Mousset, 1970).

*Fruit*

In this paper "fruit" refers to fruits, flowers, bud scales, and seeds. Flowers, which may be observed throughout the year (Malaisse, 1974), contribute very little to miombo productivity (Figure 12–1). Bud scales are also insignificant in terms of litter. On the other hand, a large number of species have heavy fruits, such as the woody pods of Caesalpiniaceae and the large berries of *Strychnos* spp.

General fruiting over a large area occurs only infrequently for one species; local fruiting of several species may be observed more commonly. The amount of fruit and seed produced by an individual also may be relatively large. For instance, one *Brachystegia spiciformis* produces 26.6 kg dry weight of seed and pods, while *Strychnos innocua*, a small tree, produces almost 10 kg (Table 12–1). Average yearly fruit fall was about 160 kg/ha for the four first years of observation; however, during 1972 values 12 times higher (2 tons/ha) were observed. This high productivity seems to be related to a wet year (1971) being followed by a dry year with high insolation.

*Wood*

Little data exist on yearly wood production in African tropical forests. We may expect that trees in miombo have growth rings since the climate is seasonal. Preliminary studies have revealed, however, a variety of rings corresponding to abnormal cold dry season flushing, fires, or caterpillar defoliation. Consequently it seems better to estimate radial changes in tree trunks by successive

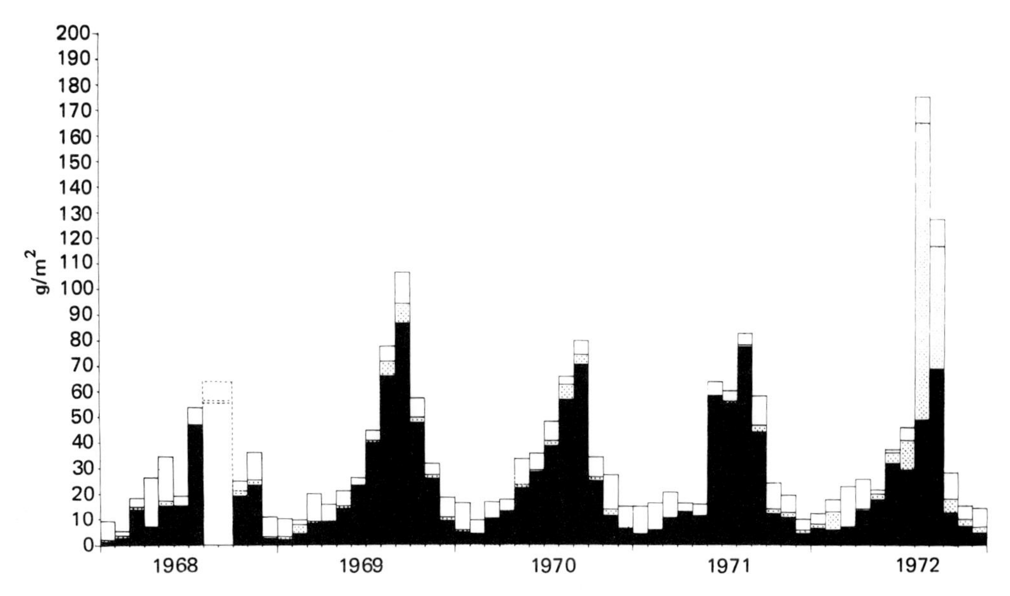


**Fig. 12–1.** Litter fall in the Kasapa miombo (1968–1972). ■ = leaves; ▩ = fruit; □ = twigs. (After Malaisse, unpublished data.)

**Table 12–1.** Parameters of total fruit or seed crops of miombo (oven-dried to constant weight for dry weight).

| Species | Circumference at 1.3 m (cm) | Total Number Fruit or Seed per Tree | $x$ Ind. Seed Dry Wt. (g) | Dry Wt. Total Seed Crop (g) | Max. Dispersal Movement (m) |
|---|---|---|---|---|---|
| *Strychnos innocua* Del. | 48.0 | 177 | 54.989 | 9,733 | 6 |
| *Combretum zeyheri* Sond. | 37.8 | 293 | 0.993 | 291 | 42 |
| *Swartzia madagascariensis* Desv. | 59.4 | 298 | 4.272 | 1,273 | 7 |
| *Pterocarpus chrysothrix* Taub. | 164.5 | 2,530 | 1.987 | 5,027 | 20 |
| *Uapaca kirkiana* Mull. Arg. | 74.7 | 3,998 | 0.517 | 2,067 | 8 |
| *Brachystegia spiciformis* Benth. (valve of pod) | 96.3 | 4,036 | 5.561 | 22,446 | 13 |
| *Brachystegia spiciformis* Benth. (seed) | 112.4 | 7,921 | 0.530 | 4,196 | 20 |
| *Hymenocardia acida* Ful. | 66.4 | 7,101 | 0.034 | 239 | 28 |
| *Albizia adianthifolia* (Schum.) W. F. Wight (seed) | 101.4 | 10,620 | 0.037 | 397 | 103 |

girth measurements. These were made during the rainy season to avoid dry season shrinkage, which occurs in some tropical forest species (Daubenmire, 1972a). Observations at two-year intervals were carried out in Kasapa miombo. The basal area on the experimental plot of ⅛ ha was 1.667 $m^2$ in 1969 (or 13.335 $m^2$/ha). For the period 1969–1973 increase in basal area was 0.194 $m^2$ (0.048 $m^2$/yr), an increase of 2.9 percent. The maximum yearly increase occurred in the 30- to 40-cm girth class (Table 12–2).

On the experimental plots recent mortality of trees represented 0.130 $m^2$ and dead trees lying on the forest floor 0.018 $m^2$ of wood. At Luiswishi miombo, there were 54 standing dead lignous plants per hectare, with a basal area of 4.95 $m^2$/ha. Volume and weight of trunk fall have not yet been measured. However, branch fall has been estimated on 16 plots of 100 $m^2$ each as about 4.4 tons/ha/yr.

Yearly fall of wood under 2 cm in diameter was 0.87 ton/ha. Twig fall fluctuates seasonally with maxima in the late rain and in the warm dry seasons. The late rainy season maximum is related to the high moisture of rotted wood, which falls abundantly. The warm dry season maximum results from the gusts of wind that frequently occur just before the first rains.

## Litter Decomposition

Litter accumulation was maximum at the end of the early rainy season (4.4 tons/ha) and minimum in the dry cold season, just after passage of fire (1.6 tons/ha). Mean yearly accumulation was 3.3 tons/ha (Malaisse et al., 1972).

**Table 12–2.** Radial and sectional area (1.5 m height) changes in a ⅛-ha plot at Kasapa miombo at these dates in the period 1969–1973.

| Diameter Classes at 1.50-m Height (cm) | No. Obser-vation | Average Radius (cm) | | | Aver. Ann. Radial Incre-ment (mm) | Total Sect. Area at 1.50-m Height (cm$^2$) | Average Basal Area (cm$^2$) | | | Aver. Ann. Sect. Incre-ment (cm$^2$) | Total Annual Sect. In-crease (cm$^2$) |
|---|---|---|---|---|---|---|---|---|---|---|---|
| | | 20.2.69 | 10.12.70 | 15.1.73 | | 20.2.69 | 20.2.69 | 10.12.70 | 15.1.73 | | |
| 0– 4.9 | 14 | 2.05 | 2.39 | 2.56 | 1.28 | 195.15 | 13.97 | 18.67 | 20.34 | 1.59 | 22.16 |
| 5.0– 9.9 | 41 | 3.47 | 3.80 | 4.06 | 1.48 | 1,621.87 | 39.56 | 47.20 | 53.73 | 3.54 | 145.27 |
| 10.0–19.9 | 32 | 6.79 | 7.07 | 7.33 | 1.35 | 4,824.31 | 150.76 | 163.34 | 170.07 | 4.83 | 154.45 |
| 20.0–29.9 | 11 | 11.49 | 11.72 | 11.90 | 1.05 | 4,597.04 | 417.91 | 436.37 | 449.98 | 8.02 | 88.20 |
| 30.0–39.9 | 2 | 15.28 | 15.74 | 16.19 | 2.27 | 1,466.97 | 733.48 | 778.50 | 823.32 | 22.46 | 44.93 |
| 40.0–49.9 | 0 | — | — | — | — | — | — | — | — | — | — |
| 50.0–59.9 | 2 | 24.68 | 25.03 | 25.05 | 0.92 | 3,963.46 | 1,981.23 | 2,035.84 | 2,040.30 | 14.77 | 29.54 |
| Total (⅛ ha) | 102 | | | | | 16,668.80 | | | | | 484.55 |

Litter decomposition in tropical regions is known to occur more rapidly than in temperate regions (Jenny et al., 1949; Laudelout and Meyer, 1954; Nye, 1961; Yoda and Kira, 1969; Bernhard-Reversat, 1972). In miombo, data on soil microbial flora (Malaisse et al., 1972) suggest that there are seasonal rhythms in decomposition. For example, aerobic digestion of cellulose is inefficient during the entire rainy season but becomes important in the dry season. Other microbial processes, such as proteolytic digestion, starch digestion, and ammonification, have a maximum activity in the main rainy season and appear to be highly correlated with soil moisture content. According to Goffinet (1973), seasonal fluctuations of Collembola populations in miombo litter are very dependent on moisture. Mite populations also fluctuate but are able to maintain a low density through the dry season.

## Tree and Shrub Strata

### *Leaves*

In order to follow leaf litter decomposition in miombo three experiments were carried out: (1) recent oven-dried mixed leaf litter was left to decay on the forest floor at the beginning of December, and five samples were taken each month during a 12-month period; (2) recent oven-dried mixed leaf litter was left to decay on the forest floor the first of each month and removed at the end of the month; (3) decay of litter of eight important species was followed during a six-month period starting in November. The results may be summarized as follows:

Leaf litter decomposition is rapid in December with activity of bacteria, mites, and Collembola (Figure 12–2). Soil microbial activity takes place on leaf litter, which is incorporated into the runoff galleries of termites. Dry litter placed for one month on the soil surface may lose between 1 and 51 percent of its weight according to season. Loss is high from January to August and is mainly correlated with termite activity.

Decomposition varies greatly between species. The ratio between fast and slow rates of decomposition is 3.7 (Figure 12–3). The most rapid rate is that of *Pterocarpus angolensis*, which lost half of its weight in 73 days. In absence of termite attack, leaf decomposition is relatively constant within a species; the observed variation was generally 0.5 percent and is always less than 4 percent variation between samples. On the other hand, termite consumption occasionally may reach 92 percent. Accordingly, species with high attraction for termites, such as *Ochna schweinfurthiana* and *Pseudolachnostylis maprouneifolia*, can be distinguished from others with low attraction, such as *Parinari mobola, Marquesia macroura*, and *Uapaca pilosa*. On the soil the main role in fruit decomposition is played by termites for lignous diaspores and ants for freshy diaspores.

### *Wood*

Wood decomposition was followed experimentally using living twigs of *Pterocarpus angolensis*, 1 to 1.5 cm diameter (10 g dry weight). The observed

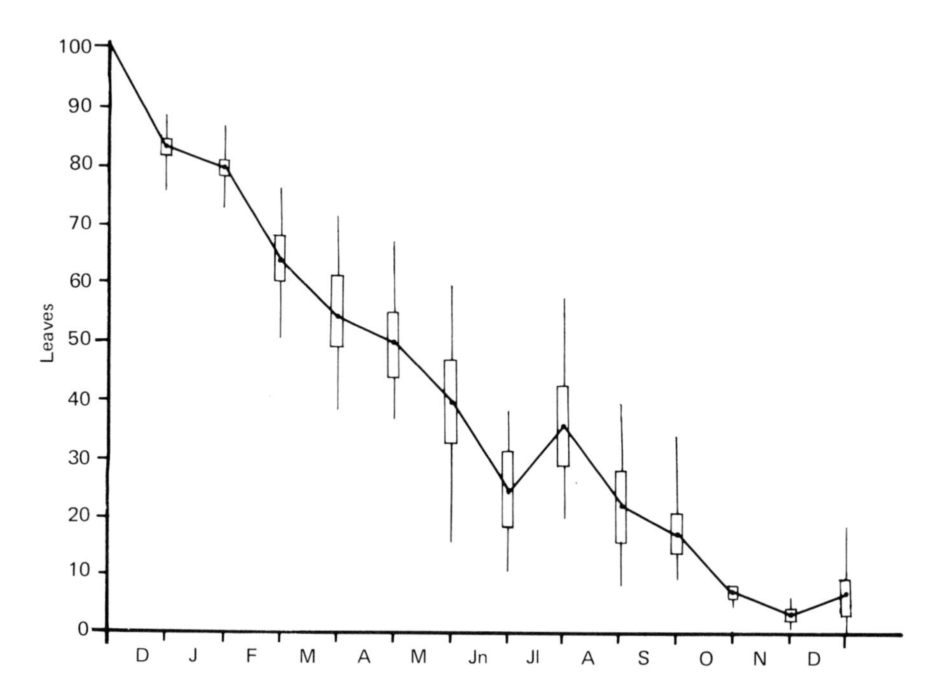


**Fig. 12–2.** Mixed leaf litter decomposition during a 12-month period. The experiment started in December 1971. Vertical line indicates range; bar indicates the standard error.

rate of decomposition was lower than that of leaves (Figure 12–4). However, the rate was even lower than expected from observation of natural litter. This was probably due to the fact that the experiment used healthy dried rather than dead wood, which was not exposed to fire nor strongly attacked by termites because of the small size of the twigs. These latter factors act to increase the rate of wood breakdown in miombo.

### Herbaceous Strata

According to Freson (1973), herbaceous litter decays more quickly than tree and shrub litter. Dead material is present as early as December (Figure 12–5), only three months after the new growth of vegetation. The maximum decomposition is reached in February to April and is principally due to termite activity. The effects of fire on the herbaceous strata will be discussed below.

### Soil Cellulolytic Activity

The activity of cellulolytic organisms was measured by a method similar to that proposed by Unger (1960). Wire netting with a 16-mm mesh, enclosing a

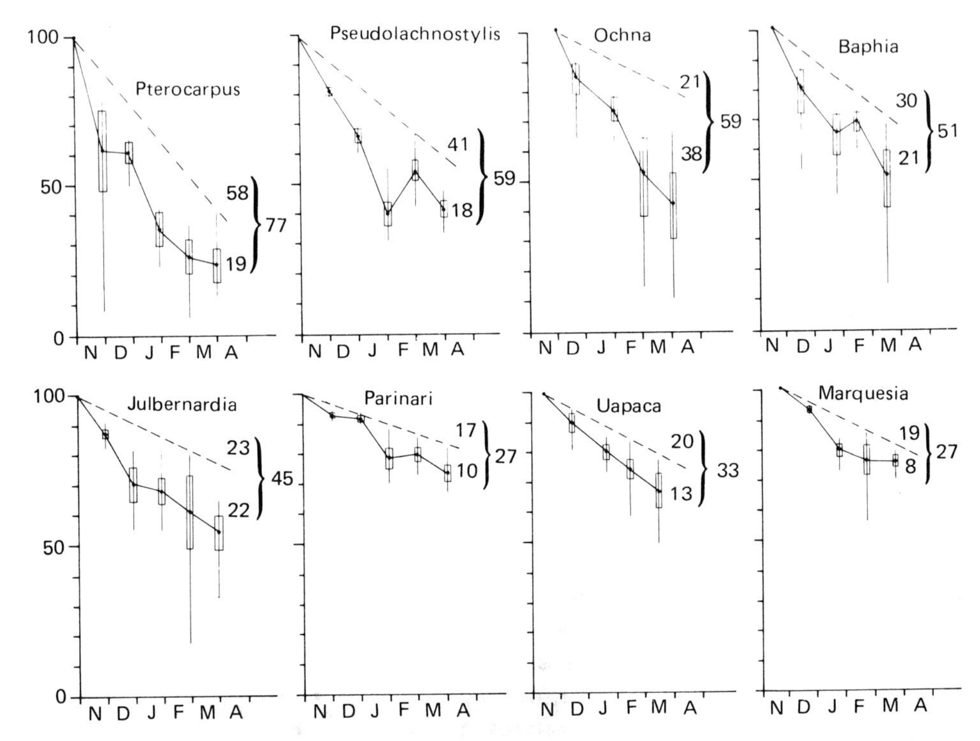


**Fig. 12–3.** Comparison of rates of leaf decomposition for different species: *Pterocarpus angolensis* DC., *Pseudolachnostylis maprouneifolia* Pax, *Ochna schweinfurthiana* F. Hoffm., *Baphia bequaertii* De Wild., *Julbernardia paniculata* (Benth.) Troupin, *Parinari curatellifolia* Planch. ex Benth., *Uapaca pilosa* Hutch. and *Marquesia macroura* Gilg. The experiment started in November 1972. Vertical line indicates total range of values; bar indicates standard error; connecting line indicates arithmetic mean. The upper number indicates the loss in percent due to nontermite activity; the lower one, the loss due to termite activity; the number to the right, the total percentage of decomposition. (Afer Malaisse-Mousset, unpublished data.)

double thickness of rectangular cellulolytic paper (5 by 10 cm) of known weight, was vertically hidden in the ground, the upper side at soil surface level. After a time, the paper was removed, cleaned, and weighed. Breakdown of cellulose is expressed in percent of dry weight loss. The method was not totally satisfactory because, due to the high clay content, fine soil particles attached to the paper, especially in the rainy season. Decomposition appeared to be higher during this last period; however termite activity also differed greatly among the samples. For instance termites never visit papers placed in areas with continuous ant activity.

## Effects of Fire

Fire is a very important factor in litter breakdown. Natural fire in savanna may occur but it has been rarely reported (Lebrun, 1947). Most of the fires are

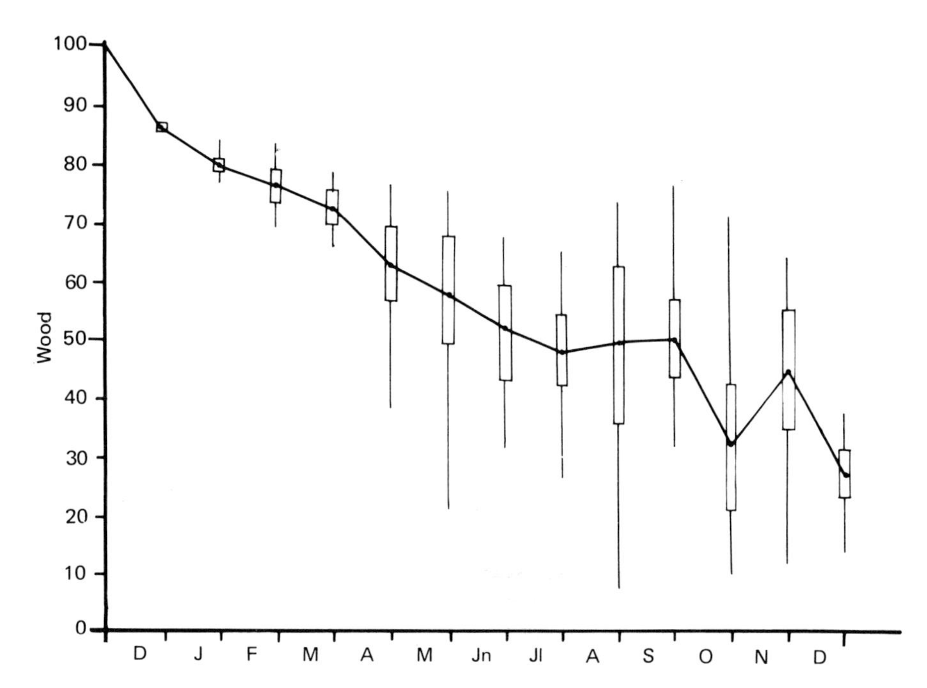


**Fig. 12–4.** Wood decomposition during a 12-month period. For explanation of symbols, see Figure 12–3.

lighted by man. Bushfires are known in savanna and woodland areas all over Africa, and their practice seems very ancient. Fire often destroys a large part of grass, tree and shrub leaves, and branches in the miombo community. However, there may be an important difference in the effect of early and late fires. Early fires tend to remain close to the ground, scorching the bases of the tall shoots, but not consuming them. Late fires consume practically all of the aerial parts or kill the aerial parts of any woody plant of small diameter (Daubenmire, 1972b). In miombo, fire destroys most of the two- to three-year-old shoots and the woody seedlings so that regeneration is very restricted.

The effect of burning on soil temperature (Masson, 1948; Pitot and Masson, 1951), on microflora (Henrard, 1939), and on small arthropods (Goffinet, 1973) has been studied. Fire causes mostly a superficial depressive effect on soil fungi and animal populations. On the other hand, its influence on soil structure and on organic matter impoverishment produced by the high rate of leaching during the first rains may be far more important.

## Effects of Termites

From the point of view of diet three groups of termites may be distinguished: the humivorous, the lignivorous, and the mixed termites (Deligne, 1966; Bouillon, 1970). Humus-feeding species, such as Cubitermes, Anoplotermes, Crene-

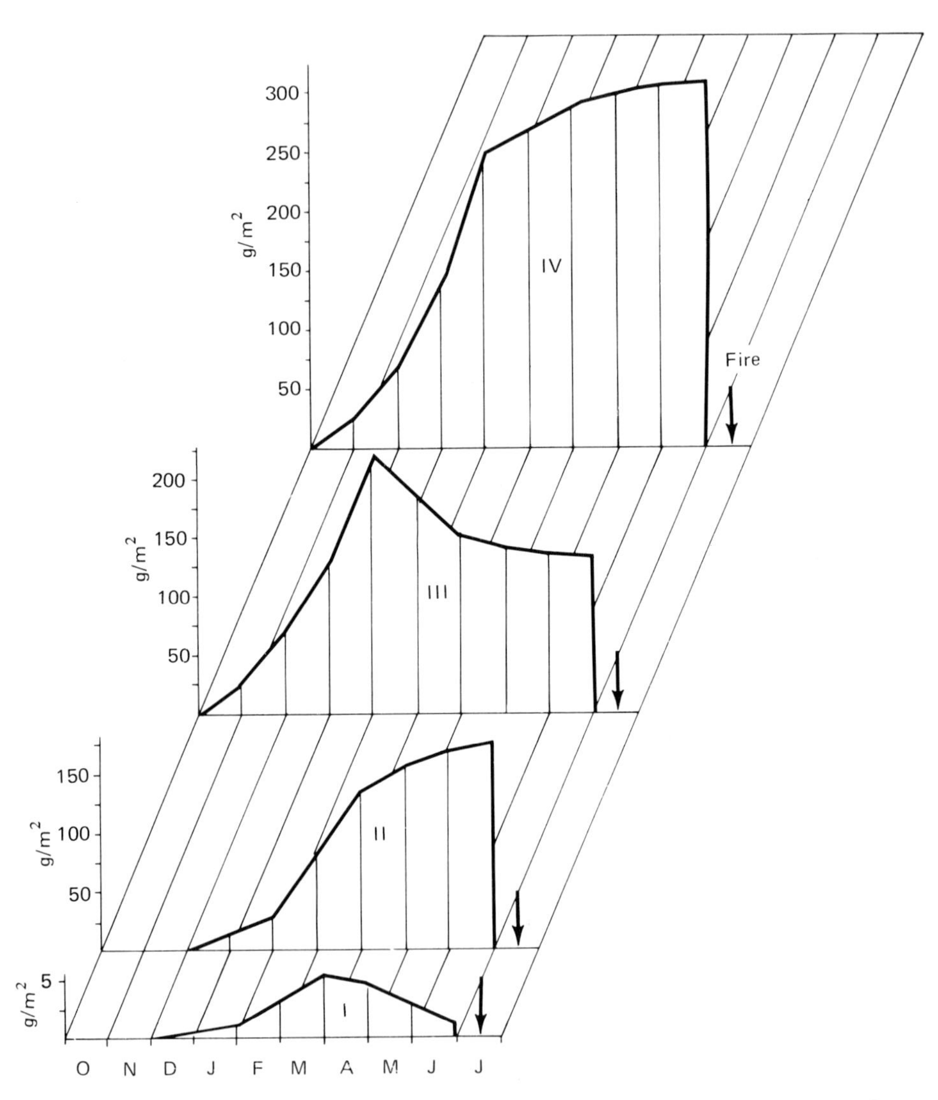


**Fig. 12–5.** Monthly variations of the herbaceous strata in Luiswishi miombo from October 1969 to July 1970 (Freson, 1973). Line I, decomposition; line II, necromass; line III, biomass; line IV, total organic matter.

termes, and Basidentitermes, ingest decaying organic matter and mineral soil. Their role in the pedogenesis is to produce, through their feces, a more stable organic matter for the physicochemical and bacterial agents of degradation (Boyer, 1971). Average abundance and standing crop of the humivorous termites point out their importance in the miombo. Total termite populations make up more than 80 percent of the total soil animal biomass. Per hectare, 1,450 calies of small and medium size occur. They contain approximately 17.5 millions of individuals, with a dry weight of 21.8 kg (Goffinet and Freson, 1972; Malaisse et al., 1972). Preliminary results suggest that Cubitermes alone constitute at least 50 percent of these forms. In the intercalic zone, Cubitermes and Anoplotermes make up more than 90 percent of the total termite biomass.

The lignivorous termites—Microcerotermes, Amitermes, and Macrotermitinae—attack intact vegetative dead material. Contrary to the humivorous termites, foraging of superficial organic matter by lignivorous termites occurs often in covered access routes. *Pseudocanthotermes spiniger, Odontotermes* sp., and *Macrotermes falciger* are the three Macrotermitinae most frequently found foraging in miombo. The abundance of *Odontotermes* can be more than 15,000 individuals per square meters, corresponding to more than 1 ton/ha fresh weight (Table 12–3). *Macrotermes falciger* is a large termite. It was observed all year, except during the warm dry season (middle July to the end of September), cutting small branches, leaves, fruits, and grasses before transforming this material into sawdust and incorporating it into its fungus gardens. In Luiswishi miombo this species is commonly found in the great mounds topping the characteristic "tumuli" that abound in the region.

**Table 12–3.** Abundance and biomass of three species of Macrotermitinae during their foraging activity (after Goffinet, unpublished data).

| | Date of Observation | Individuals/m² | | | Fresh weight g/m² | | |
|---|---|---|---|---|---|---|---|
| | | Workers | Soldiers | Total | Workers | Soldiers | Total |
| *Pseudocanthotermes spiniger* (Sjöstedt) | 13.1.70 | 2425 | 3650 | 6075 | 10.83 | 7.84 | 18.67 |
| *Odontotermes* sp. | 15.1.70 | 1550 | 1500 | 3050 | 8.96 | 35.79 | 44.75 |
| | 2.3.71 | 13175 | 2275 | 15450 | 66.37 | 41.87 | 108.24 |
| | 10.5.71 | 1200 | 704 | 1904 | 4.56 | 11.91 | 16.47 |
| *Macrotermes falciger* (Gerstäcker) | 12.1.72 | 1500 | 100 | 1600 | 51.50 | 8.00 | 59.50 |
| | 13.1.72 | 1000 | 50 | 1050 | 36.50 | 5.75 | 42.25 |

As Bodot (1967) has shown for *Macrotermes* and *Ancistrotermes* in a savanna of Ivory Coast, the foraging activity of lignivorous termites has a seasonal periodicity depending on climate and abundance of nutrients. In miombo, foraging activity was observed from November to June (*Odontotermes* and *Macrotermes*) with maximal intensity in the late rainy season. Observation of very reduced fungus gardens of *Microtermes* at the end of the dry season strongly suggests that this material constitutes reserves for the late dry months.

## Annual Rate of Litter Decomposition

The annual rate of litter decomposition may be expressed as the ratio of litter fall to litter accumulation (Nye, 1961). Miombo, in absence of fire, has an annual rate of tree litter decomposition of 111 percent. This value is lower than that for other African tropical forests (Table 12–4), but miombo is

**Table 12–4.** Litter decomposition in African forests.

| Vegetation Types | Mixed Rain Forest | Brachystegia Rain Forest | Macrolobium Rain Forest | Musanga Rain Forest | Subequator Rain Forest | Moist Semideciduous Forest | Mixed Dry Lowland Forest | Dry Evergreen Forest (Muhulu) | Woodland (Miombo) | Riparian Forest (Mushitu) |
|---|---|---|---|---|---|---|---|---|---|---|
| References | Bartholomew et al. (1963), Laudelout and Meyer (1954) | | | | Bernard (1950), Bernard-Reversat et al. (1972) | Greenland and Kowal (1960), Lawson et al. (1970) | Madge (1965) | Malaisse et al. (1970), Malaisse (unpublished data) | Malaisse et al. (1972), Goffinet and Freson (1972) | Freson (unpublished data) |
| Places | Yangambi | | | | Banco | Kade | Ibadan | Lubumbashi | | |
| Country | Zaïre | | | | Ivory Coast | Ghana | Nigeria | Zaïre | | |
| Geographical coordinates | 0°48′N 24°24′E | | | | 5°21′N 4°05′W | 6°09′N 0°55′W | ±7°N 4°W | 11°29′S 27°36′E | 11°29′S 27°36′E | 11°43′S 27°25′E |
| Altitude (m) | 360 | | | | 10–30 | 170 | 250 | 1,208 | 1,208 | 1,190 |
| Mean annual rainfall (mm) | 1700 | | | | 2,100 | 1,630 | ±1,200 | 1,273 | 1,273 | 1,279 |
| Average annual temp. (°C) | 24.5 | | | | 26 | 28 | 27 | 20°3 19°1[c] | 20°3 21°6[c] | 20°3 17°2[c] |

| Dry season period | Absent | | | | I, somet. (VIII, II) | I, somet. (VIII, XII, II) | XI–III | V–X | V–X | V–X |
|---|---|---|---|---|---|---|---|---|---|---|
| No. species/ha | | | | | | ±270 | | ±37 | ±160 | ±52 |
| No. tree species/ha | | | | | 65 | ±125 | | 9 | 21 | 22 |
| No. stem/ha | | | | | | | | 2,340 | 540–880 | 1693 |
| Dominant tree height (m) | | | | | 29–49 | | | 22 | 17 | 32 |
| Sectional area at 1.3 m ($m^2$/ha) | | | | | 35 | 56 | | 40 | 13–22 | 52 |
| Litter on the ground (kg/ha): | | | | | | | | | | |
| Dry months | | | | | | | 2,450 | 7,913 | 1,670 | |
| Wet months | | | | | | | 1,700 | 6,090 | 4,420 | |
| Annual mean | | | | | | 2,260 | | 6,532 | 3,270 | |
| Litter fall (kg/ha/yr): | | | | | | | | | | |
| Leaf (lignous plants) | | | | | 8,190 | 9,400 | 3,400 | 4,694 | 2,915 | 4,512 |
| Flowers + fruits | | | | | 1,100 | | | 1,489 | 537 | 179 |
| Branches | | | | | 2,580 | | | 2,962 | 874 | 1,201 |
| Total | 12,400 | 12,300 | 15,300 | 14,900 | 11,870 | | 5,600 | 9,145 | 4,325 | 5,897 |
| Annual rate of litter decomposition %[a] | 316[b] | | | | 330–420 | 465[b] | 200–485 | 140 | 111–132[d] | |

[a] Nye method.
[b] After Dommergues (1963).
[c] On the ground, in the vegetation type.
[d] In presence of fire.

adapted to a longer dry season. For the herbaceous strata of miombo, no litter decomposition rate is available, as experiments in the absence of fire have not been carried out.

## Discussion

Miombo presents two well-defined synusiae: the tree and shrub strata and the herb strata. The latter produces yearly 3.2 tons/ha, of which the greatest part is destroyed by fire. The former has a litter fall of 2.9 tons/ha of leaves, 0.5 ton/ha of fruit, and about 4.5 tons/ha of wood. The processes of litter decomposition may be grouped into three main categories: microflora and microfauna, termites, and fire. The first category presents a relative spatial homogeneity and a well-defined seasonal periodicity. However, great differences in rate of decomposition may exist in relation to species of plants, so that it is possible to distinguish species with leaves having a fast and others a slow decomposition. These insure some soil protection against the first showers of the early rainy season. This last group seems to be better represented in well-developed miombo than in open woodland.

Effects of termites and fire on litter decomposition have in common a spatial heterogeneity and a certain complementarity in time. Termites are highly active all year long, except during the warm dry season when fire is the most efficient. Early fire is mainly related to herb layer density, which varies from place to place, while late fire depends more on the amount of litter available. The cumulative effect of termites and fire accelerates the speed of litter decomposition by a factor of about 2. The irregularity in the timing and space of these three factors emphasizes the heterogeneity of miombo and makes more difficult a fine approach to litter breakdown.

## Acknowledgment

The writers are indebted to Professor F. Golley and Dr. E. Medina for reviewing the manuscript and helpful discussions. They thank Professor A. Bouillon for termite determination. L. Lemaire insured the photographic documentation. For their assistance in the field, Kisimba, Bulaimu, Kabiombwe, and Kapongo are gratefully acknowledged.

## References

Bartholomew, M. V., J. Meyer, and H. Laudelout. 1953. Mineral nutrient immobilization under forest and grass fallow in de Yangambi (Belgian Congo) region. *Publ. INEAC Bruxelles Ser. Sci.* 57:1–27.

Bernard, E. 1950. Aperçus fondamentaux sur la climatologie du Katanga. *Comp. Rend. Congr. Sci. Elisabethville, C. S. K., Bruxelles* 4(2):56–70.
Bernhard-Reversat, F. 1972. Décomposition de la litière de feuilles en forêt ombrophile de Basse Cote d'Ivoire. *Oecol. Plant.* 7(3):279–300.
———, C. Huttel, and G. Lemee. 1972. Quelques aspects de la périodicité écologique et de l'activité végétale saisonnière en forêt ombrophile sempervirente de Côte-d'Ivoire. In *Tropical Ecology*, P. Golley and F. Golley, eds. pp. 217–234. Athens.
Bodot, P. 1967. Cycles saisonniers d'activité collective des Termites des savanes de Basse-Côte d'Ivoire. *Insectes Sociaux* 14:359–388.
Bouillon, A. 1970. Termites of the Ethiopian region. In *Termites.* Vol. 2, pp. 153–280. New York: Academic Press.
Boyer, P. 1971. Les différents aspects de l'action des Termites sur les sols tropicaux. In *La vie dans les sols.* pp. 279–334. Paris: Gauthier Villars.
Daubenmire, R. 1972a. Phenology and other characteristics of tropical semideciduous forest in north-western Costa-Rica. *J. Ecol.* 60(1):147–170.
——— .1972b. Ecology of *Hyparrhenia rufa* (Nees) in derived savanna in north-western Costa-Rica. *J. Appl. Ecol.* 9:11–23.
Deligne, J. 1966. Caractères adaptatifs au régime alimentaire dans la mandibule des Termites (Insectes Isoptères). *Compt. Rend.* 263:1323.
Dommergues, Y. 1963. Les cycles biogéochimiques des éléments minéraux dans les formationes tropicales. *Bois et Forêts des Tropiques* 87:9–25.
Fanshawe, D. 1971. The vegetation of Zambia. *For. Res. Bull. Min. Rural Dev. Zambia* :1–67.
Freson, R. 1971. Ecologie et production primaire d'une phytoplanctocénose tropicale: le lac de retenue de la Lubumbashi. Ph.D. thesis, Univ. Louvain.
———. 1973. Aperçu sur la biomasse et la productivité de la strate herbacée au miombo de la Luiswishi. *Ann. Univ. Abidjan Ser. E. Ecol.* 6:265–277.
Goffinet, G. 1973. Contribution à l'étude de l'écosystéme forêt claire (miombo). Note 12: Recherches préliminaires sur les fluctuations saisonnières des peuplements en Acariens et en Collembols au niveau de la litiere du Miombo. *Ann. Univ. Abidjan Ser. E. Ecol.* 6:257–263.
———, and R. Freson. 1972. Recherches synécologiques sur la pédofaune de l'écosystème Forêt claire (Miombo). (Contribution No. 4 à l'étude de l'écosystème Forêt claire.) *Bull. Soc Ecol.* 3(2):138–150.
Greenland, D., and J. Kowal. 1960. Nutrient content of the moist tropical forest of Ghana. *Plant Soil* 12(2):154–174.
Henrard, F. 1939. Réaction de la microflore du sol aux feux de brousse. Essai préliminaire exécuté dans la région de Kisantu. *Publ. INEAC Ser. Sci. 20.*
Jenny, H., S. Gessel, and C. Bingham. 1949. Comparative studies of decomposition rates of organic matter in temperate and tropical regions. *Soil Sci.* 68:419–432.
Kirita, H., and K. Hozomi. 1969. Loss of weight of leaf litter caught in litter trays during period between successive collections—A proposed correction for litterfall data to account for the loss. *Jap. J. Ecol.* 19(6):243–246.
Laudelout, H., and J. Meyer. Les cycles d'éléments minéraux et de matiere organique en forêt équitoriale congolaise. *Comp. Rend. 5e Congr. Intern. Sci. Sol Léopoldville* 2:267–272.
Lawson, G. W., K. O. Armstrong-Mensah, and J. B. Hall. 1970. A catena in tropical moist semi-deciduous forest near Kade, Ghana. *J. Ecol.* 58:371–398.
Lebrun, J. 1947. La végétation de la plaine alluviale au Sud du Lac Edward. *Inst. Paris Nat. Congo Belge, Expl. Parc Nat. Albert* 1:1–800.

———, and G. Gilbert. 1954. Une classification écologique des forêts du Congo. Bruxelles: INEAC, Ser. Sci. 63:1–89.

Madge, D. 1965. Leaf fall and litter disappearance in a tropical forest. *Pedobiologia* 5(4):273–288.

Malaisse, F. 1973. Contribution à l'étude de l'écosystème forêt claire (Miombo). Note 8: Le projet Miombo. *Ann. Fac. Sci. Abidjan Ser. Ecol.* 6:227–250.

———. 1974. Phenology of the zambezian woodland area with an emphasis on miombo ecosystem in phenology and modeling seasonality. In *Ecological Studies.* 8:269–286. Berlin-Heidelberg-New York: Springer-Verlag.

———, J. Alexandre, R. Freson, G. Goffinet, and M. Malaisse-Mousset. 1972. The miombo ecosystem: A preliminary study. In *Tropical Ecology*, P. and F. Golley, eds. pp. 363–405. Athens: Univ. Georgia Press.

———, and M. Malaisse-Mousset. 1970a. Contribution à l'étude de l'écosystème forêt claire (Miombo): Phénologie de la défoliation. *Bull. Soc. Roy. Bot. Belgique* 103:115–124.

Malaisse, F., M. Malaisse-Mousset, and J. Bulaimu. 1970b. Contribution à l'étude le l'écosystème forêt dense sèche (muhulu). Note 1: Phenologie de la défoliation. *Trav. Serv. Sylv. Pisc. Univ. Off. Congo Lubumbashi*, 9:1–11.

Malaisse-Mousset, M., F. Malaisse, and C. Watula. 1970. Contribution à l'étude de l'écosystème forêt claire (Miombo). Note 2: Le cycle biologique à Elaphrodes lactea (Gaede) Notodontidae et son influence sur l'écosystème miombo. *Rev. Univ. Lubumbashi Ser. B Sci.* 25:75–85.

Masson, H. 1948. La température du sol au cours d'un feu de brousse au Sénégal. *Agron. Trop.* 3(3–4):174–179.

Newbould, P. 1970. Methods for estimating the primary production of forests. *I.B.P. Handbook No. 2*, pp. 1–62. Oxford: Blackwell.

Nye, P. 1961. Organic matter and nutrient cycles under moist tropical forest. *Plant Soil* 13(4):333–346.

Pitot, A., and H. Masson. 1951. Quelques données sur la température au cours des feux de brousse aux environs de Dakar. *Bull IFAN* 13(3):710–732.

Schmitz, A. 1963. Apercu sur les groupements végétaux du Katanga. *Bull. Soc. Roy. Bot. Belgique* 96:233–447.

Unger, H. 1960. Der Zellulosetest, eine Methode zur Ermittlung der zellulolytischen Aktivität des Bodens in Feldversuchen. *Z. Pflanzenernahr. Düng. Bodenk. NF* 91:44–52.

White, F. 1965. The savanna woodland of the zambezian and sudanian domains. An ecological and phytogeographical comparison. *Webbia* 19(2):651–681.

Yoda, K., and T. Kira. 1969. Comparative ecological studies on three main types of forest vegetation in Thailand. V. Accumulation and turnover of soil organic matter, with notes on the altitudinal soil sequence on Khao (Mt.) Laang, peninsular Thailand. *Nature and Life in Southeast Asia.* 6th edit., pp. 83–110. Tokyo: Jap. Soc. Promotion Science.

# CHAPTER 13

# Nutrients in Throughfall and Their Quantitative Importance in Rain Forest Mineral Cycles

FRANCE BERNHARD-REVERSAT

## Introduction

This study is a part of a larger study on nutrients and organic matter in the moist evergreen forest of Ivory Coast, which is being carried out at the Adiopodoume station of the Office de la Recherche Scientifique et technique Outre Mer.

Two types of forest were studied. The Banco forest grows on soils lying on tertiary sands, which are very poor in nutrients. The Yapo forest grows on soils lying on schists, which are richer in nutrients but contain 50 to 80 percent gravel. In the Banco forest two sites were studied, one on the plateau and the other in a thalweg (valley) where the phosphorus, potassium, and calcium content of vegetation and soil were higher.

Mean annual rainfall in the Banco forest is 2,100 mm, but during the period of our investigation the rainfall was only 1,400 and 1,800 mm per year. Rainfall distribution over the year is very unequal. On the average there are two rainy seasons, the main one in June and another in October. In the Yapo forest, mean annual rainfall is about 1,800 mm, with the June and October peaks not as sharp as at the previous site.

## Nutrient Content of Rainwater in the Open

The nutrient content of rainwater was was studied by Roose (1972), 15 km from the Banco at Adiopodoumé. Mean contents were 1.58 ppm nitrogen, 0.10 ppm phosphorus, 0.36 ppm potassium, 1.48 ppm calcium, and 0.27 ppm magnesium. The amounts of nutrients corresponding to 1,650 mm of rain (throughfall 1,450 mm plus etsimated interception) calculated from the data of Roose are (in kilogram per hectare per year) 25 nitrogen, 1.6 phosphorus, 5.8 potassium, 24 calcium, and 4.3 magnesium.

## Nutrients in Throughfall Water

On each plot the water was collected weekly by 12 rain gauges. Samples of four consecutive weeks were merged proportionately to corresponding rainfalls. These samples were kept in a cold room until analysis.

The nutrient contents of throughfall given in Table 13–1 are calculated over one year for the Banco forest and over seven months for the Yapo forest (mean content weighted according to the amount of water). Comparing rainfall and throughfall, the "atmospheric" portion of calcium in throughfall water is high; the nitrogen is appreciable but the amount brought by leaching is greater than by the rain. The atmospheric parts of phosphorus, potassium, and magnesium in throughfall water can almost be neglected.

**Table 13–1.** Nutrient concentration in throughfall waters (mg per liter).

| Site | N | P | K | Ca | Mg |
|---|---|---|---|---|---|
| Banco plateau | 6.1 | 0.13 | 4.0 | 2.5 | 2.8 |
| Banco thalweg | 5.8 | 0.65 | 12.2 | 3.3 | 3.6 |
| Yapo | 2.5 | 0.36 | 5.1 | 2.2 | 1.6 |

In the Banco forest the results were obtained by pairs of samples made on the same date on the two plots, corresponding to the same rainfall. This enabled us to examine the statistical difference between the plateau and the thalweg. The throughfall waters collected in the thalweg site are significantly richer in phosphorus and potassium (with a 99 percent confidence). We were not able to compare the Yapo and Banco waters in the same way but it seems that nitrogen and magnesium contents are lower at Yapo.

The nutrients found in the throughfall waters come chiefly from the leaves. We have no leaf analysis, but average content of fresh leaf litter is available and given in Table 13–2 (Bernhard, 1970).

**Table 13–2.** Annual mean of nutrient concentration in fresh leaf litter (percent of dry weight).

| Site | N | P | K | Ca | Mg |
|---|---|---|---|---|---|
| Banco talweg | 1.54 | 0.069 | 0.22 | 0.56 | 0.46 |
| Banco thalweg | 1.80 | 0.158 | 0.91 | 0.95 | 0.41 |
| Yapo | 1.40 | 0.050 | 0.28 | 1.32 | 0.29 |

As far as phosphorus and potassium are concerned, the differences observed in throughfall waters reflect those observed in the leaves. Higher values were observed at the thalweg of Banco forest than at the other sites, and at Yapo forest, the low magnesium content of the leaves seems to reappear in the waters. The throughfall waters, however, do not reflect the important differences found in calcium content of the leaves. This may be explained by the accumulation of unleachable calcium in cell wall structures.

### Relation Between Rainfall and Nutrients

The relation between the amount of rainfall over four weeks and its nutrient content was studied in the Banco forest (Figure 13–1). Three types of curves are obtained. The magnesium curve is a plateau, indicating that the amount of nutrients brought to the leaf surface and weathered per time unit is constant. The amount leached does not increase with an increase in rainfall over 200 mm/ 4 weeks.

Another type curve is that showing a continuous increase, which may be due to increasing amount of nutrient brought to the leaf surface with greater weathering or, in the case of calcium and nitrogen, to the high content of the rainwater itself as measured in the open.

The third type of curve is that of potassium at the thalweg site: The amount leached is at its maximum with a rainfall about 200 to 250 mm/4 weeks, and then decreases with increasing rainfall. This might be due to a secondary effect of the rainy season during which the labile potassium can be washed down in the soil and become less available for the vegetation. This is only an hypothesis and probably does not apply to phosphorus in the thalweg site, since this nutrient is less weatherable than potassium

## Nutrient Amounts Brought to the Soil

The annual amounts of throughfall over two years at Banco are compared with other tropical forests in Table 13–3. In the Banco forest the values over two years are similar. There is a substantially larger amount of phosphorus and potassium in the thalweg as compared to the plateau. Other elements are relatively similar at the two sites. Furthermore, the overall data are similar to those obtained in two other tropical forests. However, temperate forests give lower values (Carlisle et al., 1967; Madgwick and Ovington, 1959; Mina, 1965). We conclude that tropical forests, which are evergreen and have a high leaf area index, have a much higher flux of minerals through the throughfall pathway.

Figure 13–2 illustrates the seasonal variation in throughfall. Nutrient leaching is great at the end of the dry season, with the appearance of the first important rain (March–May).Increase in rainfall in June and July has little influence on the phenomenon. This is particularly noticeable in the case of potassium: its highest leaching occurs before the highest rainfall.

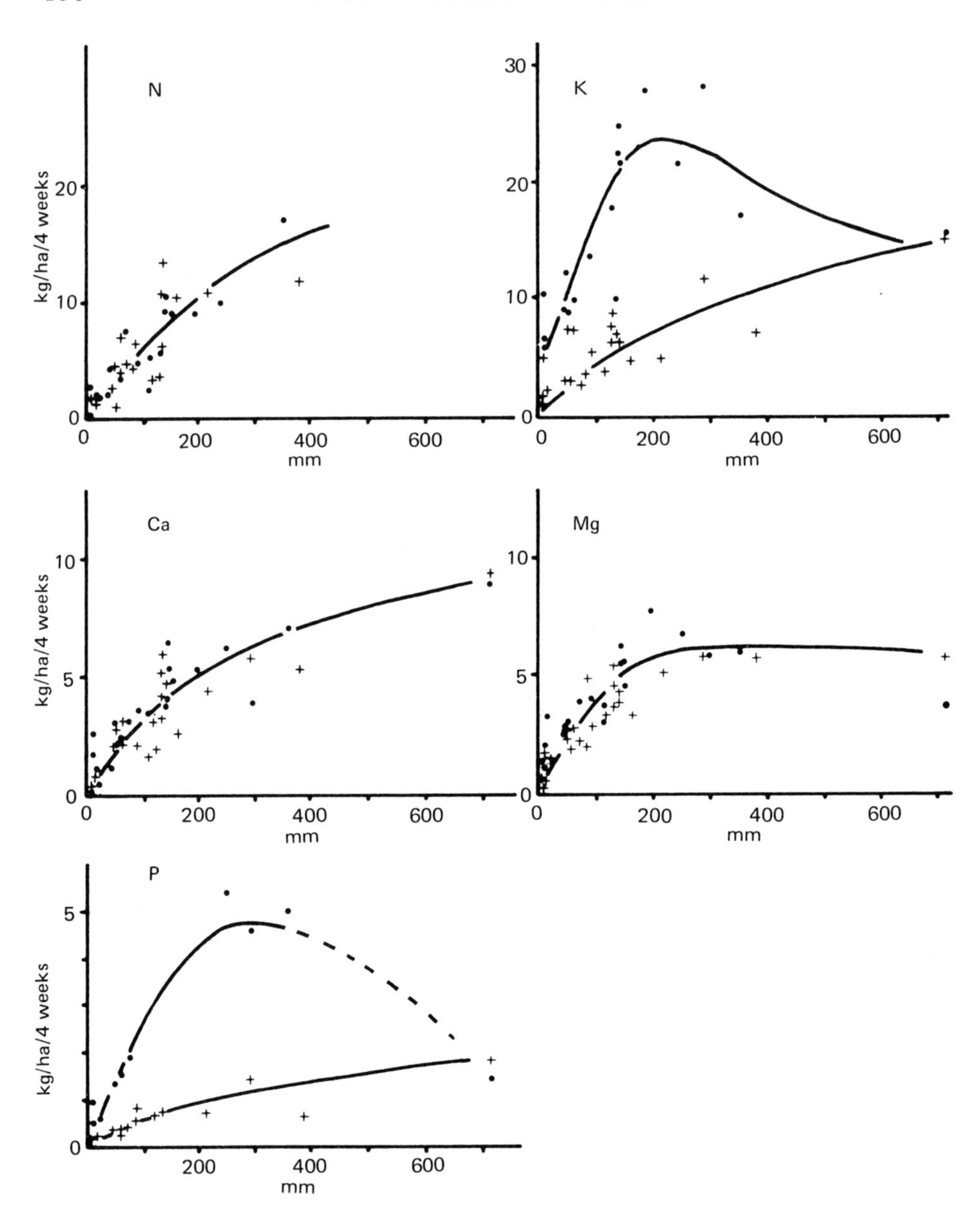


**Fig. 13–1.** Amount of nutrients in throughfall versus amount of precipitation. Crosses indicate data from the plateau; points indicate data from the thalweg.

## Leaching and Mineral Cycling in the Banco Forest

The annual amount of nutrients moving to the forest floor is calculated by adding the nutrients of the litter to those of the throughfall (leaves, flowers, fruits, twigs, branches). The stem flow is neglected. Table 13–4 gives the global amount and the percentage of each part.

**Table 13–3.** Annual amounts of nutrients in throughfall (kg/ha).

| Site or Author | Rainfall (mm) | N | P | K | Ca | Mg |
|---|---|---|---|---|---|---|
| Banco plateau | 1,430 | 88 | 1.8 | 57 | 36 | 40 |
| 1970, 1971 | 1,750 | 71 | 2.6 | 73 | 43 | 42 |
| Banco thalweg | 1,500 | 88 | 10.1 | 183 | 50 | 54 |
| 1970, 1971 | 1,780 | 74 | 9.4 | 166 | 43 | 43 |
| Nye (1961) | 1,651 | 26.4[a] | 4.1 | 237 | 41.5 | 29.0 |
| Roose (1972) | 2,100 | 51.2 | 4.3 | 67.1 | 65.4 | 34.8 |

[a] Mineral nitrogen only.

The leached part is most important for the potassium; it represents more than 60 percent of the amount annually recycled at both sites. Leaching of phosphorus is negligible at the plateau site where the vegetation is poor in this element and where nearly all of the phosphorus is recycled through the litter fall. At the thalweg site, leaching represents a larger part of this nutrient cycle. For calcium, magnesium, and nitrogen, the leached fraction ranges from 15 to 56 percent.

Water was collected from a spring downstream of the thalweg station. Its nutrient content is assumed to represent the output from the ecosystem (McColl, 1970). There were 1.33 ppm nitrogen, 0.015 ppm phosphorus, 0.46 ppm potassium, 1.77 ppm calcium, and 0.89 ppm magnesium in the spring waters. These values are much greater than those found by McColl except for potassium, which was essentially the same.

Comparison of the spring water with the throughfall water at the thalweg site shows that the spring water has a low nutrient content. Apparently the soil-root system efficiently retains the nutrients brought to it in litter and throughfall. In the case of cations, this efficiency increases in the order calcium-potassium-magnesium. The highest output concentrations observed are for those nutrients that have the highest concentration in the input (rainfall); calcium and nitrogen.

**Table 13–4.** Annual amounts of nutrients brought to the soil and percentage of different parts.

| Site | | N | P | K | Ca | Mg |
|---|---|---|---|---|---|---|
| Plateau | Total kg/ha/yr | 258 | 9.8 | 85 | 97 | 91 |
| | Rainfall % | 9 | 14 | 6 | 22 | 4 |
| | Leaching % | 25 | 4 | 61 | 15 | 40 |
| | Litter % | 66 | 82 | 33 | 63 | 56 |
| Thalweg | Total kg/ha/yr | 246 | 24 | 264 | 135 | 90 |
| | Rainfall % | 10 | 6 | 2 | 16 | 4 |
| | Leaching % | 26 | 38 | 67 | 21 | 56 |
| | Litter % | 64 | 56 | 31 | 63 | 40 |

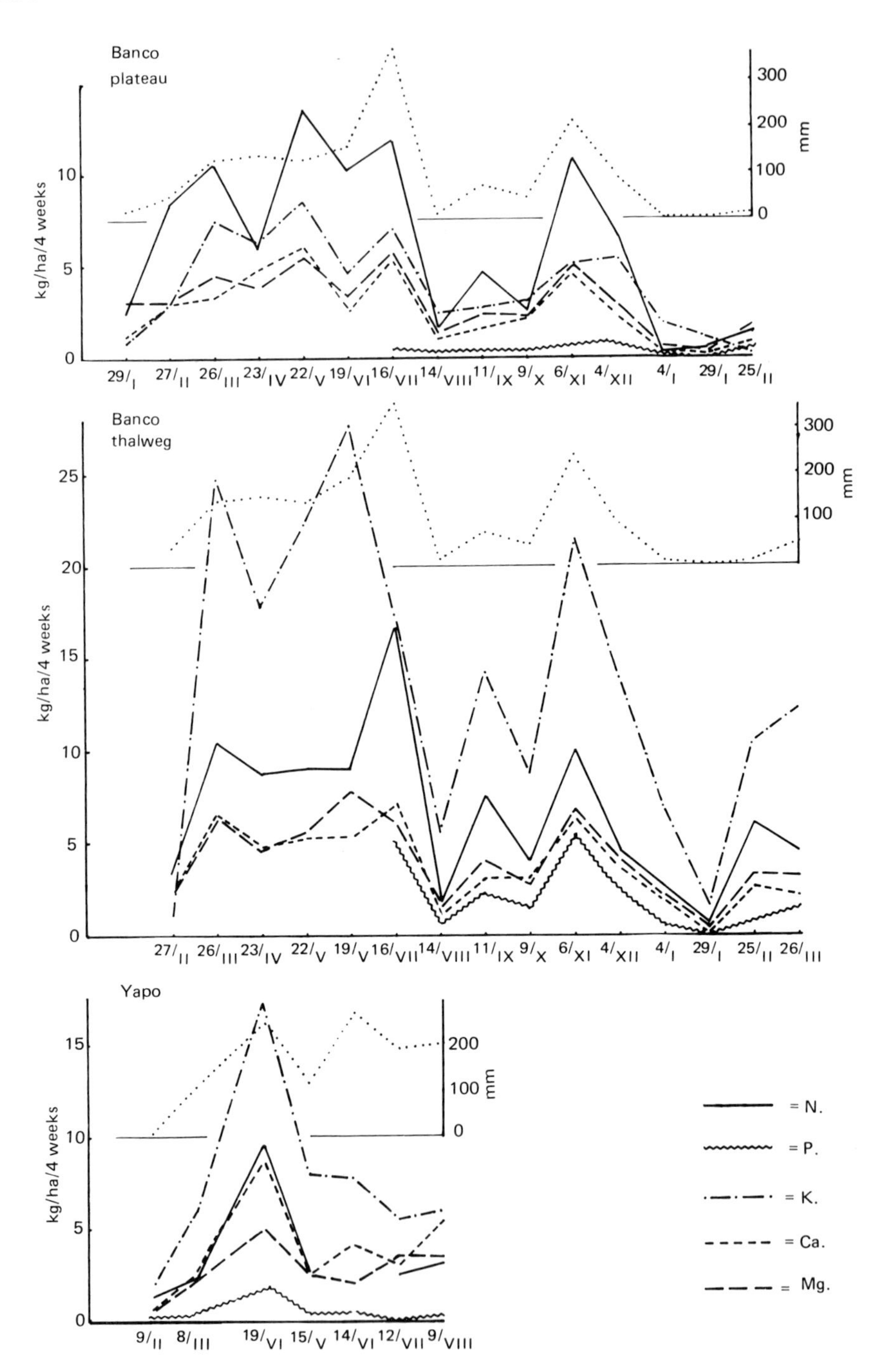


**Fig. 13–2.** Seasonal pattern of nutrients amounts in throughfall. Rainfall is shown by the dotted line above the figure for nutrients.

## Acknowledgment

Our program was carried out with the help of the water and soil analysis laboratory of Orstom, Adiopodoume, and we thank Mr. Gouzy who directed the chemical analysis. We acknowledge the collaboration of J. Delaunay and the help of J. Tcheoulou in our work.

## References

Bernhard, F. 1970. Etude de la litiere et de sa contribution au cycle des elements mineraux en foret ombrophile de Cote-d'Ivoire. *Oecol. Plant.* 5:247–266.

Carlisle, A., A. H. F. Brown, and E. J. White, 1967. The nutrient content of tree stem-flow and ground flora litter and litter and leachates in a sessile Oak (*Quercus petrae*) woodland. *J. Ecol.* 55:615–627.

McColl, J. G., 1970. Properties of some natural waters in a tropical wet forest of Costa Rica. *BioSci.* 20:1096–1100.

Madgwick, H. A., and J. D. Ovington. 1959. The chemical composition of precipitations in adjacent forest and open plots. *Forestry* 32:14–22.

Mina, V. N. 1965. Leaching of certain substances by precipitations from woody plants and its importance in the biological cycle. *Soviet Soil Sci.* 609–617.

Nye, P. H. 1961. Organic matter and nutrient cycles under moist tropical forest. *Plant Soil* 13:333–346.

Roose, E. J. 1972. Quelques effets des pluies sur la mise en valeur des sols ferralitiques et ferrugineux tropicaux. *Rapport Orstom.*

# CHAPTER 14

## Production, Energy Pathways, and Community Diversity in Forest Birds

JAMES R. KARR

### Introduction

Recent studies on a variety of organisms have demonstrated that no single hypothesis accounts for all observed patterns in biotic diversity. Relative influence of factors varies among environments and differs among taxonomic groups due to differences in morphology, population dynamics, biology, and complexity of interspecific interactions.

Geographic patterns in faunal diversity of birds have been studied extensively (MacArthur and MacArthur, 1961; MacArthur et al., 1966; Karr, 1968; Recher, 1969; Orians, 1969; Cody, 1970; Karr, 1971; Karr and Roth, 1971). From these studies two major conclusions have emerged to explain the phenomenon of latitudinal gradients in avian community diversity at microgeographic scales: (1) A variety of new resources is available above minimal threshold levels in tropical habitats compared to similar temperate habitats; and (2) there is more precise resource subdivision, in space and possibly also in foraging tactics, in tropical areas. The new resources argument has been developed by Orians (1969) and Karr (1971). Fruits, large insects (Schoener, 1971), and nectar are the resources most frequently discussed. The resource subdivision argument was first proposed on theoretical grounds by MacArthur and Pianka (1966) and MacArthur and Wilson (1967). Subsequent studies have substantiated these conclusions in a variety of neotropical avifaunas (Orians, 1969; Terborgh and Weske, 1969; Karr, 1971; Karr and Roth, 1971; Terborgh, 1971, among others), and also in New Guinea for spatial segregation on an altitudinal basis (Diamond, 1973).

For simplicity, wherever I refer to the "new resources," it will imply resources that are exploitable in tropical forest but not temperate forest, because of their availability above threshold levels. This must be explicitly stated to avoid the inference that the newly exploited resources are unavailable in temperate forest. Because of temporal or spatial patterns of distribution and abundance, they usually are unreliable enough to preclude any bird species exploiting those resources to a significant degree. Temperate fruits are exploited opportunistically by a variety of species, especially during migration and winter, but the present discussion is concerned only with the breeding season.

There have been no previous attempts to utilize data from studies of avian communities to evaluate the productivity hypothesis that states that increased productivities will result in increased species diversities (Connell and Orias, 1964). Attempts to relate increases in productivity with increases in species diversity have failed in the case of lizards (Pianka, 1967); in fact, a negative relation between productivity and species diversity of chydorid cladocera was shown for lakes in Denmark and Indiana (Whiteside and Harmsworth, 1967).

In this paper I shall explore the following two hypotheses: (1) Diversity of a community is proportional to the number of kinds of packages of resources available in that community. (2) Community diversity is proportional to production and/or rate of energy flow in the community—the productivity hypothesis. Communities will be the resident species during the season of highest breeding activity in forest study areas where a resident is a species "that could be seen or netted almost daily" (Karr, 1971). Some of the problems of this definition of community will be discussed later.

The data for this analysis have been published elsewhere (Karr, 1968, 1971; Karr and Roth, 1971; MacArthur et al., 1972). Discussion will be based on two tropical mainland forests in the lowlands (Liberia, Panama), a tropical island (Puercos Island), and a temperate mainland (Illinois). A summary of the conditions in each study area is given in Table 14–1. Annual rainfall, mean annual temperature, and mean breeding season temperature are significantly higher in tropical than temperate areas. The number of resident species and individuals is greatest in the tropical mainland forests. By comparison, the number of individuals is greater but number of species is smaller in the tropical island fauna. Similar foliage height diversities (Table 14–1) among the areas reduces the effects of variation in vegetation structure.

## Density Compensation and Community Energetics

There are several general patterns of the number of species of birds over the world. For example, greater species diversity is related to decreasing latitude, decreasing elevation, and increasing size of islands in island archipelagos (MacArthur and Wilson, 1967; MacArthur, 1972; Diamond, 1973).

The density of individuals of all species in species-depauperate island faunas is poorly correlated with total density in nearby mainland areas (MacArthur et al., 1972). In a variety of studies of birds and mammals cited by MacArthur et al. and in recent work by Karr (unpublished) in Madagascar it has been shown that total densities on islands may be equal to, greater than,

**Table 14–1.** Location, climatic conditions, number of resident species, and size of resident bird populations for forest study areas.

| Study Area | Location | Elevation (m) | Annual Rainfall (mm) | Mean Temperature (°C) | | Number of Resident Bird Species | Resident Population (pairs/40 ha) | Foliage Height Diversity |
|---|---|---|---|---|---|---|---|---|
| | | | | Breeding Season | Annual | | | |
| 1. Illinois | 40°6′39″N, 87°44′18″W. | 170 | 960 | 19 | 11 | 32 | 489 | 1.07 |
| 2. Liberia | 7°29′N, 8°35′W. | 550 | 2,840 | 23 | 23 | 50 | 712 | 1.03 |
| 3. Panama | 9°9′35″N, 79°44′36″W | 60 | 2,600 | 27 | 27 | 56 | 728 | 1.04 |
| 4. Puercos Island | 8°21′49″N, 78°48′40″W | 5 | — | — | — | 16 | 864 | 0.90 |

or less than nearby mainland densities. MacArthur et al. (1972) suggest that the "appropriateness" of successful island colonists may in large part determine the variation in patterns of density compensation. Finally, they suggest that slightly larger bird populations on islands may be due to underrepresentation of large birds in island faunas and consequent overcompensation of individuals, which result in more precise correspondence among habitats in total biomass per unit area.

The possibility that energy compensation might occur was suggested earlier (Karr, 1971), when it was found that Panama and Illinois forest avifaunas had similar energy requirements per-day for existence ($16.9 \times 10^3$ and $16.4 \times 10^3$ kcal/40 ha/day, respectively). Existence energy is the energy required by a caged bird to maintain a constant weight, and is determined by subtracting energy excreted in feces from energy consumed. This calculation, of course, does not include energy metabolized for such activities as nesting, foraging, and flight. It does, however, facilitate comparison of various areas.

The similarity in existence energy requirements was true despite the number of species being 75 percent higher and the number of individuals being 49 percent higher in the tropical forest. Perhaps within more or less narrowly defined groups such as Brown's mammals (1971) or the birds of Panamanian mainland and island forest, density compensation occurs with relatively minor adjustments to account for differences in size distribution, while on a global scale energy compensation is a more striking phenomenon than density compensation.

Logically, since energy may be the ultimate limiting factor in biological systems, an hypothesis predicting energy compensation is at least plausible.

Ideally, we should test such an hypothesis only after many data points are available from a variety of habitats and geographic areas: tropical versus temperate, island versus mainland, forest edge (late shrub) versus forest. Since data on species composition and relative abundance can only be obtained with considerable outlay of time and energy, it is not likely that such a large number of data points will become available in the near future. In addition, there is some advantage in having data collected in such a way as to reduce inter-individual bias in techniques and procedures. The data from four forests discussed here have that advantage.

Examination of Figure 14–1 shows that total densities vary less than number of resident species. However, the greatest reduction in coefficient of variation among the study areas is for the energy requirements of the avifauna. On a per-day basis, energy compensation may be more important among widely scattered areas than density compensation.

The patterns of species densities, density per species, and energy requirements of these faunas are due to a complex of factors, including especially bird size and temperature variation among the areas. Since the average bird in the tropical avifaunas is smaller (Table 14–2; see also Karr, 1971 and Moreau, 1972) and the mean environmental temperature is higher (Table 14-1), the energy requirement per bird is lower. With reduced energy requirements per bird the total energy requirements of the tropical avifaunas approxi-

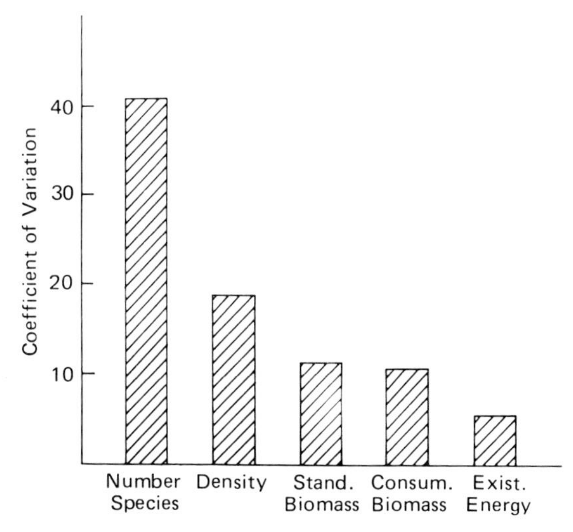


**Fig. 14–1.** Coefficients of variation of selected parameters for resident avifaunas in four forest study areas.

**Table 14–2.** Biomass and energy relations of resident forest avifaunas.

| Study Area | Biomass (g/40 ha) Standing Crop[a] | Biomass (g/40 ha) Consuming[b] | Existence Energy[c] (kcal/40 ha/day) | Mean Weight (g) |
|---|---|---|---|---|
| Illinois | 48,412 | 9,862 | 16,451 | 49.5 |
| Liberia | 49,713 | 11,929 | 18,945 | 34.9 |
| Panama | 52,682 | 13,492 | 16,950 | 36.2 |
| Puercos Island | 38,227 | 11,872 | 17,530 | 22.1 |

[a] Standing crop biomass $= \sum_{i=1}^{s} (A_i)(W_i)$

[b] Consuming biomass $= \sum_{i=1}^{s} (A_i)(W_i^b)$

[c] Existence energy $= \sum_{i=1}^{s} (a)(A_i)(W_i^b)$

where $A_i$ = abundance of species $i$; $W_i$ = individual live weight of species $i$; $b$ = slope of weight metabolism regression; $a$ = $y$ intercept of weight metabolism regression.

mate the temperate energy requirements, despite the larger number of individuals per area in the tropical fauna. The same patterns can be observed in comparisons among the three tropical forest areas, but they are less striking.

Since the variation of daily existence energy requirements is significantly below the variation in number of species within our set of communities, we reject the hypothesis that the production or rate of energy flow in a community determines the species diversity of that community.

## Patterns of Resource Exploitation

Island-mainland patterns have been eloquently discussed by MacArthur and Wilson (1967) and MacArthur (1972); so I shall not devote further discussion to patterns between the Panama mainland and Puercos Island (also see MacArthur et al., 1972). Instead, I shall concentrate on patterns among the forests representing three different continents.

Earlier studies (Salt, 1953; Karr, 1971) have compared the number of species with different food habits (such as insectivore, frugivore, omnivore). This study will follow Moreau (1966) by considering the resource exploitation patterns from the viewpoint of resource types rather than bird food habits. This may seem a trivial distinction but it has new potentials for data analysis. For example, earlier a hummingbird that exploited both nectar and insect resources was placed in the food habits category "omnivore" along with other species that may or may not take nectar but certainly exploited two kinds of resources. A motmot, which eats both large insects and fruits, might be lumped as an omnivore with the hummingbird. This is true despite the fact that their foods differ along any resource spectra dividing resources by size and type. In the new system seven categories of food resources are recognized: nectar, fruit, seeds, vertebrates, carrion, and large and small insects. In this system species that feed primarily on small insects add one unit to the small insects category, while the hummingbird adds one-half each to the small insect and nectar categories.

The results of the analysis are summarized in Figure 14–2. Of the seven

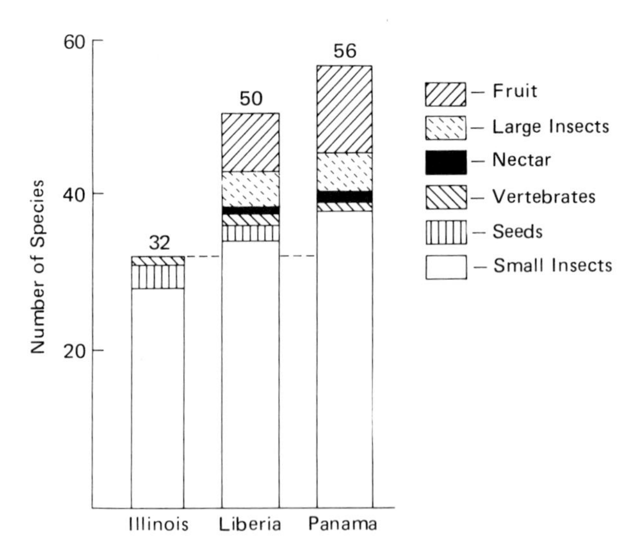


**Fig. 14–2.** Number of resident species exploiting each of several resources in three mainland forest study areas.

food resources available, one (carrion) is not exploited by any resident species in the three forest areas. Three of the food types—seeds, vertebrates, and small insects—show no obvious pattern of change among the study areas, although seeds are not exploited in Panama forest. Several food resources (fruit, large insects, and nectar), which are not exploited in the temperate forest, are exploited to varying degrees in the tropical forests.

Since we are trying to determine why there are more species in tropical than in the temperate forest, one might ask how many "excess" resident species do the more diverse faunas contain. Panama forest has 24 excess species when compared to Illinois (56 versus 32) and Liberia has 18 excess species (50 versus 32). What proportion of the excess species exploits new resources? A total of 17.5 species exploit the new resources in Panama and 12.5 species in Liberia. Thus approximately 70 percent of the excess tropical species exploit resources that are unavailable and/or unreliable in temperate forest.

The remaining 6.5 excess Panamanian species and 5.5 excess Liberian species are still unaccounted for. They come especially from species that exploit the same kinds of resources as their temperate counterparts, but also do an extensive amount of feeding on new resources.

## Resource Availability and Evolutionary Strategies

Three interacting properties of resource availability are important from the standpoint of developing evolutionary strategies: types of resources; amount of energy available in each resource, including abundance relative to some minimal threshold; and temporal pattern of availability. We have already seen that the first two account for a significant proportion of the excess tropical species with little or no change in amount of energy consumed on a per-day basis. Also some tropical species may maintain themselves by opportunistically switching foods on a daily and/or seasonal basis.

Energy packages can be thought of as pulses of resources with three more or less distinct temporal and spatial patterns of availability. Clearly, some packages are available all the time and will be exploited by species that specialize on that resource. This is the case for many tropical insectivores and also for a number of bark gleaners and bark drillers in temperate forest.

Alternatively, if a resource does not exist in a particular environment or is below some minimal threshold level throughout the year, no species will be able to depend on such a resource. This is likely the case for large insects and nectar in temperate forest.

Finally, certain resources are reliably available on a seasonal or other temporally intermittent basis. Dormancy, such as hibernation or estivation, is a common response to this circumstance. This strategy is rarely used by birds, but many plants, insects, mammals, and other groups make use of this alternative. Hummingbirds and a few other birds may reduce metabolic rates during the night (Pearson, 1960), presumably to conserve energy during a time when foraging is not possible and maintenance of normal metabolic rates would quickly exhaust energy reserves.

The majority of birds that breed in the temperate zone, especially those feeding on insects, migrate in response to seasonal variation in availability of their food rather than having a dormant period. Most migrants have obligate migration patterns (warblers, vireos) but a few seem to be facultative migrants. An example of the latter is the red-headed woodpecker (*Melanerpes erythrocephalus*), which winters in much of north-central United States in years of good acorn crops. In poor mast years they migrate southward (Bent, 1939).

On the other hand, tropical forest species rarely have long-distance migration patterns. Instead, many switch foods opportunistically on a day-to-day or seasonal basis. Problems of maintenance are adequately resolved by this opportunism, and each species gears its breeding season to coincide with the optimal availability of its food resources (Skutch, 1950). For example, insectivores and frugivores usually breed in late dry and early wet season when their foods are most available, and nectar feeders nest in early dry season to coincide with a flush of flowering in many plants.

Temperate visitors to the tropics migrate from their breeding grounds when foods are no longer available above threshold levels. It is significant that many of these species that exploit only insects on their breeding grounds (nothing else is available) become very opportunistic (including frugivory) on their wintering grounds (Willis, 1966; Leck, 1972).

## Available Energy and New Resources

What proportion of the daily energy requirements of the avifaunas derives from fruits, large insects, and nectar and from seeds, vertebrates, and small insects? How do the relative proportions of energy from these two categories of resources compare to increases in species numbers? The daily energy requirements of species feeding on the first set of resources in tropical forest vary from 25 to 37 percent of total energy requirements in the same avifaunas (Table 14–3). Comparison of percent of species and percent of energy on the new food resources indicates striking similarities between these two measures. On a per-day basis during the breeding season, energy requirements of birds seem to be very similar in both tropical and temperate forest (Table 14–2). This leads to the conclusion that community diversity does not vary in proportion to amount of energy used, but does vary with diversity of resource packaging. The increase in avian diversity is in direct proportion to the amount of energy coming from new resources (Table 14–3); that is, a shift of 20 percent of the energy to new resources results in an increase of diversity of about 20 percent.

This simplistic result must be looked at very skeptically. A major problem has not been adequately discussed: A very large number of birds are classed as rare and/or irregular in tropical forest and cannot be ignored. Addition of

**Table 14–3.** Percent of resident species feeding on "new" (fruits, large insects, and nectar) and "other" (seeds, vertebrates, and small insects) food resources and percent of energy of energy derived from "new" and "other" food resources in three forest study areas.

| Study Area | | Food Resources | |
|---|---|---|---|
| | | Other | New |
| Illinois | Species | 100 | 0 |
| | Energy | 100 | 0 |
| Liberia | Species | 75 | 25 |
| | Energy | 75 | 25 |
| Panama | Species | 69 | 31 |
| | Energy | 63 | 37 |

these rare species to the resident fauna will add little to population sizes, biomass, and energy requirements (Karr, 1971), but an explanation of patterns of avian diversity must not ignore the rare species.

## Resources and Nonresident Species

Analysis of the nonresident species was carried out by dividing them into irregular and regular species. A species was classed as regular if it was seen on the study area during many months, but generally on less than about 50 percent of the surveys in any specific month. Species that were seen only once or twice or, at most, a few times during the study were classed as irregular. Species in the irregular category could be very rare in the vicinity of the study area, nomadic, or accidental wanderers from the other habitats. The number of species is clearly higher in the tropical forests (Table 14–4), especially in the Panama forest where nearly 90 species, more than half of the total species, were in one of these two categories. By contrast, only 20 percent of the species observed on the Illinois study area were classed as either regular or irregular. Liberia also has fewer species than Panama in these two classes.

Of greater significance than the absolute number of species is the variation in number of species among the food resources. This varies from continent to continent but the greatest increases are invariably in the number of species that feed on small insects.

Theoretical considerations suggest that species segregating on a single resource dimension will maintain minimum niche differences (May and MacArthur, 1972) approximating spacing of 2$\times$ by weight (MacArthur, 1972). As the complexity of resource dimensions being subdivided increases, the complexity of interspecific interactions will increase, resulting in divergence from the predicted spacings of species.

We might be able to predict qualitatively which types of resource are more

**Table 14–4.** Number of nonresident species feeding on each food resource (see text).

| | Food Resource | | | | | | | | |
|---|---|---|---|---|---|---|---|---|---|
| | Large Insects | Small Insects | Nectar | Fruit | Seeds | Vertebrates | Fish | Carrion | Totals |
| Regular: | | | | | | | | | |
| Illinois | — | 1 | — | — | 1 | 1 | — | — | 3 |
| Liberia | 1 | 2 | — | 1 | — | — | — | — | 4 |
| Panama | 1 | 7 | — | 4 | — | — | — | — | 12 |
| Irregular: | | | | | | | | | |
| Illinois | — | 6.5 | — | — | 0.5 | 1 | 2 | — | 10 |
| Liberia | 2 | 31 | 1 | 7 | 1 | 1 | — | — | 43 |
| Panama | 1 | 38 | 4 | 17 | 2 | 9 | 2 | 3 | 75 |

likely to be divided on a single resource dimension and which will be subdivided in more complex ways. The potential complexity for subdivision of insect resources seems larger than for fruits or nectar: Fruits can only be harvested from the canopy or as fallen fruit lying on the ground, but birds can foliage glean, hover glean, bark glean, flycatch, or ground glean for insects. Each of these foraging strategies will have a set of insect sizes to be subdvided among sequences of bird species. Far fewer opportunities are available for subdivision of fruit resources. Consequently, the relative increase in small insectivore diversity might be greater than for species feeding on fruit or nectar. Variation in diversity of species that feed on large insects is likely to more closely parallel that for fruits due to the distribution and abundance of large insects. How do these predictions correlate with the real world as represented by tropical forest birds?

There are rarely more than a few species of forest floor frugivores. In Panama there is usually one (rarely two) quail doves and one tinamou, while in South American forest there will commonly be several species of tinamous (Terborgh, personal communication). These usually occur in size sequences much as the size sequences observed in New Guinea fruit pigeons (Diamond 1972, 1973).

Even a cursory review of Table 14–4 supports the hypothesis that the relative increase in species exploiting small insects is much higher than for any other energy source. The difference between Panama and Liberia is not very great for small insects but is relatively large for nectar, fruit, and vertebrates. In the case of fruits and nectar this supports my general impression that fruits and flowers were reduced in abundance in Liberian forest compared to what one would expect in similar Panama forests at an equivalent time of the year (late dry and early wet season). This may be due to any of several factors. For example, rainfall and, hence, water for production of fruits may be less reliable from year to year in Africa. This would produce a counterselective effect, reducing the likelihood of selection for seed dispersal by fruit-eating birds and/or bats, and biomass of fruit-eaters within the resident faunas are lower in Liberia than in Panama (11,000 versus 18,000 g/40 ha, respectively).

There is also a striking reduction in the number of "flesh-eating" species in the African forest. This is, I believe, due to changes in resource availability, especially in the form of lizards and snakes. In the neotropics I have been impressed by the abundance of lizards of a wide variety of sizes, especially of the genus *Anolis*. Many of the hawks (especially *Leucopternis*) seem to specialize in feeding on these and other reptiles. Even some of the bird hawks of the genus *Micrastur* may not uncommonly feed on reptiles. For example, I have seen *Micrastur mirandollei* successfully capture a snake about three feet long in Panama.

By contrast, lizards seem less abundant and less diverse in African forest, and to my knowledge no forest hawk specializes on reptiles. The abundance of large hawks and eagles may be unusually low in the vicinity of Mt. Nimba because of hunting pressure, but even if all of the potential species are considered, Liberia is still obviously short of forest raptors relative to Panama.

It appears that variation in community diversity does not occur as a result of consistent community-wide variation. Rather, certain groups have changed markedly between tropical and temperate forest (small insectivores), while others have been added in the tropics (large insectivores, frugivores, and nectarivores). In addition, in tropical forests there is an increase in the subdivision of the resources.

## Annual Productivity in Tropical and Temperate Forest

The numerous attempts to summarize the voluminous data on productivity of terrestrial environments have been only moderately successful (Whittaker, 1970; Jordan, 1971a, 1971b; Golley, 1972; Golley and Lieth, 1972). These attempts have been hampered by inconsistencies among the studies in such important characteristics as field techniques, plant parts collected, and time sequences of data collection relative to seasonality in plants and climatic patterns.

However, Golley (1972), summarizing data from Whittaker (1970), has provided at least a tentative summary of production in several habitats in both grams per square meter and kilocalories per square meter (Table 14–5). The present discussion will be restricted to consideration of kilocalories per unit area as the energy requirements for the bird communities are determined in these units.

Three general conclusions on productivity in tropical and temperate forests are particularly relevant to subsequent discussion.

1. Caloric value (kilocalorie per gram) of tropical trees tends to be lower than in temperate trees (Golley 1969, Jordan 1971). This is a significant argument for restricting the present discussion to consideration of productivities by energy and not weight.

2. A very significant conclusion of Jordan (1971a, 1971b) is that annual wood production in tropical and temperate forest is approximately the same, while annual leaf and fruit production in tropical forest is approximately three times the temperate

**Table 14–5.** Annual net primary productivity (after Golley, 1972), annual and daily energy requirements of resident forest birds, and existence energy as proportion of net annual primary production for tropical and temperate forest.

| Geographic Region | Net Annual Primary Production (kcal/m²/yr) | Annual Energy Requirements | | Existence Energy as Proportion of Net Annual Primary Production |
|---|---|---|---|---|
| | | kcal/40 ha/yr | kcal/m²/yr | |
| Tropical | 8,400 | $6.19 \times 10^6$ | 15.5 | 0.18 |
| Temperate | 5,980 | $4.17 \times 10^6$ | 10.4 | 0.17 |
| Tropical:temperate ratio | 1.48 | 1.40 | 1.40 | 1.06 |

values. Annual litter production of tropical forest, which may be more important than wood production to most animals (Willson, 1973), will exceed that of temperate forest.

3. However, total production per day of the growing season is similar in both geographic areas, as is daily production during the growing season for leaves and fruits (Willson, 1973). Therefore, on any given day in the breeding season there is about the same amount of production in temperate and tropical forest. This supports the earlier procedure of looking at energy requirements on a per-day basis for the avian communities.

What are the annual energy requirements of resident birds and how do they compare with total net productivities? It is relatively easy to determine existence energy requirements for the Panama forest as temperatures remain constant throughout the year, and I was not able to discern any significant seasonal variation in species composition and abundances of resident species. Consequently, to determine annual existence energy requirements, the per-day energy requirements were multiplied by 365, yielding a figure of 15.5 kcal/m$^2$/y (Table 14–5).

Since seasonal changes in resident avifaunas are extreme in temperate forest and I do not have a full year's data, several assumptions are necessary to determine the annual energy requirements of temperate forest residents. In a comparison of several hundreds of Audubon bird censuses for a variety of habitats, Webster (1966) found that 183 eastern deciduous forest areas in North America had winter bird populations about 25 percent of their summer (breeding) densities. Since winter temperatures are significantly below the summer temperatures, weight metabolism equations were determined for the mean temperature of 3°C. The assumption was also made that the resident faunas at breeding season and winter densities were each present for periods of six months. With these assumptions, the annual energy requirements were about 10.4 kcal/m$^2$/day. Ratios for net annual production of plants and energy requirements of birds are surprisingly similar, 1.40 and 1.48, respectively (Table 14–5). This suggests that on an annual basis, energy requirements of forest avifaunas increase in proportion to net annual primary production in the same habitats. One might then ask, what proportion of the annual energy requirements is taken by species exploiting the new food resources discussed earlier? Recall that 25 and 27 percent of the daily energy requirements are taken by the species exploiting new resources in tropical forest. That is approximately one-third of the energy, or 5.2 kcal/m$^2$/yr. This leaves 10.3 kcal/m$^2$/yr for the remaining species in the tropical forest as compared to 10.4 kcal/m$^2$/yr in the temperate forest.

## Energy Flow and Significance to Community

Not uncommonly, one hears hears the comment, "Why study a group of organisms that are relatively unimportant to the total energy dynamics of an ecosystem?" On the surface this seems a significant question with respect to birds when we consider that the annual existence energy requirements for the avifauna account for only 0.17 and 0.18 percent (Table 14–5) of the total net production in temperate and tropical forest.

Existence energy requirements are minimal estimates, as they do not include energy required for such activities as foraging, reproduction, or other maintenance activities of residents, nor do they include the energy requirements of stragglers, transients, or winter visitors. However, even if these figures increased the energy requirements by an order of magnitude, it would still be less than 2 percent of the net production.

These arguments fail to consider several very important functions provided by birds (and the same argument holds for almost any group). As demonstrated by several papers in this volume, such essential functions as pollination, seed dispersal, and so on are served by a large variety of species that contribute minimal amounts to the total energy flow. There can be little doubt that virtually any terrestrial community would be very different if these and other types of interspecific interactions were not maintained.

## Conclusions

An attempt to account for variation in avian community diversity with the productivity hypothesis gives unsatisfactory results. An alternative hypothesis that stresses the number of types of harvestable resources seems more plausible for resident bird communities. The increase in species diversities of tropical forest avifaunas over temperate forest avifaunas seems to be in direct proportion to the amount of energy passing through a variety of newly available resources in tropical communities. Since daily existence energy requirements of resident avifaunas vary much less than numbers of species or total species densities, it is suggested that energy compensation may be a relatively general phenomenon in a wide variety of similar communities.

Rare species make up a greater proportion of the avifauna in tropical than in temperate forest. Heretofore, rare species have been more or less ignored in studies of avian community organization. Theoretical considerations on the subdivision of resources suggest that species packing on relatively simple, single-dimension resources will maintain fewer species than resources that can be subdivided on several axes of exploitation. In support of this theory it is shown that the number of rare species exploiting small insects increases more rapidly than the number of rare species that exploit fruit resources. Variation in nature, abundance, and temporal availability of resources account for some anomalies in species densities and resource exploitation patterns between African and neotropical forest birds.

Ratios of net annual plant production and existence energy requirements of Panama and Illinois forest avifaunas are strikingly similar. When tropical species that exploit the "newly available" resources—fruits, large insects, and nectar—are excluded, annual energy requirements per unit area are very similar when all resident temperate species are compared with resident tropical species that exploit the same set of resources.

Three major factors influencing diversity of birds are the temporal distribution of resource (good) availability, number of types of resources (food) available, and the complexity of interactions among resource dimensions (including food types and spatial dimensions).

## Acknowledgment

I thank the LAMCO Joint-Venture Mining Company at Mt. Nimba, Liberia, for making the facilities of the Nimba Research Laboratory available; Alec Forbes-Watson and Stuart Keith for introducing me to the African avifauna; and Helen Lapham for help with fieldwork in Liberia. The Smithsonian Institution provided continuing support in the form of a postdoctoral fellowship through its bureau, the Smithsonian Tropical Research Institute, and a PL-480 Travel Grant from the Office of International Activities of the Smithsonian Institution. Travel support to attend the 2nd International Symposium in Tropical Ecology was provided by the Purdue Research Foundation and the School of Science at Purdue University. J. M. Diamond, F. B. Golley, D. H. Janzen, R. R. Roth, R. S. Wilcox, E. O. Willis, M. F. Willson, and A. T. Winfree constructively criticized the thoughts presented here.

## References

Bent, A. C. 1939. Life histories of North American woodpeckers. *U.S. Nat. Mus. Bull.* 174.

Brown, J. 1971. Mammals on mountaintops: Nonequilibrium insular biogeography. *Am. Naturalist* 105:467–478.

Cody, M. L. 1970. Chilean bird distribution. *Ecology* 455–464.

Connell, J. H., and E. Orias. 1964. The ecological regulation of species diversity. *Am. Naturalist* 98:399–414.

Diamond, J. M. 1972. Avifauna of the eastern highlands of New Guinea. *Publ. Nuttall Ornith. Club No. 12* 438 pp.

———. 1973. Distributional ecology of New Guinea birds. *Science* 179:759–769.

Golley, F. B. 1969. Caloric value of wet tropical forest vegetation. *Ecology* 50:517–519.

———. 1972. Energy flux in ecosystems. In *Ecosystem Structure and Function*, J. A. Wiens, ed. pp. 69–88. Corvallis: Oregon State Univ. Press.

———, and H. C. Lieth. 1972. Bases of organic production in the tropics. In *Tropical Ecology with an Emphasis on Organic Production.* pp. 1–26. Athens: Univ. Georgia Press.

Jordan, C. F. 1971a. Productivity of a tropical forest and its relation to a world pattern of energy storage. *J. Ecol.* 59:127–142.

———. 1971b. A world pattern in plant energetics. *Am. Sci.* 59:425–433.

Karr, J. R. 1968. Habitat and avian diversity on strip-mined land in east-central Illinois. *Condor* 70:348–357.

———. 1971. Structure of avian communities in selected Panama and Illinois habitats. *Ecol. Monographs* 41:207–233.

———, and R. R. Roth. 1971. Vegetation structure and avian diversity in several new world areas. *Am. Naturalist* 105:423–435.

Lack, C. F. 1972. Seasonal changes in feeding pressures of fruit- and nectar-eating birds in Panama. *Condor* 74:54–60.

MacArthur, R. H. 1972. *Geographical Ecology*. 269 pp. New York: Harper and Row.

———, and J. W. MacArthur. 1961. On bird species diversity. *Ecology* 42:594–598.

———, and E. R. Pianka. 1966. On optimal use of a patchy environment. *Am. Naturalist* 100:603–609.

———, and E. O. Wilson, 1967. *The Theory of Island Biogeography*. Princeton, N.J.: Princeton Univ. Press.

———, J. M. Diamond, and J. R. Karr. 1972. Density compensation in island faunas. *Ecology* 53:330–342.

———, H. Recher, and M. Cody. 1966. On the relation between habitat selection and species diversity. *Am. Naturalist* 100:319–332.

May, R. M., and R. H. MacArthur. 1972. Niche overlap as a function of environmental variability. *Proc. Nat. Acad. Sci. U.S.A.* 69:1109–1113.

Moreau, R. E. 1966. *The Bird Faunas of Africa and Its Islands*. 424 pp. New York: Academic Press.

———. 1972. *The Palaearctic-African Bird Migration Systems*. 384 pp. New York: Academic Press.

Orians, G. H. 1969. The number of bird species in some tropical forests. *Ecology* 50:783–801.

Pearson, O. P. 1960. Torpidity in birds. *Bull. Mus. Comp. Zool. Harvard* 124:93–103.

Pianka, E. R. 1967. On lizard species diversity: North American flatland deserts. *Ecology* 48:333–351.

Recher, H. 1969. Bird species diversity and habitat diversity in Australia and North America. *Am. Naturalist* 103:75–80.

Salt, G. W. 1953. An ecologic analysis of three California avifaunas. *Condor* 55: 258–273.

Schoener, T. W. 1971. Large billed insectivorous birds: A precipitous diversity gradient. *Condor* 73:154–161.

Skutch, A. F. 1950. The nesting seasons of Central American birds in relation to climate and food supply. *Ibis* 92:185–222.

Terborgh, J. 1971. Distribution on environmental gradients: Theory and a preliminary interpretation of distributional patterns in the avifauna of the Cordillera Vilcabamba, Peru. *Ecology* 52:23–40.

———, and J. S. Weske. 1969. Colonization of secondary habitats by Peruvian birds. Ecology 50:765–782.

Webster, J. D. 1966. An analysis of winter bird-population studies. *Wils. Bull.* 78: 456–461.

Whiteside, M. C., and R. V. Harmsworth. 1967. Species diversity in Chydorid (Cladocera) communities. *Ecology* 48:664–667.

Whittaker, R. H. 1970. *Communities and Ecosystems*. New York: Macmillan.

Willis, E. O. 1966. The role of migrant birds at swarms of army ants. *Living Bird* 5:187–231.

Willson, M. F. 1973. Tropical plant production and animal species diversity. *Trop. Ecol.* 14:62–65.

# Savannah and Grassland Systems

Ecologists are well aware that a major part of the tropics is covered by savannah, dry forest, or grassland vegetation. In many countries the rain forest, which has attracted so much of the attention of temperate visitors to the tropics, may comprise relatively little of the landscape. Savannahs and grasslands are important in all three tropical continents, and the following papers provide information for Africa, the Americas, and India.

Lamotte describes the long and intensive studies made in the savannah of Ivory Coast. This investigation has focused on the ecosystem, and it comprises one of the few IBP tropical ecosystem studies in the world. Sarmiento and Monasterio develop an explanation for savannah vegetation in the tropics, especially discussing the American example, with which they have extensive experience. Their approach is broad or, in their terms, holocoenotic, focusing on the interaction of soil, water, vegetation, and biotic and human influences as they interact to result in savannah communities. San Jose and Medina present some detailed data on production and transpiration in one of the Venezuelan savannahs discussed by Sarmiento and Monasterio, providing an example of the data supporting their conclusions. Shrimal and Vyas also present detailed information on productivity of a dry Indian grassland, following the methodology and approach of R. Misra.

# CHAPTER 15

## The Structure and Function of a Tropical Savannah Ecosystem

MAXIME LAMOTTE

### Introduction

The higher levels of biological organization, called ecosystems, are of growing interest to ecologists. Ecosystems are defined as systems of plants and animals interacting together in a given environment. Research on ecosystems can be quite complicated since even the simplest systems have numerous plant and animal species, which show great diversity of size, distribution, rate of biological increase, diet, and so on. These parameters also vary with time, so that it is necessary to account for their heterogeneity in time and space.

This report provides preliminary information on a ten-year study of a *Borassus* palm savannah ecosystem at Lamto, Ivory Coast (Figure 15–1). Although the report is necessarily incomplete, it is designed to provide the elements necessary for a synthesis of trophic relationships and the flow of energy in the savannah ecosystem. Data will be presented on the general physical environment, the structure of the ecosystem, and the functioning of the primary producers, consumers, and decomposers.

### Climatic Factors[1]

The Lamto savannah (6° 13′ N; 5° 02′ W) is about 100 km north of the Gulf of Guinea, at the edge of the rain forest (Figure 15–2).

In this region the mean monthly temperatures are remarkably constant, as shown by the average monthly values obtained over a ten-year period (in °C):

[1] Data from J.-L. Tournier.

| Jan. | Feb. | March | April | May | June |
|---|---|---|---|---|---|
| 26.82 | 28.21 | 28.28 | 27.79 | 27.26 | 26.91 |
| July | Aug. | Sept. | Oct. | Nov. | Dec. |
| 25.27 | 25.14 | 25.70 | 26.19 | 26.86 | 26.59 |

**Fig. 15–1.** Aerial view of a section of *Borassus* palm savannah at Lamto. One can also observe gallery forest in the photograph and the remains of quantitative sample plots.

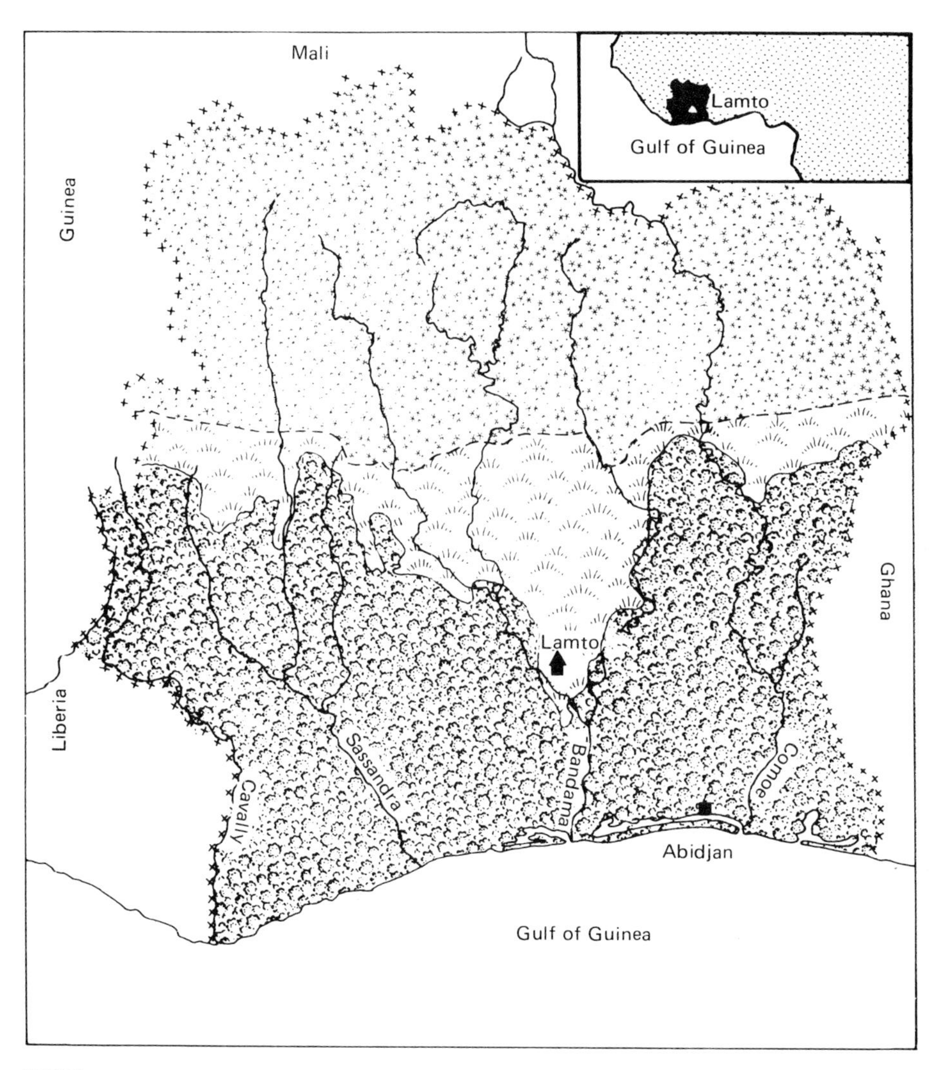


Dense Forest (forêt dense)

Guinean Savannah (savanes guinéenes)

Northern Savannah (savanes septentrionales)

**Fig. 15–2.** Major types of vegetation in the Ivory Coast and the location of Lamto in the savannah.

There is a similar lack of variation in the mean daily temperature, which fluctuates between 24° and 30°C in the dry season and between 25° and 28°C in the wet season. The diurnal variation in temperature is high, but the minimum temperature does not go below 16°C nor the maximum above 34°C. The maximum daily variation (about 18°C) occurs during the dry season (December, January); the minimum variation (8°C) occurs during the wet season (July, August).

As far as the biota is concerned, rain is the predominant environmental factor. Annual rainfall averages 1,300 mm but may be as low as 900 mm and as high as 1,700 mm.

In contrast to temperature, rainfall measurements exhibit a definite seasonal pattern. A strong difference exists between the dry season, which generally lasts from November or December to the end of February, and the rainy season of more than eight months, which may be interrupted in August by a short dry season. The ombrothermic diagram (Figure 15–3), illustrating the balance between precipitation and evaporation and the mean monthly temperature, shows clearly the alternation of seasons. As the diagram is based upon the mean of data from ten years, this obscures the fact that there is some variability from year to year.

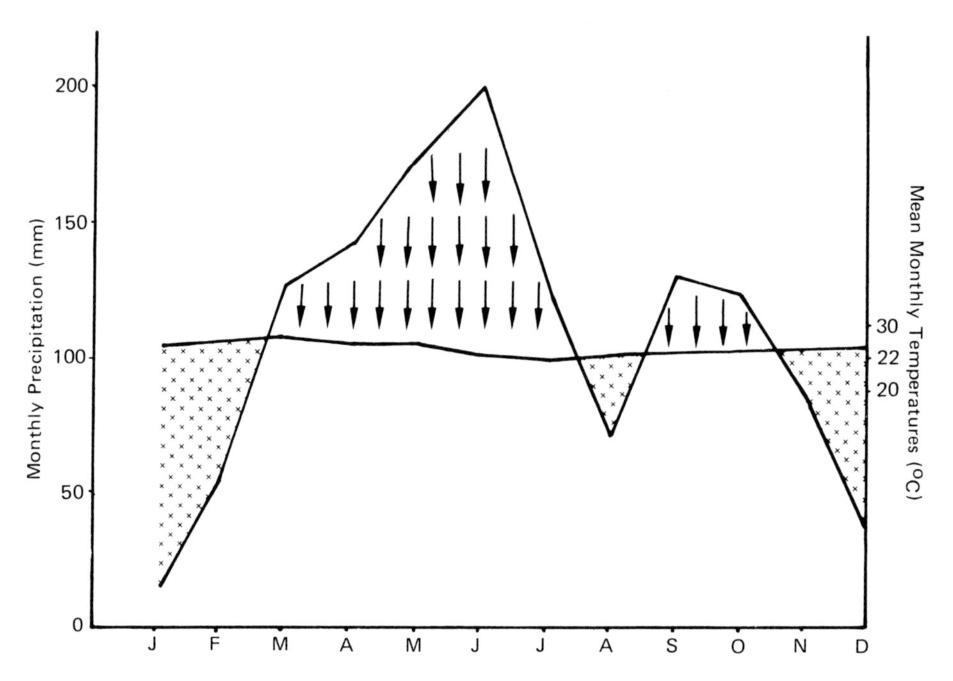


**Fig. 15–3.** Ombrothermic diagram based on ten years of observations in the Lamto savannah. (After Lachaise (1972) and unpublished data of J.-L. Tournier.)

Photoperiod varies little, from a minimum in January (11 hr, 49 min) to a maximum in June–July (12 hr, 26 min). In contrast, seasonal insolation varies markedly, with the minimal values between June and September, in the extent of cloudiness. Insolation varies from 346 cal/cm$^2$/day in the wet season (July) to 544 cal/cm$^2$/day at the end of the dry season (March). This represents about 1,686 $\times$ 10$^6$ cal/m$^2$/yr, or a mean of 462 cal/cm$^2$/day. The percentages of the input in the visible, infrared, and ultraviolet are of the order of 71, 20, and 9 percent of the annual total, respectively.

## Primary Producers

The Lamto savannah may be divided into two major types, between which exist a number of intermediates:

1. Herbaceous savannah is characterized by a grass layer of *Loudetia simplex* dominated by a tree stratum, composed of one species, the *Borassus* palm (*Borassus aethiopum*). The herbaceous stratum also contains about 300 stalks per hectare of woody hemicryptophytes and chamaephytes.

2. Shrub savannah is characterized by an important shrub stratum and by the development of a more or less open tree stratum. The grass layer, dominated by Andropogoneae (*Andropogon schirensis, Hyparrhenia* spp.), is composed of numerous species of grasses, as well as a number of small woody species. From the density of trees and shrubs several facies may be identified, which merge imperceptibly into one another.

The structure of both types of savannah is greatly influenced each year in January by fire. Clearly, the annual bush fire plays a greater role in preventing any reforestation of the savannah than the climatic or edaphic factors.

Because of fundamental differences in their structure and biology, the grass and shrub layers require relatively independent study with distinct techniques for each.

### The Herbaceous Stratum[2]

The herb layer contains a relatively high number of species, most of which are grasses (Figure 15–4). Only some are very abundant and represent a notable fraction of the vegetation. These are: Gramineae—*Hyparrhenia diplandra, Hyparrhenia chrysargyrea, Hyparrhenia rufa, Hyparrhenia dissoluta, Imperata cylindrica, Andropogon schirensis, Schizachirium platyphyllum, Brachiaria brachylopha;* Cyperaeceae—*Bulbostylis aphyllanthoides, Fimbristylis ferruginea, Cyperus obtusiflorus, Fimbristylis monostachya*; Leguminosae—*Tephrosia bracteolata, Indigofera polysphaera*; Compositae—*Vernonia guineensis, Aspilia bussei.*

The grasses are distributed in dense clumps. The vertical structure has also been defined (Figure 15–5) and varies with fire.

The morphology of all the savannah species has been studied in detail. Figure 15–6 shows the large development of subterraneous organs that represent a major fraction of the biomass.

Changes in biomass have been followed seasonally, considering both the above- and belowground portions. The aboveground portion of the herbs varies greatly over a year, dropping to zero immediately after passage of fire. In the *Loudetia* savannah the aboveground living biomass reaches a maximum of 5.5

[2] Work of J. César and J.-Cl. Roland.

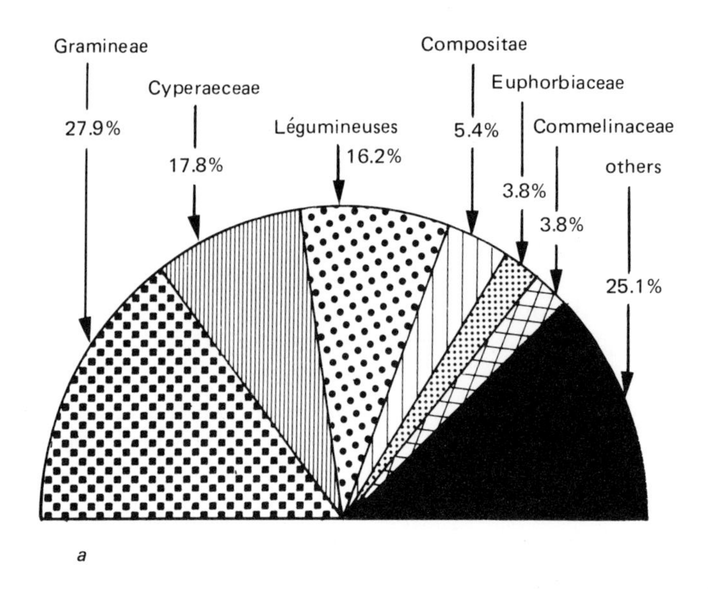


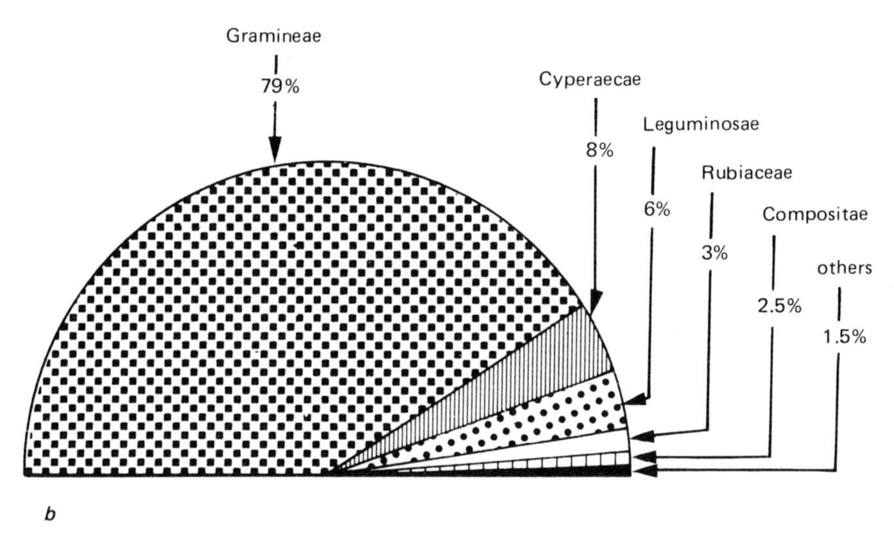


**Fig. 15–4.** Specific composition of the herb stratum. Numbers of species (*a*) and biomass (*b*) corresponding to the different families. (After Roland, 1967.)

tons/ha (dry weight) (Figure 15–7); the total biomass, living or dead, reaches 7 tons/ha in November. In the *Andropogon* savannah the aboveground biomass attains 2.3 tons/ha two months after the fire, and reaches a maximum in October of 8 tons/ha of living material. After June the dead matter begins to accumulate, attaining almost 3.3 tons/ha just before the fire; the maximum total standing crop is 10 tons/ha, in November. In the wooded *Andropogon*

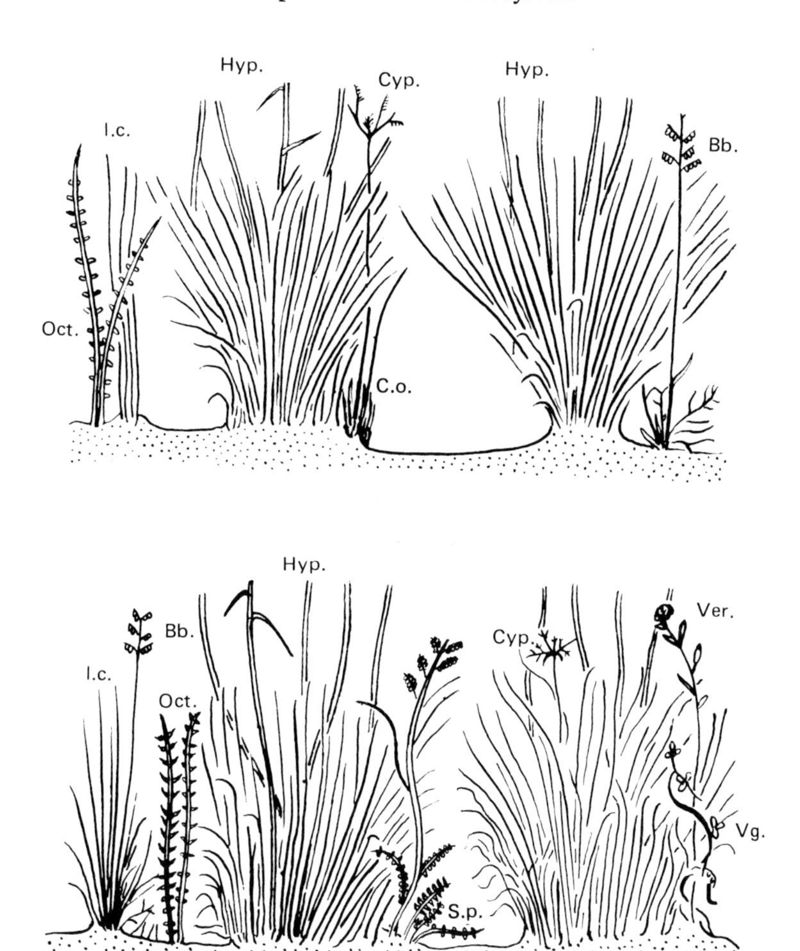


**Fig. 15–5.** Vertical structure of the herb stratum in the *Andropogon* savannah. At the top is burned savannah; at the bottom, unburned. Hyp: Tufts of different species of *Hyparrhenia*; Bb.: *Brachiaria brachylopha*; Ic.: *Imperata cylindrica*; Sp.: *Schizachyrium platyphyllum*; Cyp.: *Cyperus schweinfurthianus*; Co.: *Cyperus obtusiflorus*; Vg.: *Vigna multinervis*; Oct.: *Octodon setosum*; Ver.: *Vernonia nigritiana*. (After Roland, 1967.)

savannah the living aboveground herb biomass only reaches 5 tons/ha in August and the total biomass 7 tons/ha in December. The situation is very different in the unburned savannahs (Figure 15–7*d*). The minimum standing crop falls to 1 to 2 tons/ha in March–April, depending upon the facies. The total biomass attains an October peak of 13.5 tons/ha, of which 10 tons/ha is living.

The seasonal fluctuations have been followed for the principal species in the phytocenosis. The peaks of the different species do not actually coincide in time. Figure 15–8 shows the seasonal changes in biomass that have been observed for some of the more typical species. The root biomass of the herb stratum is always maximum in the upper 10 cm of soil and decreases exponentially

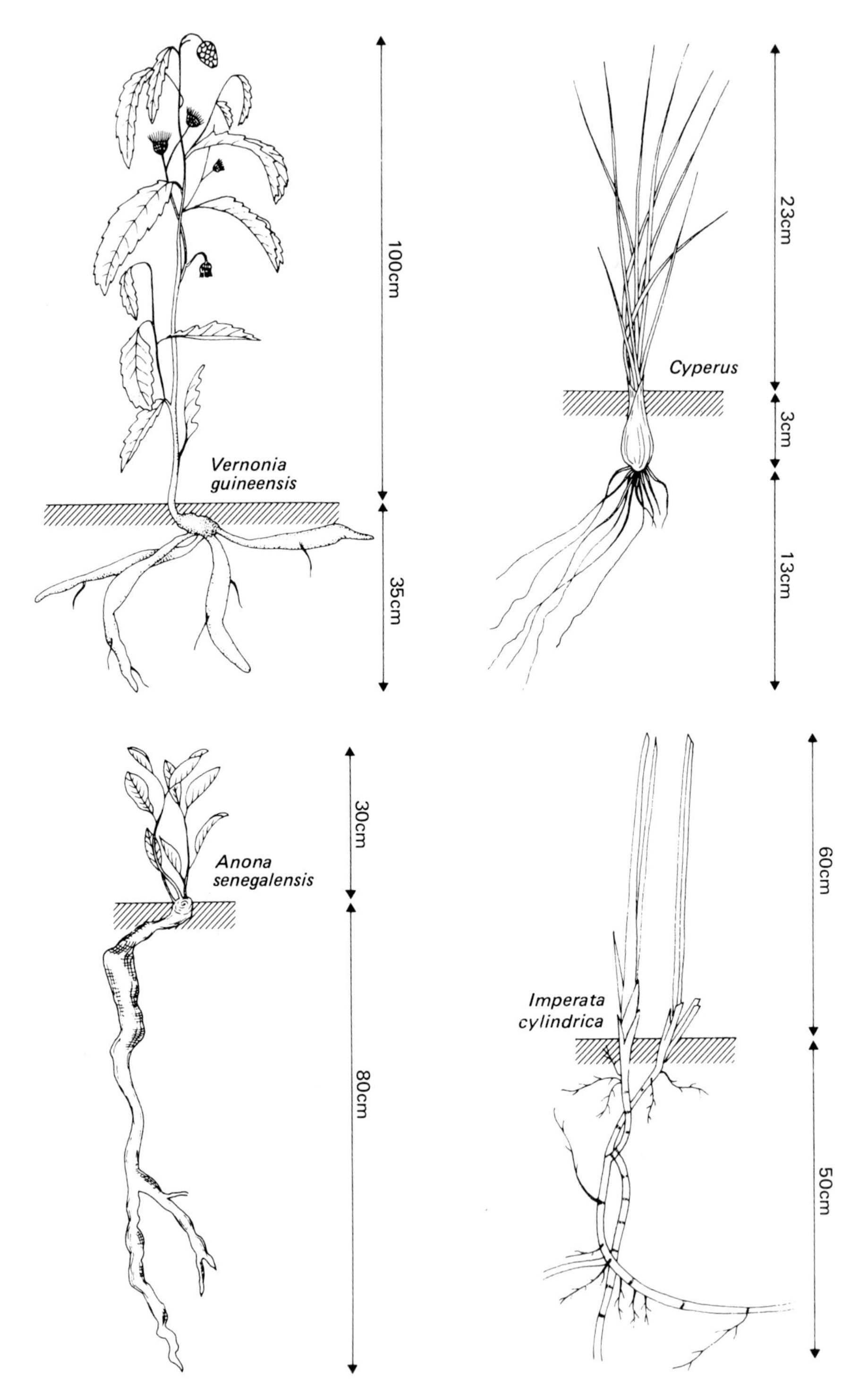


**Fig. 15–6.** Structure of some species of herbs in the savannah. (After Monnier, 1968.)

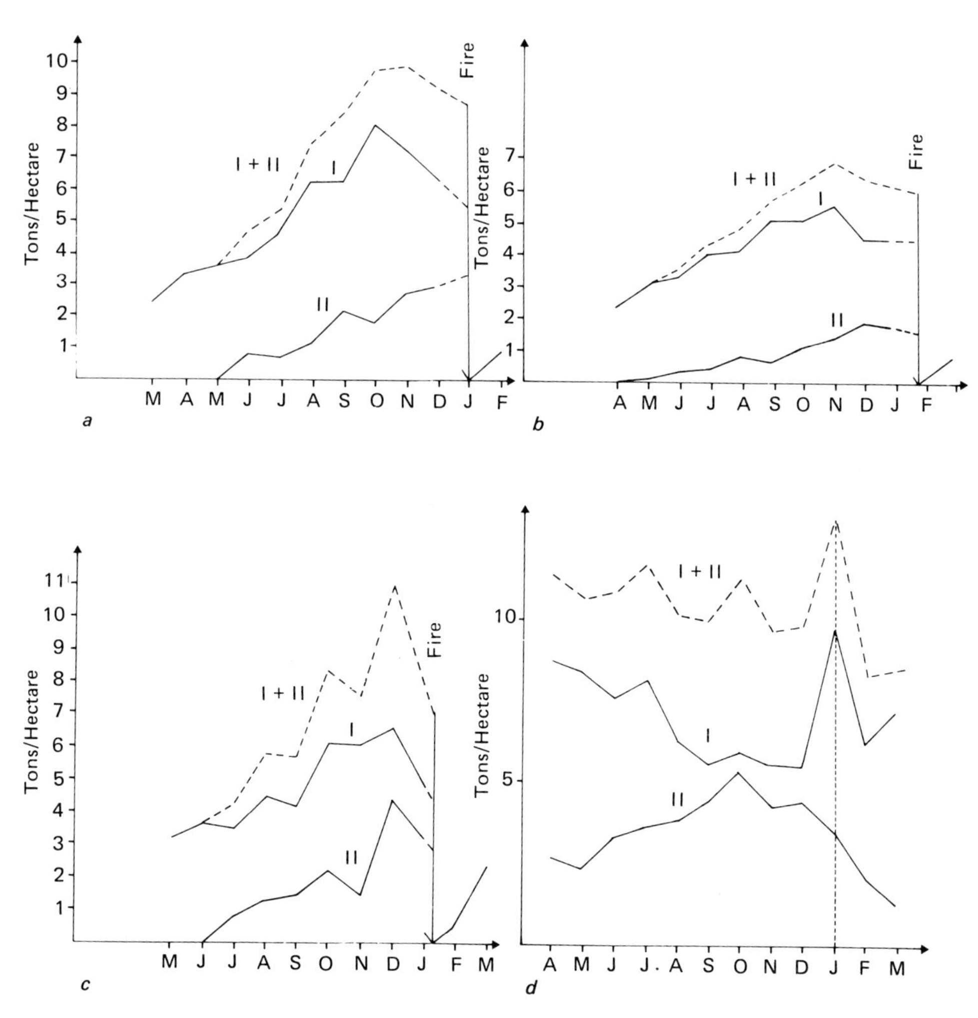


**Fig. 15–7.** Seasonal variation in aboveground biomass of the herb strata. (After Cesar, 1971.) *I*, Live matter. *II*, Dead matter. *a*, Grass *Andropogon* savannah. *b*, Grass *Loudetia* savannah. *c*, Wooded, *Andropogon* savannah. *d*, Unburned wooded savannah.

with depth. On the average, the belowground biomass of the herb stratum is 19 tons/ha for the *Loudetia* savannah, 14 tons/ha for the *Andropogon* savannah, and 10 tons/ha for the *Andropogon* shrub-savannah.

The seasonal fluctuations of the underground parts are very irregular and vary from one station to another. Over the year biomass fluctuates between 9 and 21 tons/ha in the *Andropogon* savannah (Figure 15–9*a*), 14 and 25 tons/ha in the *Loudetia* savannah, 6 and 17 tons/ha in the *Andropogon* shrub-savannah, and finally 5 and 20 tons/ha in the unburned savannah (Figure 15/9*b*). These data give a minimum estimate of the underground production of the herb stratum.

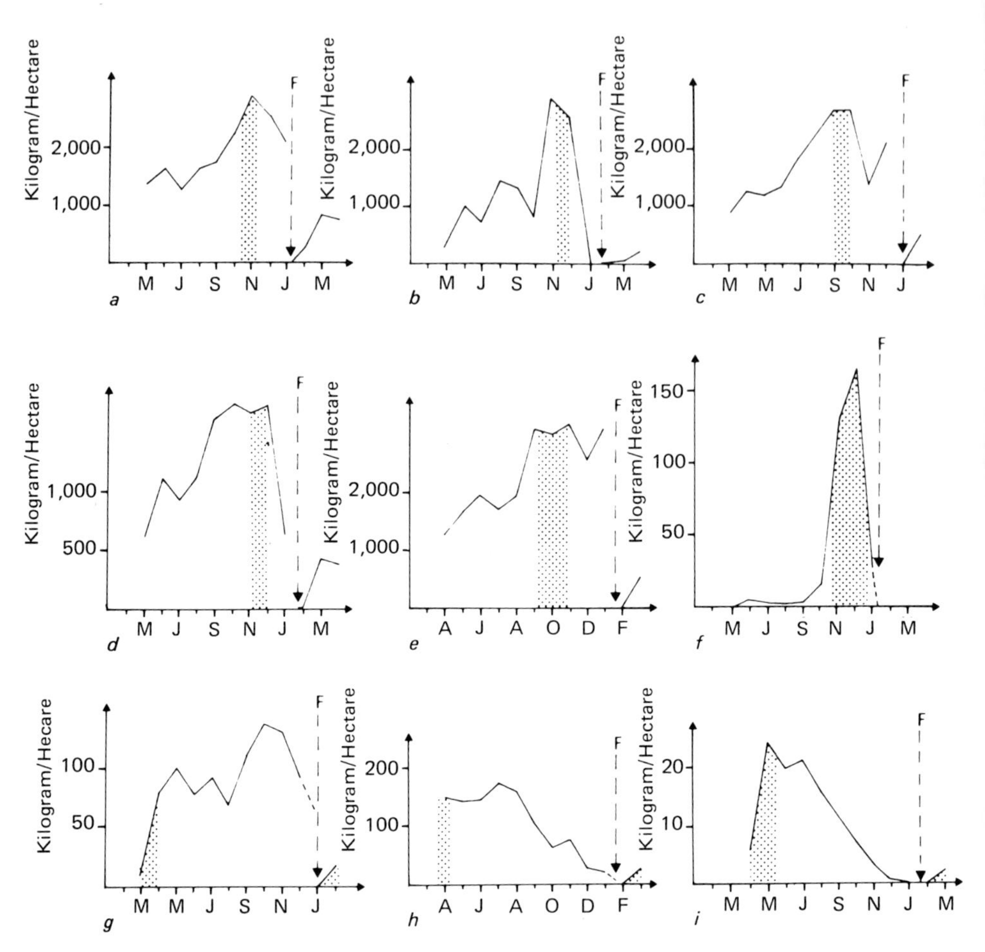


**Fig. 15–8.** Seasonal variation of biomass of selected species in burned savannah. (After Cesar, 1971.) a, *Hyparrhenia diplandra*; b, *Sorghastrum bipennatum*; c, *Andropogon schirensis*; d, *Hyparrhenia chrysargyrea*; e, *Loudetia simplex*; f, *Tephrosia elegans*; g, *Brachiaria brachylopha*; h, *Bulbostylis aphyllantoides*; i, *Cyperus schweinfurthianus*. Grey area represents the period of reproduction.

## The Woody Elements[3]

Besides the *Borassus* palms, the woody elements are essentially represented by four species: *Crossopterix febrifuga* (*Rubiaceae*), *Piliostigma thonningii* (Cesalpiniaceae), *Bridelia ferruginea* (Euphorbiaceae), and *Cussonia barteri* (Araliaceae). In addition, one may encounter *Annona senegalensis* (Annonaceae), of smaller size, and *Ficus capensis* (Moraceae), mostly located near the gallery forests.

On the study plots in different types of savannah, protected or not from fire, the spatial distribution of species was studied as a whole and by stratum.

[3] Work of J.-Cl. Menaut and J.-Cl. Roland.

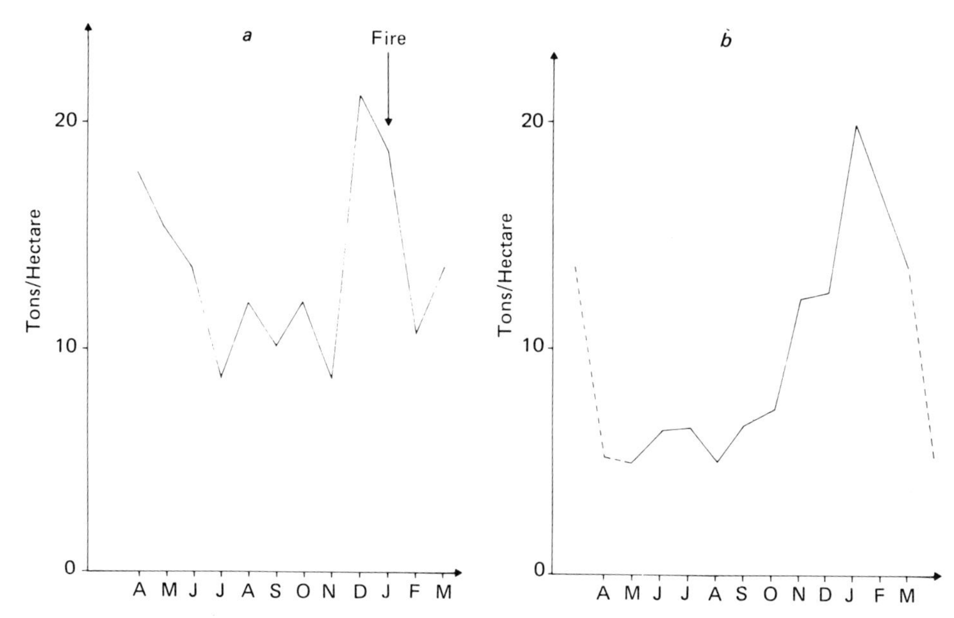


**Fig. 15–9.** Seasonal variation of root biomass. (After Cesar, 1971.) *a*, In a section of regularly burned savannah. *b*, In unburned savannah.

For each tree or shrub, the height, circumference at different levels, and form and surface of the crown were measured.

From measurements of a great many of individuals it is possible to correlate weight and some other simple measurements (height, circumference, surface of the crown) (Figure 15–10). As for the herbs, the importance of the roots is an ecological characteristic of populations in savannah regions, where there is a seasonal water deficit and where there are regular fires. From these kinds of data it was possible to estimate the biomass of each element on the plots to evaluate the total biomass of wood in the various facies. Table 15–1 gives an example of these results.

Graphs of correlation between weight and other measures of vegetation permit an estimate of the linear and weight growth of trees and shrubs and the productivity of the woody fraction. The first series of measurements covers three years of growth. The estimate of total production of trees and shrubs had been completed by measuring annual litter production per year.

The phenology of the woody species has been followed in detail. Several days after the passage of fire, the leaves drop, followed for the most part by the shoots of the preceding year. Soon the new buds develop, and two months after the appearance of the first shoots, the trees have completed the development of their foliage. A true state of rest does not exist over the year.

To these seasonal variations may be added noncyclic changes that result in a progressive modification of the vegetation. It appears that the *Borassus* savannah at Lamto is maintained in its present state only by fires that occur annually at the beginning of the dry season. In the absence of annual burning,

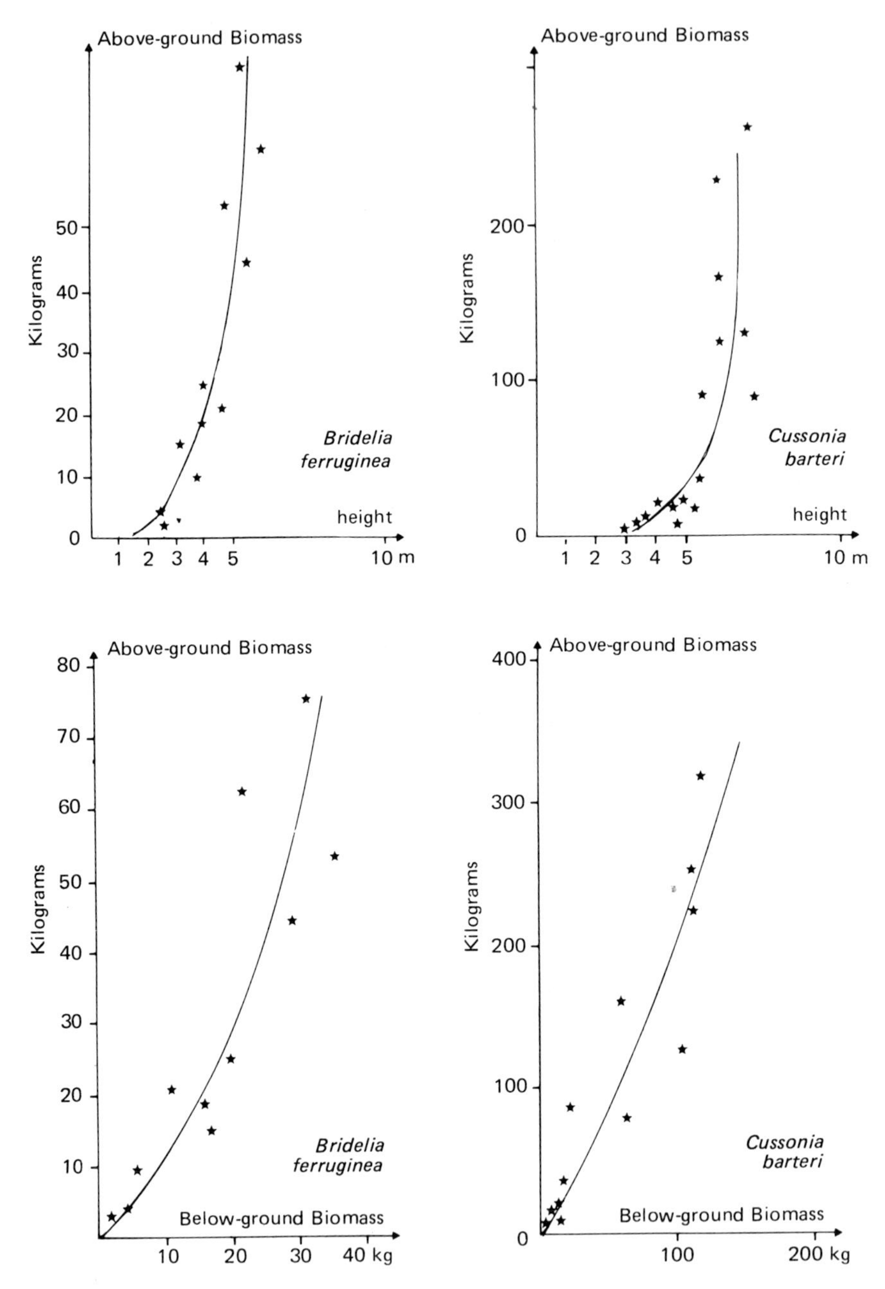


**Fig. 15–10.** Relative variation of aboveground biomass and height (top) and above- and belowground biomass (bottom) in *Bridelia ferruginea* and *Cussonia barteri*. (After Menaut, 1972.)

**Table 15–1.** Biomass of leaves, wood, roots of trees, shrubs, and palms in three facies of savannah. (After J. C. Menaut.)

| | Open Shrub Savannah | | | Dense Shrub Savannah | | | Tree Savannah | | |
|---|---|---|---|---|---|---|---|---|---|
| | Leaves | Woods | Roots | Leaves | Woods | Roots | Leaves | Woods | Roots |
| Trees and Shrubs | | | | | | | | | |
| 0–2 m | 5 | 10 | 30 | 16 | 25 | 260 | 50 | 80 | 380 |
| 2–8 m | 300 | 7,400 | 3,600 | 1,600 | 28,600 | 12,800 | 2,100 | 25,600 | 14,800 |
| over 8 m | 0 | 0 | 0 | 100 | 4,000 | 1,200 | 1,700 | 28,600 | 11,100 |
| Palms | 900 | 12,000 | 10,100 | 900 | 12,00 | 10,100 | 1,200 | 16,000 | 13,400 |
| Total in tons/ha | 1.2 | 19.4 | 13.7 | 2.6 | 44.6 | 24.4 | 5.1 | 70.3 | 39.7 |

the shrub and tree strata develop rapidly, and the woody forest species appear. Some observations suggest that in 30 years an open forest devoid of grasses may develop in place of the savannah. Thus in the studied area the savannah appears to be a "fire climax."

In arboreal facies of the savannah, the woody elements represent a large part of the total biomass, but their photosynthetic organs are of much less importance (Figure 15–11). They play a relatively modest role in ecosystem production, which is directly tied to the quantity of chlorophyll.

The relation between biomass and annual production may be graphically represented by rectangles with the width proportional to the production and surface of the biomass. The height of the rectangles shows the time of renewal of the biomass and gives a minimal estimate of the turnover rate (Lamotte, 1969). The set of rectangles constitutes a synthetic representation that could be called a "spectrum of ecosystem primary production." Figure 15–12, based on our preliminary results, represents the ecological spectrum of the shrub savannah.

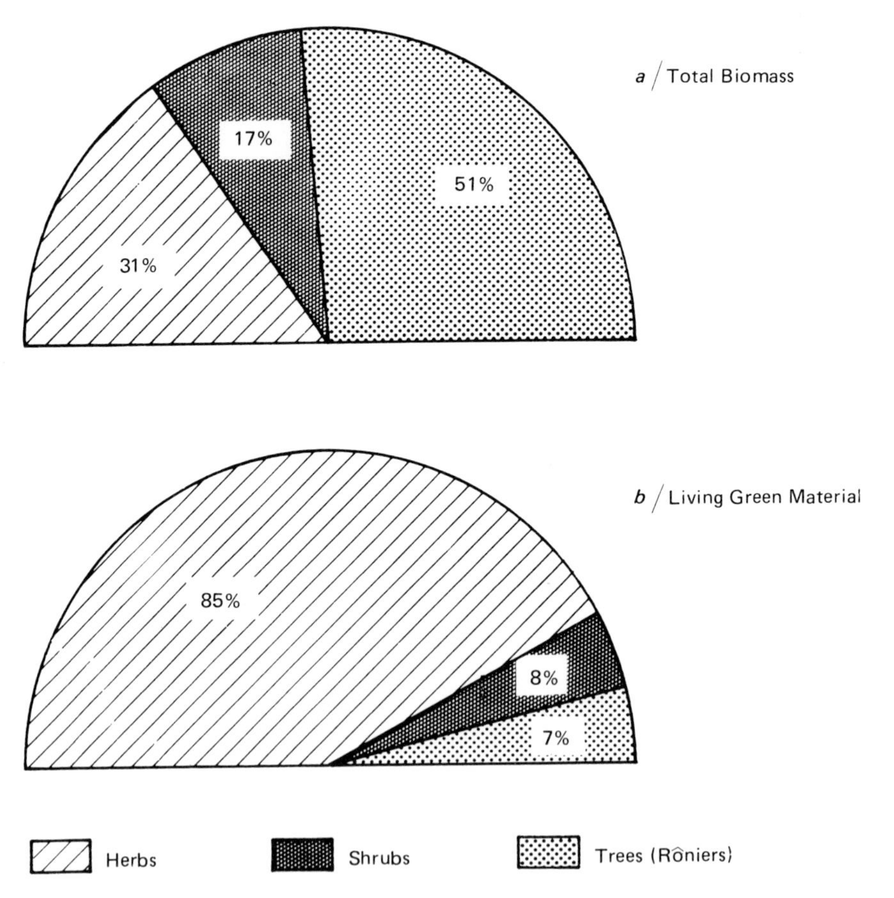


**Fig. 15–11.** Vegetation biomass (in fresh weight) in the Lamto savannah at the end of the growth period showing the relative importance of the herb, shrub and tree strata. (After Roland, 1967).

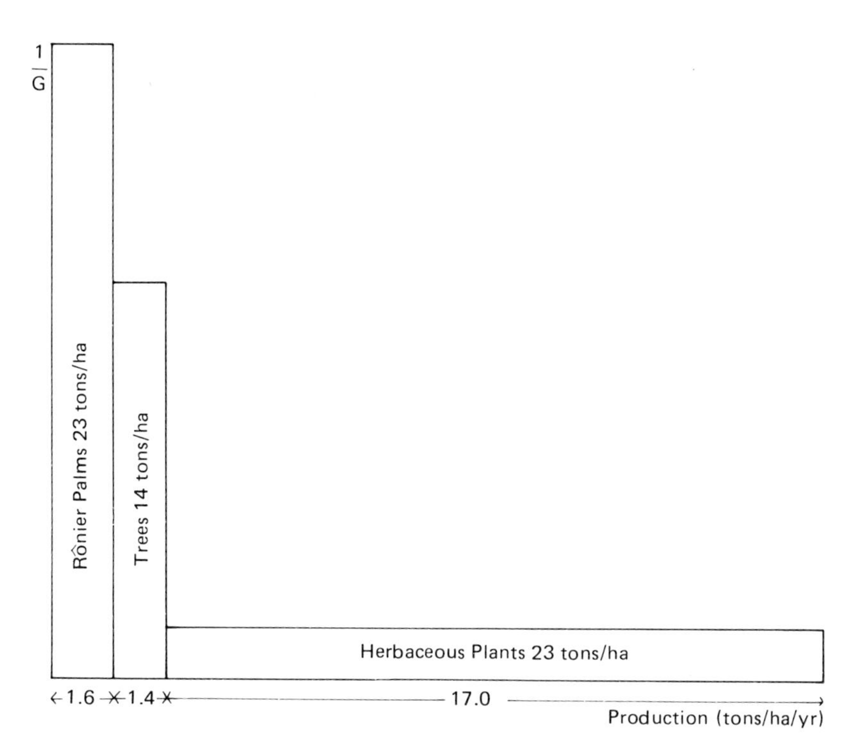


**Fig. 15–12.** Synthetic diagram of the production structure of the Lamto savannah. The surface of the rectangles represent biomass, the width production and the height is the inverse of the turnover rate.

## Animal Life in the Aboveground Strata

Contrary to the primary producers, the other organisms of the biocenosis have such a varied diet that it is almost impossible to define the different trophic levels satisfactorily. Considering the size of the organisms, as well as their extreme diversity, we have considered mainly small mammals, reptiles, and amphibians, nonsocial arthropods of the herb stratum, ants, termites, earthworms, soil microarthropods, actinomycetes, and bacteria.

For each group quantitative samples were taken to show (1) the density and biomass of the different species, (2) the structure of their populations, and (3) variations during the year. This sampling chiefly involved exhaustive hand-collecting on a predetermined and sometimes fenced surface, which, depending upon the group involved, ranged from 1 $cm^2$ to 2,500 $m^2$ (Figures 15–13 to 15–16). For aboveground arthropods the quadrats were of 1, 4, 10, 25, and 100 $m^2$; for amphibians, reptiles, and small mammals they were of 100, 400, 900 and 2,500 $m^2$.

In addition to collections, the abundant species were also raised in the

**Fig. 15–13.** Collection cage of 1 m². (Photo by Y. Gillon.)

laboratory or in field enclosures. These experiments permitted us to obtain a measure of the speed and duration of development, fecundity, individual growth and reproduction, and food consumption.

Interpretation of the results from field studies and experiments permit us, in the best case, to measure net production and food consumption of the population of the abundant species, and, by extrapolation, those of the numerically less important groups. From these data an energy budget can be calculated.

The following gives examples of some of the results.

## The Vertebrates

Birds have an important place in the savannah ecosystem since they are represented by a number of species occupying very diverse ecological niches. The most abundant birds are granivorous or frugivorous species, but many are insectivorous. Their numbers are minimal in December in spite of the presence of migrants, but the seasonal variation is of low amplitude: their density varies from a mean of about 9 to 12 individuals per ha and their biomass

**Fig. 15–14.** Placement of a collection cage (biocenometer) for sampling 10 m². (Photo by Y. Gillon.)

**Fig. 15–15.** Quantitative sampling of invertebrate fauna of the herbs strata on a surface of 25 m² surrounded by a tarpaulin. (Photo by Y. Gillon.)

**Fig. 15–16.** Quantitative sampling of fauna on 25 $m^2$ (left) and on 100 $m^2$ (right). (Photo by Y. Gillon.)

from 850 to 1,300 g/ha. We lack the precise information needed to estimate the biological production of the bird populations; 0.2 to 0.5 kg/ha (live weight) constitutes a provisional estimate.

Among mammals, the rodents[4] with about 12 species are the best represented group; these include *Lemniscomys striatus*, *Uranomys ruddi, Pelomys dybowskii*, and *Dasymys incomtus*. The mean biomass of rodents varies from year to year as a function of climatic factors. For the year 1964–1965, biomass was of about 1,200 to 1,600 g/ha (live weight). This biomass corresponds to a production of about 4 kg/ha/yr live weight, or 1,200 g/ha/yr dry weight. The insectivora *Crocidura* represents a smaller biomass of about 85 g/ha in unburned savannah and less than 50 g/ha in burned savannah.

Large mammals, notably antelopes and monkeys, presently occur in such low numbers that they probably play no significant role in the ecosystem. This is in contrast to the situation in similar habitats elsewhere where buffalo, hippopotami, and elephants play a major role as consumers of herbage.

The most numerous vertebrates on the savannah are reptiles and amphibians[5] but their densities depend mainly upon the alternating wet and dry seasons characteristic of the regional climate (Figure 15–17). Roughly, numbers and biomass increase to a maximum at the end of the rainy season and diminish in the dry season. In the burned savannah with few trees, for example, the ranges in seasonal mean values are:

[4] Data of L. Bellier.
[5] Data of R. Barbault.

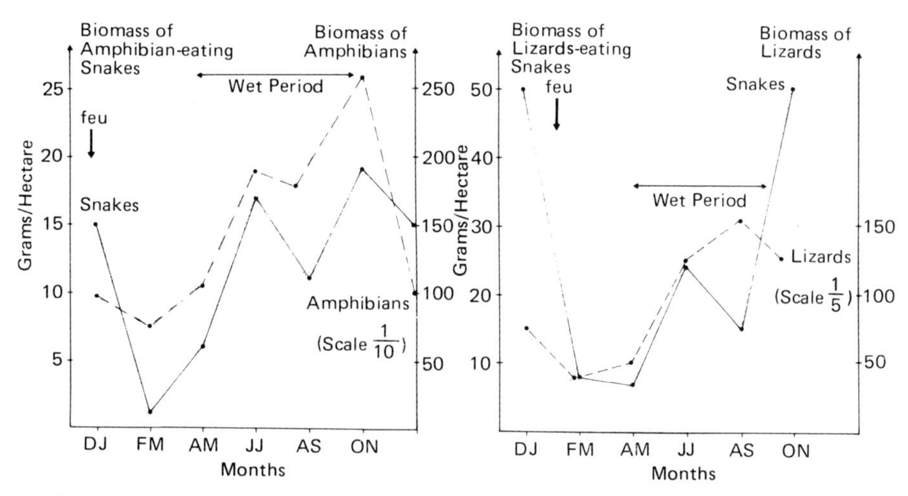


**Fig. 15–17.** Seasonal changes in the evolution of the biomass of amphibian feeding snakes and amphibians (left) and the sauriphage snakes and lizards (right). (After Barbault, 1967.)

| | *In Fresh Weight* | *In Dry Weight* |
|---|---|---|
| Snakes | 75 to 270 g/ha | 22 to 71 g/ha |
| Amphibians | 77 to 258 g/ha | 15 to 52 g/ha |
| Lizards | 39 to 153 g/ha | 12 to 46 g/ha |

To these surface-dwelling species must be added the populations living in *Borassus* palms (0 to 110 g/ha of amphibians and 100 g/ha of lizards, fresh weight).

Amphibians and lizards feed on invertebrates and are essentially opportunists. As they are of different sizes and occupy space and time differently, all of the species do not feed on the same prey. The lizards eat arachnids and orthoptera; only one species (*Panaspis nimbaensis*) eats termites. All species, except geckos and myrmecophageous *Agama agama*, neglect ants and coleoptera, which constitute the staple diet of the amphibians. The snakes stand at a higher level of the food web; with the exception of termite feeders and rare unspecialized insect feeders, all consume vertebrates. Most of the species are specialized, feeding on amphibians (*Natriceteres olivaceus* and *Causus rhombeatus*), eggs (*Dasypeltis scabra*), or small mammals (*Bitis arietans* and *Boaedon lineatus*). Nevertheless, two of the commonest species are euryphagous, with a preference for lizards in the case of *Psammophis sibilans* and rodents in the case of *Echis carinatus*.

An analysis of the seasonal cycle in population structure was conducted for each dominant species to estimate the average growth rate of individuals and the annual population production. Figures 15–18 and 15–19 represent, for example, the survival and growth curves of populations of *Mabuya buettneri*, a lizard that hatches at the end of the dry season, and of which most individuals die a short time before the annual fire. In this case, the population is limited to a unique annual cohort.

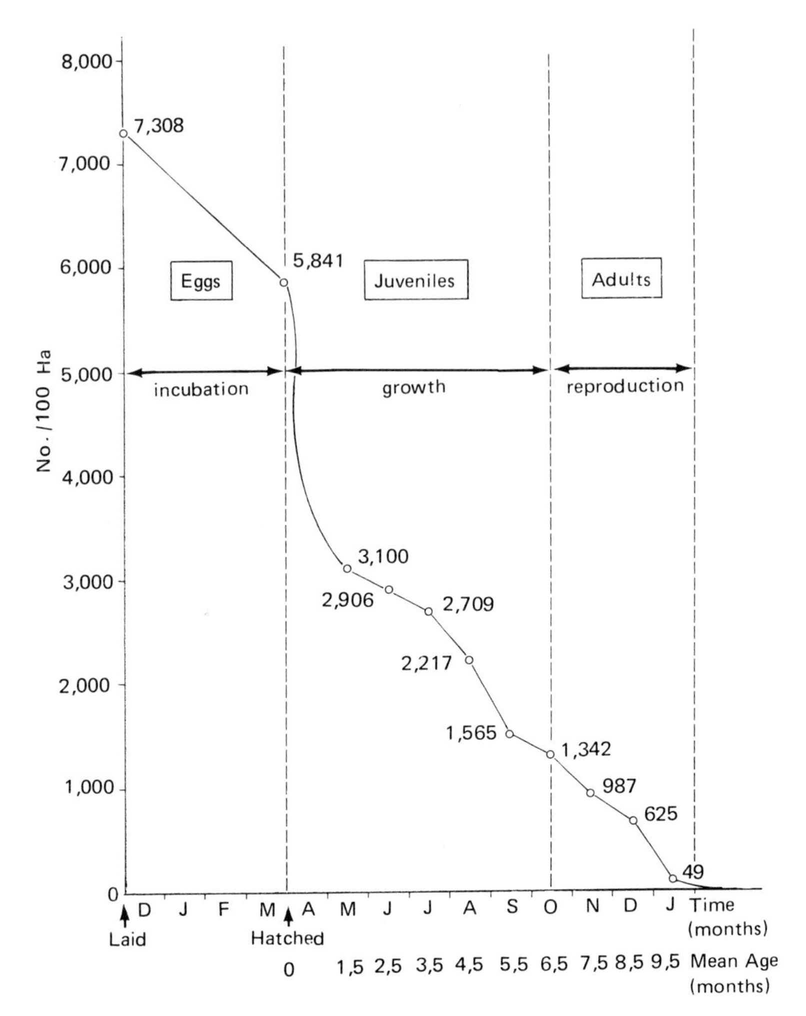


**Fig. 15–18.** Survival curve of the lizard *Mabuya buettneri* in burned savannah. (After Barbault, 1971b.)

The ground-living lizards of the burned grass savannah have an average net production of 291 g/ha live weight, or 87 g/ha dry weight. Higher values may be observed (670 g/ha live weight) in shrub savannah. Their production rate, that is, the ratio of production to average biomass ranges from 2.7 in *Mabuya maculilabris*, where adult life is relatively long, to 3.3 in *Mabuya buettneri* where adults have a very short life span after reproduction.

The overall consumption of lizard populations averages 3.1 kg/ha/yr live weight, or 900 g dry weight, representing 6 to 12 percent of the animals weight daily. The ratio of production to consumption is about 11 percent in *Mabuya buettneri* and 10.5 percent in *Mabuya maculilabris*.

Amphibians are represented by a number of species. Many of them live only slightly more than a year. This is the case for the small *Phrynobatrachus* and

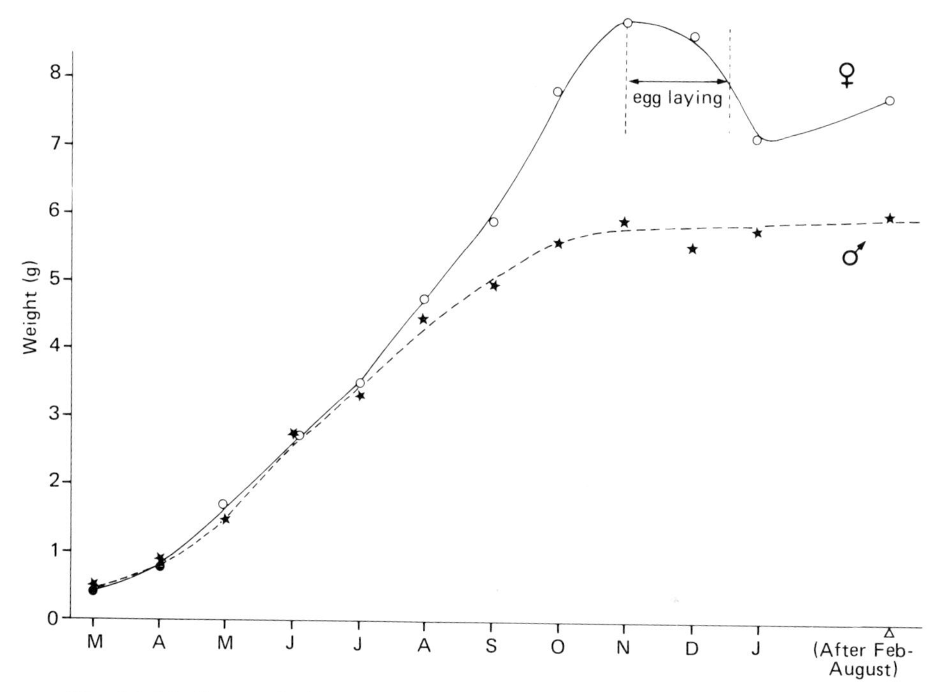


**Fig. 15–19.** Seasonal changes in fresh weight of *Mabuya buettneri* in the savannah. (After Barbault, 1971b.)

*Arthroleptis*, which constitute the majority of individuals. The larger species such as *Bufo regularis* and some *Ptychadena* have a longer life span.

For an average year and an "average" savannah, the total amphibian production is about 450 g/ha/yr live weight (or 100 g dry weight), with a production rate close to 3. Considering that the ratio of production to consumption in amphibians is, as in lizards, about 10 percent, the consumption has been estimated at 1,000 g/ha/yr dry weight, or about 5000 kcal/ha/yr.

The major characteristic of these populations is their great qualitative and quantitative variability in time, depending on the amount and distribution of rainfall. Maximum density may vary by a factor of 4. This variability is superimposed on the seasonal variation which produces the diversity and heterogeneity of the amphibian facies.

## The Invertebrates

It is not possible to summarize in a few pages the accumulated results on the populations of Lamto savannah invertebrates. Hundreds of species coexist, with very different densities that vary widely seasonally. Diverse sampling techniques have been used to collect animals, depending upon their size, density, and agility. The populations of the grass layer (not including social forms such as ants, nor very mobile forms such as Diptera) will be described by minimal density estimates of the major groups that have been collected in quadrats by D. and Y. Gillon (Table 15–2).

The densities of the various insects are extremely variable according to season, whether one considers numbers of individuals or biomass. The graphs of Figure 15–20 show some features of the seasonal fluctuations over an annual cycle in the burned savannah.

In all the burned formations, which form the greatest part of the Lamto savannah, fire, together with climatic factors, induces an annual rhythm in the plant and animal life. D. and Y. Gillon have analyzed the effect of fire immediately after its passage and its repercussions in the following months. Burning the savannah is not an always consistent phenomenon. The intensity and speed of fire are functions of a large number of climatic factors, such as wind, temperature, rainfall during the week and month preceding the fire, as well as local factors, such as slope of the ground, quantity of combustibles, which in turn depend upon the vegetation and the data of preceding fires. In addition, the temperature during the course of fire is variable, depending upon the stratum above the ground. Attaining elevated values above the soil (600°C at 20 cm, 300°C at 1.5 m, the upper limit of the vegetation, 75°C at 4m), temperatures are lower under the soil surface (65°C) or in the interior of a

**Table 15–2.** Numbers and biomass (March to December 1965) for 100 m$^2$ of burned (SB) and unburned savannah (SNB) for diverse groups of arthopods (percentage is of all arthropods [TA]).

| | Numbers | | | | Biomass | | | |
|---|---|---|---|---|---|---|---|---|
| | Number/ 100 m$^2$ | | % TA | | % mg 100 m$^2$ | | % TA | |
| | SB | SNB | SB | SNB | SB | SNB | SB | SNB |
| Arachnids | 1,107 | 1,274 | 47.8 | 38.6 | 12,722 | 19,128 | 17.5 | 20.3 |
| Myriapoda | 78 | 98 | 3.4 | 3.0 | 8,193 | 6,417 | 11.3 | 6.8 |
| Caterpillars | 54 | 89 | 2.3 | 2.7 | 6,527 | 8,406 | 9.0 | 8.9 |
| Blattids | 132 | 438 | 5.7 | 13.3 | 3,133 | 8,774 | 4.3 | 9.3 |
| Mantids | 48 | 71 | 2.1 | 2.2 | 3,056 | 5,296 | 4.2 | 5.6 |
| Acridids | 221 | 182 | 9.6 | 5.5 | 19,023 | 16,571 | 26.2 | 17.6 |
| Tetrigids | 27 | 31 | 1.2 | 0.9 | 1,160 | 576 | 1.6 | 0.6 |
| Grillids | 152 | 171 | 6.5 | 5.2 | 5,112 | 6,625 | 7.0 | 7.0 |
| Grasshoppers | 83 | 58 | 3.6 | 1.8 | 5,446 | 6,065 | 7.5 | 6.4 |
| Pentatomids | 46 | 128 | 2.0 | 3.9 | 1,735 | 5,216 | 2.4 | 5.5 |
| Coreids | 17 | 38 | 0.7 | 1.1 | 340 | 613 | 0.5 | 0.7 |
| Lygeids | 35 | 203 | 1.5 | 6.2 | 253 | 1,831 | 0.3 | 1.9 |
| Reduvids | 58 | 64 | 2.5 | 1.9 | 1,646 | 1,604 | 2.3 | 1.7 |
| Homoptera | 135 | 205 | 5.8 | 6.2 | 669 | 960 | 9.9 | 1.0 |
| Carabids | 31 | 86 | 1.3 | 2.6 | 1,353 | 3,885 | 1.9 | 4.1 |
| Other Coloeoptera | 92 | 168 | 4.0 | 5.1 | 2,297 | 2,285 | 3.3 | 2.4 |
| Total arthropods | 2,316 | 3,304 | 100.0 | 100.0 | 72,663 | 94,251 | 100.0 | 100.0 |
| Total insects | 1,131 | 1,932 | | | 51,748 | 68,707 | | |

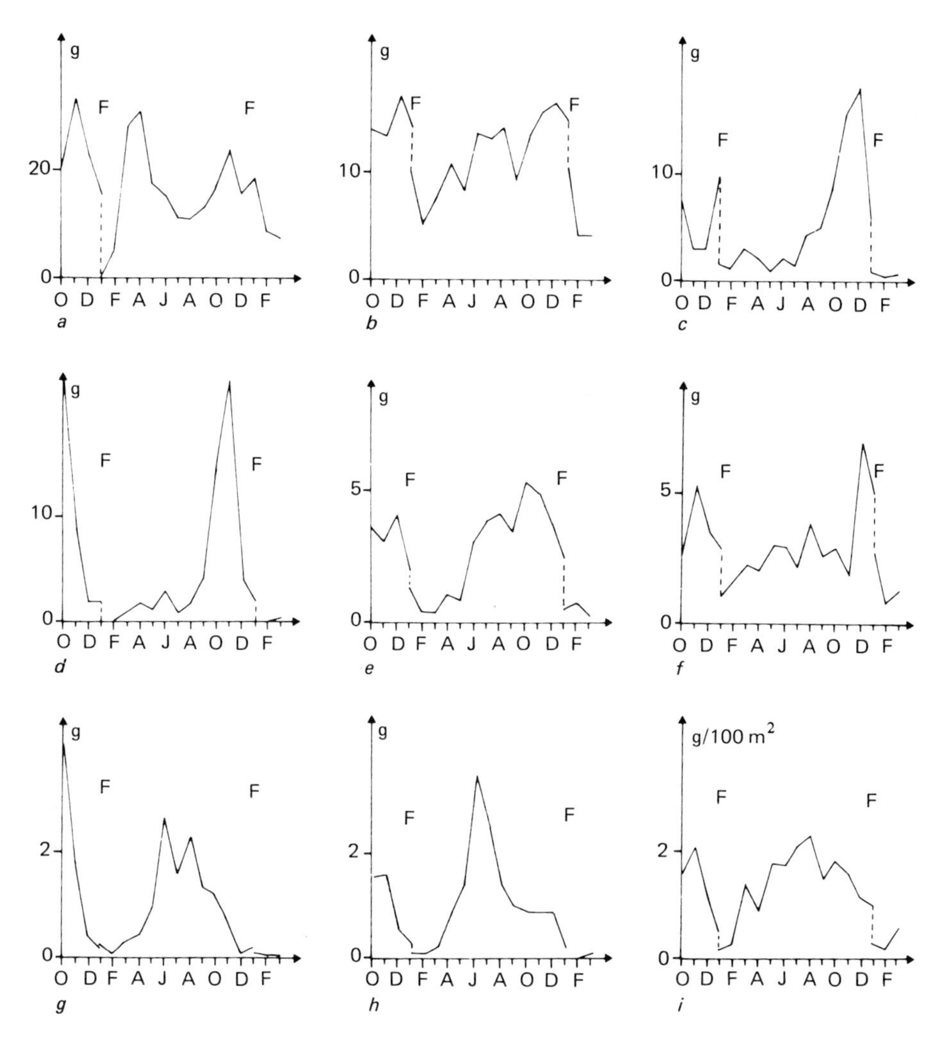


**Fig. 15–20.** Seasonal variation in fresh weight biomass of major groups of arthropods in Lamto savannah. (After Gillon and Gillon, 1967.) a, Acridids; b, Spiders; c, Caterpillars; d, Grasshoppers; e, Blattids; f, Mantids; g, Tetrigids; h, Carabids (imagos); i, Reduvids.

bunch of grass (65°C) and do not vary at all below 5 cm. of soil. At soil level maximum temperatures may range from 75° to 350°C at two points no farther than 2 m apart.

## Immediate Effect of Fire

Fire alarms insects even at a great distance (Figure 15–21). Before the flames arrive, animals become very agitated in the herbage. This is probably the reason why hundreds of kites fly just before the fire and dive in front of

**Fig. 15–21.** Passage of fire in the *Loudetia* savannah.

the flames for insects. The disappearance of strong flying groups after fire indicates that they escaped or were killed by fire and birds of prey. The weak effect of fire on the fauna of the soil surface is explained by the low temperatures observed in the center of grass tufts, under the soil surface or, similarly, the irregularity of the terrain. Finally, the decrease in the mean size, observed in almost all groups, shows that the larger animals on the soil surface are the most vulnerable to the flames and that the good fliers of larger size escape more efficiently.

## Repercussions of Fire in the Following Month

Fire transforms the herb stratum of the savannah into a vast field of ash from which emerge blackened stubbles (Figure 15–22). In the following days small green leaves grow at the base of grass tufts and in a month the burned savanna is green again. At this time grasses reach 10 cm in height but cover only 20 percent of the soil surface.

Although the vegetation then vigorously grows, and in a unburned

**Fig. 15–22.** View of the savannah just after passage of fire.

savannah the arthropod populations remain relatively stable with normally only a slight decrease in the short dry season, in a burned savannah the arthropod fauna decreases 39 percent in numbers. This is not reflected at the biomass level, however, because the average individual weight increases. Table 15–3 shows these variations in numbers and in biomass resulting from fire. Figure 15–23 shows the relative importance of these variations in the seasonal cycle and a comparison with the unburned savannah.

The groups that are almost unaffected by the immediate effect of fire show the strongest decline: More than 50 percent of the arachnids, blattids, tetrigids, gryllids, and carabids disappear in a month (Table 15–4). With the exception of lygeids, the good flying groups are the only ones to increase in importance. The acridids are even more numerous than before the fire or in the unburned savannah at the same time, whereas the tettigonids, reduvids, and leafhoppers are more numerous than just after the fire, as are the caterpillars marking the return of the adult butterflies and moths to oviposit in burned savannah.

**Table 15–3.** Numbers and biomass in grams per 100 m² of the arthropod assemblage the day before the fire, the next day, and a month after in burned savannah, and the percent of the fauna the day before the fire.

| | | Day Before Fire | Next Day | | After One Month | |
|---|---|---|---|---|---|---|
| January fire | Number | 2,688 | 1,710 | 64% | 1,044 | 38% |
| | Weight | 60.1 | 19.5 | 32% | 19.8 | 33% |
| April fire | Number | 3,007 | 2,853 | 88% | 1,044 | 35% |
| | Weight | 78.4 | 45.9 | 59% | 25.2 | 32% |
| January fire | Number | 1,848 | 1,243 | 67% | 716 | 38% |
| | Weight | 67.7 | 40.4 | 60% | 26.6 | 39% |

These late upheavals within the arthropod populations illustrate two phenomena:

1. The return of strongly flying insects, inhabiting herbs, hence heliophilous, which come from refuges to recolonize the new environment
2. The disappearance, either by flight or death, of the arthropods of the lower level, probably because of the complete change in their environment

Indeed, before a fire, the savannah consisted of several different strata: herbs, a mat of dead grasses, and the soil surface. After fire there is only one level. For the arthopods this means a lack of cover, a reduction in living space, and a simplification of the environment. This probably results in strong competition for shelter, intense predation among arthropods, and especially a greater vulnerability to large insectivores such as birds. The spectacular change in color of some species of acridids, mantids, and lepidopterans is related to these environmental changes.

On the other hand, the microclimatic conditions, which were particularly stable within the grass layer itself, are now totally changed. In the absence of grass cover the sunlight strikes the naked ground directly. This results in strong daily variations in temperature, humidity, and luminosity. Finally, the primary trophic level is very different after the passage of fire: If regrowth is rapid and fresh nutrients are not lacking, the litter or food accumulated for a year is lacking for the detritivores that all live on the soil surface.

Fire induces, therefore, two major changes in the arthropod fauna: (1) during its passage, it destroys some of the animals and drives out the strong fliers but does not greatly influence the soil surface and its fauna; (2) the new environment created permits a sunloving and mobile fauna of grasses to appear, but the groups spared during its passage are either eliminated or decimated.

Notwithstanding the differences observed under different fire conditions, about 32 to 39 percent of numbers and 35 to 36 percent of the biomass of arthropods living in the savannah before the fire may be found there a month

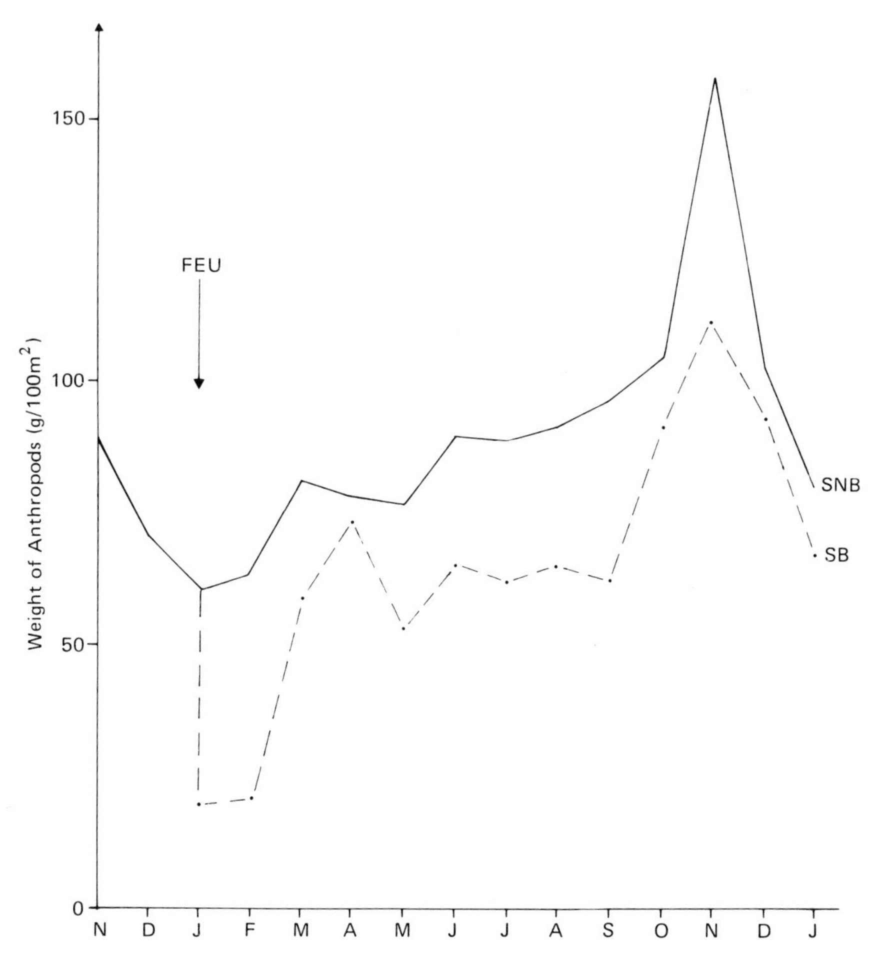


**Fig. 15–23.** Change in biomass of nonsocial arthropods in the herb strata of savannah in 1965. (After Gillon and Gillon, 1967.) *SB* = burned savannah; *SNB* = unburned savannah.

later (Table 15–3). Fire is thus responsible for the disappearance of about 60 percent of the fauna.

## Long-Term Influence of Fire on the Arthropod Fauna

Barely two months after the passage of fire of January 1965 the numbers and biomass of arthropods increased again in burned savannah (Fig. 15–23). However, regardless of the time of the year, there are fewer arthropods in the burned than in the unburned savannah. In the course of one annual cycle (1965–1966), an average of 330,000 arthropods per hectare, weighing 9,200 g, were captured in unburned savannah compared to 230,000, weighing 6,200 g, in the burned savannah in April. The fire had thus reduced the numbers and

**Table 15–4.** Numbers per 100 m² of the main groups of arthropods the day before the January 1965 fire, the next day, and the following month (percentage is of the numbers before the fire).

| | Day Before Fire | Next Day | | After One Month | |
|---|---|---|---|---|---|
| Arachnids | 1,341 | 1,273 | 95% | 533 | 39% |
| Myriapoda | 10 | 7 | 70% | 4 | 40% |
| Caterpillars | 83 | 23 | 28% | 30 | 36% |
| Blattids | 104 | 86 | 83% | 29 | 28% |
| Mantids | 84 | 39 | 46% | 36 | 43% |
| Acridids | 135 | 9 | 7% | 200 | 148% |
| Tetrigids | 12 | 8 | 67% | 2 | 17% |
| Grillids | 144 | 64 | 44% | 19 | 13% |
| Grasshoppers | 22 | 0 | 0% | 1 | 5% |
| Pentatomids | 82 | 21 | 26% | 18 | 22% |
| Coreids | 42 | 1 | 2% | 1 | 2% |
| Lygeids | 40 | 24 | 60% | 32 | 80% |
| Reduvids | 37 | 7 | 19% | 13 | 35% |
| Homoptera | 331 | 11 | 3% | 17 | 5% |
| Carabids | 30 | | 43% | 2 | 7% |

the biomass of arthropods in the herb stratum by about 30 percent over the entire year.

Moreover, in the absence of fire, the fauna evolves into a different type. The blattids, a moisture-loving group, decrease in density from 55,500/ha in unburned savannah to 16,100/ha in burned savannah. Inversely, the orthopterans increase from 48,100 in unburned to 67,100 in burned savannah.

Eleven months after the passage of fire, the population of the burned savannah was still 30 percent less than that of the unburned savannah; yet, for most species, one or two generations already had developed. This long-term influence of fire can probably be attributed to the differences between burned and unburned savannah. In fact, one can always recognize a savannah that had burned during the year by the distribution of its grass clumps with the bare soil unprotected by the matted carpet of dead leaves present (Figure 15–5). The burned savannah constitutes an unfavorable environment for moisture-loving arthropods, which are partially eliminated. As a whole, it offers to various groups fewer ecological possibilities than the unburned savannah.

## Production and Energy Balance

Here we concentrate especially on the acridids,[6] which are most abundantly represented in the savannah, with an average population density of 16,000 individuals per hectare (554 g dry weight). Juveniles make up 80 percent

[6] Work of Y. Gillon.

of the population, but only 131 g as compared to 368 g of imagos. Considering the seasonal variation in numbers, the maximal density is reached in June (25,000/ha, with 21,700 young) and the lowest in January (6,900/ha, with 2,000 young). By contrast, the highest mean biomass is 1,060 g/ha in December and the lowest is 275 g/ha in August, when adults are at their minimum density (see Figure 15–20). Within each species the increase in individual weight results in an increase in total biomass, despite the mortality that decimates the cohort. The standing crop biomass reaches its maximum when imagos appear in the population.

Growth production has been estimated from the development of annual species in the field and from laboratory rearing of polyvoltine species. The two methods were compared for the species *Orthochtha brachycnemis*. In this case, some maturation production continues to take place in the imagos after the last molt. Production of new individuals has been estimated on the basis of the average number of eggs laid per day in the breeding cages of different species and of the number of females present at each instant in nature. Total production is the sum of the productions of the different species.

The growth production of young has been estimated at 1,242 g/ha/yr, the production of maturation at 77 g/ha/yr for males and 326 g/ha/yr for females, the production of eggs at 1,944 g/ha/yr, approximating 3,600 g/ha/yr dry weight. Converted to calories, these productions are 618 cal/m$^2$/yr for young, 209 cal/m$^2$/yr for the maturation of imagos, and 1,147 cal/m$^2$/yr for eggs. Production due to reproduction constitutes therefore the greatest part—58 percent—of the total production of the acridid population.

The daily food consumption for the life span was determined in the laboratory for *Anablepia granulata, Catantopsilus taeniolatus, Orthochtha brachycnemis*, and *Rhabdoplea munda*. Ratios were established between the daily food intake and the weight of the individual young on the one hand, and of adult males and females on the other. Monthly consumption of different species is therefore estimated as a function of their biomass. The total annual consumption from these specific consumptions is, in dry weight, 27.5 kg of herbage per hectare for young, 32.9 kg/ha for females, and 9.2 kg/ha for males, making a total of 69.6 kg/ha, or 29.4 kcal/m$^2$/yr, that is, 15 times the acrid production. Figure 15–24 shows graphically the distribution of energy flow in acridid populations in the Lamto savannah.

The mean ratio of production to consumption (P/C), is 6.7 percent, the specific values ranging from 5.3 (*Dnopherula bifoveolata*) to 9.5 percent (*Anablepia granulata*). Other elements of the energy budget have been calculated but with less precision because of the difficulty of determining exactly the assimilation rate, which is probably around 20 percent.

The production turnover ($P/\overline{W}$) is about 6.5. For different species this ratio depends basically on the number of generations per year. It is clearly higher—over 11—in species with three generations such as *Rhabdoplea munda* than in species with two generations such as *Orthochtha brachycnemis* or *Anablepis granulata*. The species with one generation have the lowest rate of production, roughly 4.

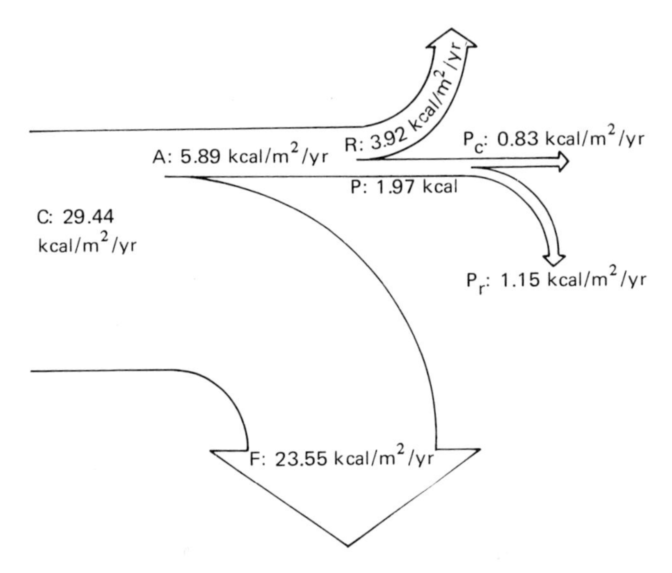


**Fig. 15–24.** Energy flow utilized annually on 100 $m^2$ by populations of acridids in savannah. (After Gillon, 1972b.) $C$ = consumption; F = feces; $A$ = assimilation; $R$ = respiration; $P_c$ = growth production; $P_r$ = production of reproduction.

Extrapolating the results obtained for acridids to all the other arthropods captured in the Gillons' quadrats, one obtains a total production of about 12 kg/ha/yr. Consumption, using a more speculative estimate, is about 240 kg/ha/yr.

It is also necessary to take into account the insects above the grass layer (which were not included in the preceding samplings) notably the diptera.[7] The biomass of these animals is not high, but their development is rapid, giving a relatively important production. A single *Borassus* palm fruit permits the development of 1,000 *Drosophila* larvae, and since each palm tree produces 100 fruits per year, there may be 100,000 larvae. The total production could thus exceed 500 g/ha/yr fresh weight, or 160 g dry weight. The production based on *Ficus* fruits, restricted to the sections of the savannah with most trees, is even greater.

Again, the ants have not been considered in the preceding paragraphs, for their study must be based on different sampling techniques than those used for nonsocial arthropods.[8] The formicids almost always have their nests underground, but exploit mostly the aboveground strata of the savannah, and play an essential role in the ecosystem. With more than 150 species living underground, on the ground, or in trees, some of which are strict carnivores and others polyphagous, the ants are ubiquitous and occupy extremely varied ecological niches.

[7] Work of D. Lachaise.
[8] Work of J. Lévieux and B. Darchen-Delage.

Numerous data have been gathered on the biology of the more abundant species, but it would be imprudent here to estimate their biomass and energy flow. For the time being, we can only say that the total density of underground nests is 7,000/ha; but, depending on the species, a nest may contain from several dozen to several thousand individuals.

## Soil Populations

The organic matter in the litter and soil provides food for a number of animals. Certain of these animals consume the already decayed material and help to complete the decomposition and mineralization of organic matter. Others behave as predators. It would be premature to describe in detail such a complex trophic network. Instead, we will attempt to give an analytical description of the soil populations, showing as far as possible, the trophic role, the food consumption and biological production of the organisms considered. Among the animal groups studied, the most abundant and most characteristic are the termites, the earthworms, and the microarthropods, but other groups such as the myriapods, the ants, or the underground larvae of coleoptera also play an important role in the functioning of the underground community.

### Termites[9]

Termites constitute an essential component of the consumer level in any tropical ecosystem. They are represented at Lamto by three groups that play very different ecological roles. The *foraging termites* of the genus *Trinervitermes* construct small aboveground nests and consume living or partly dead herbage that they harvest at night. They are represented by five species. Their density is on the average about 60 nests per hectare containing 1.3 million individuals, which represent 6 kg of live biomass, 1.2 kg dry weight, or 5,800 kcal. These termites consume from 6 to 44 kg/ha of dry grass per year.

The *humivorous termites* are near the end of the chain of decomposition since they mineralize an important part of organic matter. At Lamto they are almost always found underground. Depending upon the type of savannah, the density may range from more than 4.5 million/ha to only 0.5 million. In the case of the highest density, the live biomass may be 12 kg/ha, or dry, 4.4 kg, which is equivalent to 7,400 kcal. Food consumption by these termites is estimated to be equivalent to 30 kg of cellulose per hectare per year, which implies rearrangement of several dozen tons of surface soil. The role of the humivorous termites, while slight from the point of view of consumption, is important for the structure and aeration of the soil.

[9] Data of G. Josens

The *fungus growing termites* consume dead wood and dry grasses from the litter. They build gardens with this litter, where the fungi grow and provide food for the ants. These insects are represented at Lamto by four species of underground *Macrotermitinae*. On the average 1 ha contains 57,000 gardens, which represents 175 kg dry weight. These gardens are tendered by 4.5 million termites, which have a fresh weight of 4.9 kg or a dry weight of 1.2 kg, which is equivalent to 5,600 kcal. Since the time of renewal of the gardens is of about two months, these termites incorporate in the termitaria 1,000 kg/ha/yr of dry litter.

**Earthworms**[10]

There are nine species of earthworms, excluding those living in trees. Examination of their diets, morphology, pigmentation, and vertical distribution permits their separation into three groups: (1) litter species that consume fragments of dead roots and especially debris of grass and tree leaves collected at the surface; (2) deep earth-feeding species with unique earth-feeding habits; and (3) intermediate species, which are the most abundant and feed on earth collected near the soil surface mixed with fragments of roots or dead leaves.

It can be shown that litter species, which represent 5 percent of the biomass savannah, exceed 10 percent in the tree savannah and 33 percent in the savannah protected from fire for ten years. This is inverse to the deep earth-feeding species, which represents 5 percent of the biomass in unburned savannah, more than 9 percent in tree savannah, and reach their maximum of 28 percent in the grass savannah.

A surface of 1 ha, in which a combination of the different types of savannah was represented, contained an average of 2,300,000 earthworms per hectare, weighing 490 kg live weight. Depending upon the type of savannah, the population may vary from 1,800,000 to 3,400,000 individuals and 390 to 570 kg/ha, respectively. The populations in the unburned savannah are the most dense but have the least biomass. It is in the tree savannah that the biomass is maximal and the grass savannah is intermediate. Dry weight represents only about 8 percent of the fresh weight.

The most abundant species is *Millsonia anomala* (Omodeo), which constitutes 40 to 80 percent of the biomass of this group of organisms in all savannah types. These nonpigmented worms are large, measuring 15 to 20 cm in length, with a diameter of 7 mm. They live most often near the ground surface.

Reproduction takes place twice a year, from May to July and from October to September. The young hatch three weeks after eggs are laid if the soil is wet, or one or two months later if drought interrupts the development of cocoons. Each year we note the appearance of two generations. As they are relatively well separated in time, they may be distinguished on the histograms

[10] Data of P. Lavelle.

illustrating the population structure (Figure 15–25). The interpretation of these monthly histograms has been aided by knowledge derived from raising individuals in natural environments. It is thus possible to trace for each generation growth and survival curves during the period of study. From these data the production of a generation can be calculated; then, by adding results from each generation, the total production of the species can be estimated.

In the grass savannah, a population of *Millsonia anomala* produced about 460 kg/ha in live weight, or about 190,000 kcal, between August 1969 and August 1970, and 405 kg (175,000 kcal) between August 1971 and August 1972. The ratios of production to weight ($P/\overline{W}$) for the two periods are 2.7 and 1.7, respectively. The difference here is correlated to important differences in the rainfall and consequently in the population structure. Most other species have a slower growth and hence a lower productivity. The total production is estimated at between 600 and 1,000 kg/ha/yr in fresh weight, 45 to 80 dry weight, and 250,000 to 450,000 kcal.

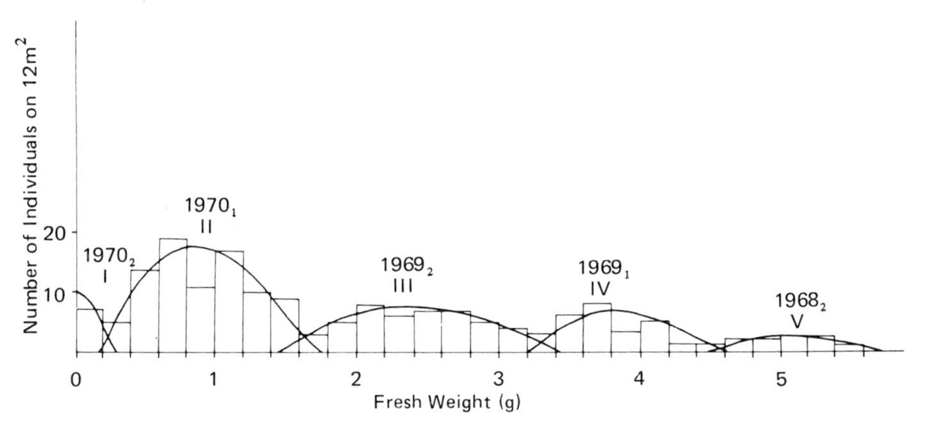


**Fig. 15–25.** Demographic composition of a population of *Millsonia anomala* in August 1970 in the herbaceous savannah with *Loudetia simplex.*

Earth consumption by *Millsonia* has been estimated by measuring the defecation of worms in the laboratory, which varies according to soil humidity and weight of worm. It is highest in the young, which at the optimum humidity (11 percent $\cong$ pF 2.5) ingest daily 24 times their weight in earth. The maximum for adults is only 9 times, at a much higher humidity (20% $<$ pF 2). From such data the consumption of field populations of *Millsonia* can be calculated, knowing the monthly weight structure and the average soil humidity. The value obtained is 500 tons of dry earth per hectare between August 1971 and August 1972 in herbaceous *Loudetia* savannah.

The rate of assimilation was found to be very low; a preliminary series of measurements made with *Millsonia anomala* showed the ratio to be about 9 percent. Knowing the mean caloric value of earth (0.056 kcal) and that of dried worms (5.3 kcal), a provisional energy budget for the population may be estimated (see Figure 15–26).

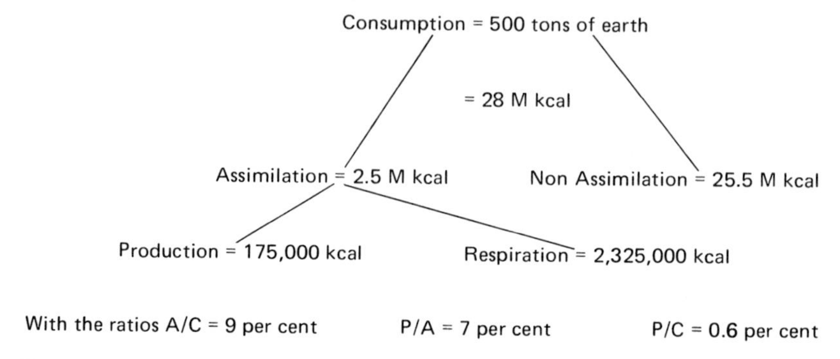


**Fig. 15–26.** Provisional energy budget for the earthworm *Millsonia anomala.*

**Microarthropods**[11]

Because of their numbers, the soil microarthropods constitute one of the most essential elements in the ecosystem. Their density is about 10,000 individuals per square meter, a very similar value to those already observed in other herbaceous communities of tropical Africa. This density is nevertheless about ten times lower than that in herbaceous biotopes in Europe. Table 15–5 indicates the seasonal development of microarthropods observed at two stations is burned savannah and in a savannah protected from fire for eight years.

The Acari constitute the most abundant group. Among these, the Oribatids (fungivorous and litter feeders) average 30 to 50 percent of the total. The Gamasida (predators) represent about 2 percent. The Collembola, which are for the most part fungivorous, are relatively uncommon in the soil. The soil populations are also characterized by an unexpectedly high relative importance of Symphilans and Pauropods. We noted also a strong contingent of wingless coccids feeding on roots. The use of some data on the mean individual biomass permits the following estimates of the fresh weight biomass (kg/ha) of microarthropods:

| | *Acari* | *Collembola* | *Other Microarthropods* | *Total* |
|---|---|---|---|---|
| FSI | 1.4 | 0.4 | 2.5 | 4.3 |
| FS2 | 2.9 | 0.8 | 8.5 | 12.2 |

Without adequate study on the annual cycles of the main species and precise demographic data, an estimate of production is very speculative. Taking into account the existence of two to four generations per year according to species, and assuming a production turnover ($P/\overline{W}$) of 3 for a cohort (similar to that of many arthropods), suggests a production of 180 kg/ha/yr (in unburned savannah) and 36 kg/ha/yr (burned savannah) in fresh weight.

[11] Data of F. Athias.

**Table 15–5.** Seasonal development of microarthropods in burned savannah with few shrubs (FS1) and in savannah protected from fire for eight years (FS2).

| | Dec. | Jan. | Feb. | Mar. | Apr. | May | June | July | Aug. | Sept. | Oct. | Nov. | Dec. | Jan. | Feb. |
|---|---|---|---|---|---|---|---|---|---|---|---|---|---|---|---|
| FS2 | | | | | | | | | | | | | | | |
| Acari | 47,425 | 12,162 | 29,500 | 925 | 9,550 | 2,725 | 8,062 | 2,387 | 3,650 | 6,500 | 8,437 | 9,375 | 56,087 | 19,050 | 2,750 |
| Collembola | 2,550 | 4,362 | 3,566 | 1,050 | 2,000 | 1,387 | 1,750 | 1,162 | 1,750 | 250 | 562 | 1,250 | 2,812 | 1,525 | 125 |
| Other Micro-arthropods | 6,100 | 2,937 | 2,133 | 2,900 | 1,100 | 1,137 | 1,312 | 3,187 | 5,800 | 3,375 | 3,437 | 5,187 | 13,837 | 5,900 | 1,000 |
| FS1 | | | | | | | | | | | | | | | |
| Acari | 15,850 | 3,300 | 2,450 | 575 | 1,850 | 2,562 | 17,087 | 3,875 | 7,000 | 18,525 | 9,812 | 6,387 | 31,325 | 24,100 | 16,500 |
| Collembola | 650 | 725 | 525 | 350 | 200 | 3,437 | 762 | 2,050 | 2,775 | 500 | 887 | 775 | 2,750 | 1,900 | 250 |
| Other Micro-arthropods | 2,075 | 1,787 | 1,525 | 1,000 | 450 | 1,437 | 3,062 | 4,250 | 5,312 | 6,750 | 3,812 | 5,440 | 6,712 | 9,250 | 5,875 |

## Microflora[12]

Determination of the quantitative importance and of the role of the soil microflora poses even more complex problems than those posed by the nematodes and microarthropods. Accurate census techniques do not exist, and, still less, methods of estimating biomass. The data obtained at Lamto have been gathered with dilution techniques, which certainly underestimate the density of organisms present but also do not show the temporarily inactive state of many of the organisms.

The Actinomycetes are remarkably numerous in the Lamto soils, especially in the burned savannah of *Hyparrhenia* (2,600,000 per gram of dry soil at Station FS 1 and 1,900,000 at FS 2). Yet the burned savannah shelters three times fewer soil fungi than savannah protected from fire (100,000 compared to 30,000 per gram of soil), due no doubt to the persistance of some litter materials on the surface of the soil.

The number of bacteria, apparently immense, is nevertheless low compared to that observed in most other soils studied elsewhere in the world. This relatively low density of the diverse groups of bacteria is evidently linked to the chemical poverty of the soil. Mineralization of nitrogen, in particular, is very slow.

The abundance of diverse functional groups varies from station to station, but the anaerobic forms are relatively more frequent everywhere. In sifted soil the aerobic nitrogen fixers are not shown by dilution technique, for the fixation of nitrogen appears to be essentially carried out in the rhizosphere of the higher plants.[13]

As an example, Table 15–6 shows the results of counts of bacteria in a soil profile during the dry season. It is very difficult to derive from such numbers a notion of biomass and, even more so, of biological production. When the physical environment is favorable and a metabolizable substrate is present, the production turnover ($P/\overline{W}$) of bacteria is very high, but these conditions only exist during short periods. So the only method to measure the production of bacteria is to measure directly their metabolic activity *in situ*.

**Table 15–6.** Counts of bacteria in savannah soil by dilution techniques.

| Depth (cm) | Total Micro-flora[a] | Anaerobic Forms | Cellulo-lytic Aerobes | Cellulo-lytic Anaerobes | Nitrogen Bacteria | Nitrite Bacteria |
|---|---|---|---|---|---|---|
| 0–10 | 40 | 4.0 | 0.17 | 0.2 | 0.06 | 4.0 |
| 10–20 | 35 | 1.4 | 0.17 | 0.14 | 0.05 | 3.4 |
| 20–30 | 30 | 0.7 | 0.17 | 0.14 | 0.04 | 3.0 |
| 40 | 25 | 0.17 | 0.08 | 0.1 | 0.02 | 2.0 |
| 75 | 30 | 0.25 | 0.1 | 0.1 | 0.03 | 2.8 |

[a] In millions per gram dry earth; the others, in thousands per gram dry earth.

[12] Data of I. Bacvarov, J. Pochon, A. Rambelli, and P. Villecourt.
[13] Work of J. Balandreau.

The rate of release of carbon dioxide gives a first approximation of this activity inasmuch as it can be attributed to soil microorganisms and not to roots. The measures have shown that, on average, respiration was slightly higher than 8 g $CO_2/m^2$/day, which corresponds to an expiration of 8 tons/ha/yr and an energy expense of $10^6$ kcal. In the burned savannah these losses are virtually equal to the supply of organic carbon from decomposition of the litter and the roots.

## Conclusions

While in many respects the analysis of the Lamto savannah ecosystem cannot yet be considered complete, results obtained so far already provide a basic idea of its general structure and particularly of the relative importance of the different groups. It is even possible to indicate in broad terms how the ecosystem functions and thus to specify the basic features of its physiognomy.

The study of the development of vegetation over the years in the presence or absence of fire shows that the *Borassus* savannah of Lamto is maintained stable only by the regular action of fire. Without the latter, there is recolonization by shrubs and trees. A series of different savannah types according to the more or less intense effect of fire is proposed in Figure 15–27.

Since fire is due mainly to man's actions, we may call the Lamto savannah *anthropomorphic*. But, in addition, if we consider the surrounding landscapes, its maintenance also appears to be favored by the great poverty of the soil and the climate where precipitations are extremely variable from year to year.

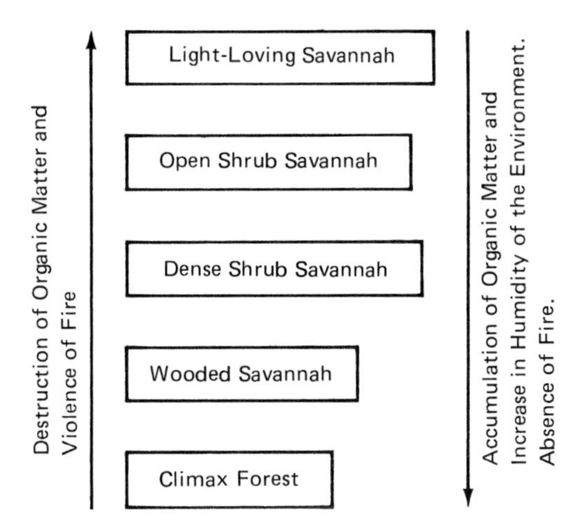


**Fig. 15–27.** Relationship between different types of savannahs and the effect of intensity of fire on organic matter.

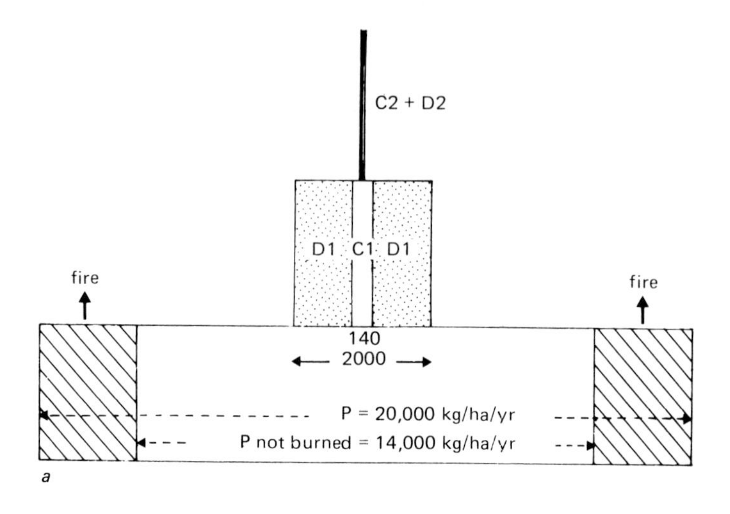


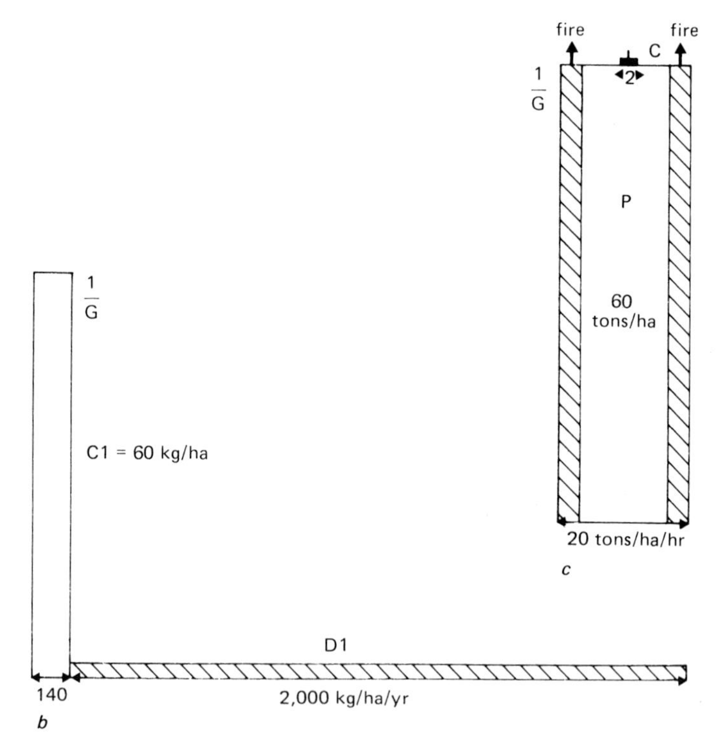


**Fig. 15–28.** Synthetic representation of the trophic structure of the Lamto savannah. *a* = Pyramid of energy transfer, *P* = production (primary), *C1* = consumers (primary), *D1* = decomposers of vegetable matter, *C2* = secondary consumers, *D2* = decomposers of animal material. *b* = Structure of primary consumers of the savannah. *C1* = primary consumers, *D1* = decomposers of vegetable material. *c* = Structure of production of the ecosystem. *P* = producers, *C* = consumers and decomposers.

In *b* and *c*, the surfaces of the rectangles represent the biomass of the trophic level or a fraction of the trophic level, the width the production, and the height (1/G) the inverse of the turnover rate.

The trophic structure of the ecosystem shows a great disparity between the biomass of primary producers and that of animal consumers, there are more than 60 tons/ha (dry weight) for the first and only 80 kg/ha for the second. Similarly, a great disparity exists between the corresponding productivities, which is accentuated by the dominant activity of decomposing microorganisms compared to that of nonmicroscopic primary consumers. This preponderance is made possible by the relatively humid climate that favors bacterial activities during a large part of the year. One must stress, moreover, the importance of detritivorous forms such as earthworms and termites (among the nonmicroscopic organisms), which are also favored by soil humidity. Inversely, the herbivores and true plant feeders consuming living vegetation do not play an important role, the large mammals not being a part of the animal community for a least four or five decades.

The scarcity of large herbivores is certainly the result of man's actions, which have virtually exterminated buffaloes, elephants, hippopotami, and even antelopes and monkeys. These forms constituted an important fraction of the primary consumer trophic level in the past and probably had an important action on the vegetation by their preference for the grass layer.

In a way, fire now replaces the action of large herbivores in the maintenance of the equilibrium of the savannah, since it destroys the surplus organic matter produced by photosynthesis, a surplus that otherwise would increase the biomass of trunks and branches that would eventually be consumed by termites, earthworms, fungi, and bacteria. The Lamto savannah (and more generally the preforest savannahs of Ivory Coast and Guinea) is fundamentally opposed to the drier savannahs of East Africa (at least in national parks). In the former, large herbivores have a reduced role, whereas in the latter they constitute an important fraction of the primary consumer trophic level. Provided humidity is sufficient, earthworms and decomposing microorganisms at Lamto occupy the role played elsewhere by large herbivorous mammals, and discreetly insure (together with fire) the mineralization of organic matter produced every year by green plants.

Utilizing the previously suggested mode of representation to show simultaneously the biomass and productivity of the various trophic levels, we may translate the energetic structure of the Lamto savannah ecosystem into the pyramid shown in Figure 15–28.

## References

Athias, F. 1971. Recherches écologiques dans la savane de Lamto (Côte d'Ivoire): Étude quantitative préliminaire des Microarthropodes du sol. *Terre Vie* 25(3): 395–409.

Barbault, R. 1967. Recherches écologiques dans la savane de Lamto (Côte d'Ivoire): Le cycle de la biomasse des Amphibiens et des Lézards. *Terre Vie* 21(3):297–318.

———. 1970. Recherches écologiques dans la savane de Lamto (Côte d'Ivoire): Les traits quantitatifs du peuplement des Ophidiens. *Terre Vie* 24(1):94–107.

———. 1971a. Les peuplements d'Ophidiens des savanes de Lamto (Côte d'Ivoire). *Ann. Univ. Abidjan Ser. E.* 4(1):133–193.

———. 1971b. Recherches écologiques dans la savane de Lamto: Production annuelle des populations naturelles du Lézard *Mabuya büttneri* (Matschie). *Terre Vie* 25(2):203–217.

———. 1972. Les peuplements d'Amphibiens des savanes de Lamto (Côte d'Ivoire). *Ann. Univ. Abidjan Ser. E.* 5(1):61–142.

Bellier, L. 1967. Recherches écologiques dans la savane de Lamto (Côte d'Ivoire): Densités et biomasses des petits Mammifères. *Terre Vie* 21(3):319–329.

———. 1968. Contribution à l'étude d'*Uranomys ruddi* Dollman. *Mammalia* 32(3): 419–446.

Bigot, L. 1967. Observations écologiques sur les Invertébrés du bosquet à *Bauhinia thonningii* Schaum dans une savane de Côte d'Ivoire. *Ann. Soc. Ent. Fr. N.S.* 3(4):1109–1128.

———. 1968. Contribution à l'étude écologique des Invertébrés de *Cussonia barteri* Harms dans la savane de Lamto (Côte d'Ivoire). *Ann. Soc. Ent. Fr. N.S.* 4(4): 875–890.

———, and F. Roux. 1966. Premières données sur l'avifaune de la savane et de la forêt-galerie de Lamto (Côte d'Ivoire). *Oiseau et RFO* 36:145–152.

Blandin, P. 1971. Recherches écologiques dans la savane de Lamto (Côte d'Ivoire): Observations préliminaires sur le peuplement aranéologique. *Terre Vie* 25(2): 218–229.

———. 1972. Recherches écologiques sur les araignées de la savane de Lamto (Côte d'Ivoire): Premières données sur les cycles des Thomisidae de la strate herbacée. *Ann. Univ. Abidjan Ser. E Ecol.* 5(1):241–264.

Bonvallot, J., M. Dugerdil, and D. Duviard. 1970. Recherches écologiques dans la savane de Lamto (Côte d'Ivoire): Répartition de la végétation dans la savane préforestière. *Terre Vie* 24(1):3–21.

Brunel, J., and J .M. Thiollay. 1969. Liste préliminaire des Oiseaux de Côte d'Ivoire. *Alauda* 37(3)230–254,(4):315–337.

Daget, J., and Ch. Lecordier. 1971a. Variabilité et structure des peuplements de Carabiques (Col.) au pied des Palmiers Rôniers dans une savane préforestière (Lamto, Côte d'Ivoire). *Rev. Biol. Ecol. Sol. VIII* 3:471–489.

———. 1971b. Structure d'un peuplement de Carabiques dans la savane de Lamto (Côte d'Ivoire). *Bull. IFAN* XXXIII, A, 2:425–435.

Cesar, J. 1971. Etude quantitative du cycle de divers groupements herbacés d'un savane preforestière de Basse Côte d'Ivoire. Thèse de 3ème cycle, Faculté des Sciences de Paris. 125 pp.

Darchen, R. 1970. Le nid de deux nouvelles espèces d'Abeilles de la Côte d'Ivoire, *Trigona (Axestotrigona) sawadogoi* Darchen et *Trigona (Axestotrigona) eburnensis* Darchen *(Hymen., Apidae). Biol. Gabon.* 6(2):139–150.

———. 1972. Ecologie de quelques Trigones (*Trigona* sp.) de la savane de Lamto (Côte d'Ivoire). Ökologie einiger Trigonen (*Trigona* sp.) in der Savanne von Lamto (Elfenbein Küste). *Apidologie* 3(4):341–367.

Delage-Darchen, B. 1971. Contribution à l'étude écologique d'une savane de Côte d'Ivoire (Lamto). Les fourmis des strates herbacée et arborée. *Biol. Gabon.* 7(4): 461–496.

———. 1972a. Une fourmi de Côte d'Ivoire, *Melissotarsus tibutans* Del. n. sp. *Ins. Soc.* 19(3):213–226.

———. 1972b. Le polymorphisme larvaire chez les fourmis *Nematocrema* d'Afrique. *Ins. Soc.* 19(3):259–278.

Delmas, J. 1967. Recherches écologiques dans la savane de Lamto (Côte d'Ivoire): premier aperçu sur les sols et leur valeur agronomique. *Terre Vie* 21(3):216–227.

Descoings, B. 1972. Note sur la structure de quelques formations herbeuses de Lamto (Côte d'Ivoire). *Ann. Univ. Abidjan Ser. E Ecol.* 5(1):7–30.

Duviard, D. 1969. Importance de *Vernonia guineensis* Benth. dans l'alimentation de quelques fourmis de savane. *Ins. Soc.* 16(2):115–134.

———. 1970a. Recherches écologiques dans la savane de Lamto (Côte d'Ivoire): l'Entomocénose de *Vernonia guineensis* Benth. (Composées). *Terre Vie* 24(1): 62–79.

———. 1970b. Place de *Vernonia guineensis Benth.* (Composées) dans la biocénose d'une savane préforestière de Côte d'Ivoire. *Ann. Univ. Abidjan Ser. E.* 3(1): 7–174.

———. 1970c. La pleurocécidie de *Piezotrachelus* sp. (*Disjunctum*) Wagner 1907 (Col. Apioninae) sur *Vernonia guineensis* Benth., Composée des savanes de Côte d'Ivoire. *Marcella* 36(4):243–256.

Gillon, D. 1970. Recherches écologiques dans la savane de Lamto (Côte d'Ivoire): Les effets du feu sur les Arthropodes de la savane. *Terre Vie* 24(1):80–93.

———. 1972. Les Hémiptères Pentatomides d'une savane préforestière de Côte d'Ivoire. *Ann. Univ. Abidjan Ser. E Ecol.* 5(1):265–371.

———, and Y. Gillon. 1965. Recherche d'une méthode quantitative d'analyse du peuplement d'un milieu herbacé. *Terre Vie* 19(4):378–391.

———, and J. Pernes. 1970. Recherches écologiques dans la savane de Lamto (Côte d'Ivoire): Comparaison entre relevés de 25 m$^2$ et 100 m$^2$ pour l'étude de la faune de la strate herbacée. *Terre Vie* 24(1):40–53.

Gillon, D., and J. Pernes. 1968. Etude de l'effet du feu de brousse sur certains groupes d'Arthropodes dans une savane guinéenne. *Ann. Univ. Abidjan Ser. E.* 1(2): 113–198.

Gillon, Y. 1968. Caractéristiques quantitatives du développement et de l'alimentation de *Rhabdoplea klaptoczi* (Karny, 1915) (Orthoptera-Acridinae). *Ann. Univ. Abidjan Ser. E* 1(2):101–112.

———. 1970. Caractéristiques quantitatives du développement et de l'alimentation de *Orthochtha brachycnemis* Karsch 1893 (Orthoptera, Acridinae). *Terre Vie* 24(3):425–448.

———. 1972. Caractéristiques quantitatives du développement et de l'alimentation de *Anablepia granulata* (Ramme 1929) (Orthoptera: Gomphocerinae). *Ann. Univ. Abidjan Ser. E.* 5(1):373–394.

———. 1973. Bilan énergétique de la population d'*Orthochtha brachycnemis* Karsch., principale espèce acridienne de la savane de Lamto (Côte d'Ivoire). *Ann. Univ. Abidjan (Colloque Lamto).* Ser. E. 6(2):105–125.

———, and D. Gillon. 1967a. Recherches écologiques dans la savane de Lamto (Côte d'Ivoire): Cycle annuel des effectifs et des biomasses d'Arthropodes de la strate herbacée. *Terre Vie* 21(3):262–277.

———. 1967b. Méthodes d'estimation des nombres et des biomasses d'Arthropodes en savane tropicale. In *Secondary Productivity of Terrestrial Ecosystems.* Vol. 2, pp. 519–543. IBP, 1966.

———, and J. Pernes. 1970. Recherches écologiques dans la savane de Lamto (Côte d'Ivoire): Comparaison de plusieurs indices de diversité d'un peuplement de Mantes. *Terre Vie* 24(1):54–61.

———, and R. Roy. 1968. Les Mantes de Lamto et des savanes de Côte d'Ivoire. *Bull. IFAN Ser. A* 30(3):1038–1151.

Girard, C. 1971. Les Elatérides de Lamto. *Bull. IFAN*, 33, A, 3, 550–651.

Godron, M. 1973. Analyse d'un échantillonnage de ligne dans la savane de Lamto (Côto d'Ivoire). *Ann. Univ. Abidjan (Colloque Lamto).* Ser. E. 6(2):25–31.

Hedin, L. 1967. Recherches écologiques dans la savane de Lamto (Côte d'Ivoire): La valeur fourragère de la savane. *Terre Vie* 21(3):249–261.

Heim de Balsac, H. 1967. Rongeurs de Lamto (Côte d'Ivoire). Faits nouveaux d'ordre anatomique, taxonomique et biogéographique. *Biol. Gabon.* 3(3):175–222.

———. 1968. Recherches sur la faune des Soricidae de l'Ouest africain (du Ghana au Sénégal). *Mammalia* 32(3):376–418.

Hummelen, P., and Y. Gillon. 1968. Etude de la nourriture des Acridiens de la savane de Lamto en Côte d'Ivoire. *Ann. Univ. Abidjan Ser. E* 1(2):199–206.

Jezequel, J. F. 1964a. Araignées de la savane de Singrobo (Côte d'Ivoire). I. Sicariidae. *Bull. Mus. Hist. Nat.* 36:185–187.

———. 1964b. Araignées de la savane de Singrobo (Côte d'Ivoire). II. Palpimanidae et Zodariidae. *Bull. Mus. Hist. Nat.* 36:320–338.

———. 1965a. Araignées de la savane de Singrobo (Côte d'Ivoire). III. Thomisidae. *Bull. IFAN Ser. A* 26:1103–1143.

———. 1965b. Araignées de la savane de Singrobo (Côte d'Ivoire). IV. Drassidae. *Bull. Mus. Hist. Nat.* 37(2):294–307.

———. 1965–1966. Araignées de la savane de Singrobo (Côte d'Ivoire). V. Note complémentaire sur les Thomisidae. *Bull. Mus. Hist. Nat.* 37(4):613–630.

Josens, G. 1971a. Recherches écologiques dans la savane de Lamto (Côte d'Ivoire): Données préliminaires sur le peuplement en Termites. *Terre Vie* 25(2):255–272.

———. 1971b. Variations thermiques dans les nids de *Trinervitermes geminatus* Wasmann, en relation avec le milieu extérieur dans la savane de Lamto (Côte d'Ivoire). *Ins. Soc.* 18(1):1–14.

———. 1971c. Le renouvellement des meules à champignons construites par 4 *Macrotermitinae* (Isoptères) des savanes de Lamto Pacobo (Côte d'Ivoire). *Comp. Rend. Acad. Sci.* 272:3329–3332.

———, and D. Corveaule. 1973. Le peuplement en Termites des savanes de Lamto (Côte d'Ivoire). Aperçu de quelques données quantitatives. *Ann. Univ. Abidjan (Colloque Lamto).* Ser. E. 6(2):99–104.

Lachaise, D. 1971a. Répartition des espèces de Drosophiles du sous genre *Sphophora* groupe *melanogaster* dans une mosaïque "savane à palmiers Rôniers-galerie forestière" (Côte d'Ivoire). *Comp. Rend. Acad. Sci.* D273:1527–1530.

———. 1971b. Répartition du complexe *D. seguyi* (Dipt. Drosophildae) au contact "savane galerie" dans une savane préforestière de Côte d'Ivoire. *Comp. Rend. Acad. Sci.* D273:1623–1924.

———. 1972. Ecologie des Drosophilidae de la savane tropicale de Lamto (Côte d'Ivoire). I. Rythmes saisonniers et circadiens des populations. Répartition des espèces. Thèse de 3ème cycle. Fac. Sci. Paris VI.

Lamotte, M. 1965. Le laboratoire d'Ecologie tropicale de la savane de Gpakobo (Côte d'Ivoire). *Notes Africaines No. 108* 131–132.

———. 1967a. Les Batraciens de la région de Gpakobo (Côte d'Ivoire). *Bull. IFAN Ser. A* 29:218–294.

———. 1967b. Recherches écologiques dans la savane de Lamto (Côte d'Ivoire): Présentation du milieu et du programme de travail. *Terre Vie* 21(3):197–212.

———. 1969. La participation au P.B.I. de la Station d'Ecologie tropicale de Lamto (Côte d'Ivoire). *Bull. Soc. Ecol.* 1(2):58–65.

———. 1973. Les travaux de la Station de recherches de Lamto. *Ann. Univ. Abidjan (Colloque Lamto).* Ser. E. 6(2):13–17.

———, D. Gillon, Y. Dillon, and G. Ricou. 1969. L'échantillonnage quantitatif des peuplements d'Invertébrés en milieux herbacés. In *Problèmes d'Ecologie: l'Échantillonnage des peuplements animaux des milieux terrestres.* Lamotte and Bourliere, ed., pp. 7–54. Masson.

Lavelle, P. 1971a. Recherches sur la démographie d'un Ver de Terre d'Afrique *Millsonia anomala* Omodeo (Oligochètes Acanthodrilidae). *Bull. Soc. Ecol.* 2(4): 302–312.

———. 1971b. Recherches écologiques dans la savane de Lamto (Côte d'Ivoire): Production annuelle d'un Ver de Terre, *Millsonia anomala* Omodeo. *Terre Vie* 2(2):240–254.

———. 1971c. Etude préliminaire de la nutrition d'un Ver de terre africain *Millsonia anomala* (Acanthodrilidae, Oligochètes). IV Coll. Int. Zool. Sol Dijon (sept 1970). *Ann. Zool. Ecol. Anim. Hors Ser.* 133–145.

———. 1973. Peuplement et production des Vers de terre dans les savanes de Lamto. *Ann. Univ. Abidjan. (Colloque Lamto).* Ser. E. 6(2):79–98.

Lecordier, C. 1971. Une nouvelle méthode de plégeage lumineux. *Bull. IFAN*, A, 33(2):481–486.

———, and Pollet A. 1971. Les Carabiques (Col.) d'une lisière forêt-galerie savane à Lamto (Côte d'Ivoire). *Ann. Univ. Abidjan E* 4(1):250–286.

———, and Vuattoux R. 1971. Les Carabiques (Col.) du palmier-rônier de la savane de Lamto (Côte d'Ivoire). *Ann. Univ. Abidjan E* 4(1):210–247.

———. C. 1972. Les Carabiques de la savane de Lamto (Côte d'Ivoire). *Bull. IFAN* Ser. A, 34(2):378–456.

Levieux, J. 1965. Description de quelques nids de Fourmis en Côte d'Ivoire. *Bull. Soc. Ent. Fr.* 70:259–266.

———. 1966. Note préliminaire sur le comportement des colonnes de chasse de *Megaponera foetens. Ins. Soc.* 13:117–126.

———. 1967a. La place de *Camponotus acvapimensis* Mayr (Hyménoptère Formicidae) dans la chaîne alimentaire d'une savane de Côte d'Ivoire). *Bull. Un. Int. et Ins. Soc.* 14(4):314–322.

———. 1967b. Recherches écologiques dans la savane de Lamto (Côte d'Ivoire): Données préliminaires sur le peuplement en Fourmis terricoles. *Terre Vie* 21(3): 278–296.

———. 1968. Influence des feux de brousse sur le peuplement en Fourmis terricoles d'une savane de Côte d'Ivoire. *Proc. 13th Congr. Entomol. Moscow.*

———. 1969. L'échantillonnage des peuplements de Fourmis terricoles. In *Problèmes d'Ecologie: l'Échantillonnage des peuplements animaux des milieux terrestres.* Lamotte et Bourliere, ed. pp. 289–300. Masson.

———, and J. K. A. Van Boven. 1969. Les Dorylinae de la savane de Lamto (Hym. Form.). *Ann. Univ. Abidjan Ser. E* 1(2):351–358.

———. 1971. Mise en évidence de la structure des nids et de l'implantation des zones de chasse de deux espèces de *Camponotus* à l'aide de radio-isotopes. *Ins. Soc.* XVIII (1):29–48.

———. 1972a. Le rôle des Fourmis dans les réseaux trophiques d'une savane préforestière de Côte d'Ivoire. *Ann. Univ. Abidjan Ser. E. Ecol.* 5(1):143–240.

———. 1972b. Les Fourmis de la savane de Lamto (Côte d'Ivoire): Éléments de taxonomie. *Bull. IFAN* A, 34(3):611–654.

———. 1972c. Le microclimat des nids et des zones de chasse de *Camponotus acvapimensis* Mayr. *Inc. Soc.* 9(2):63–79.

Menaut, J.-Cl. 1971. Etude de quelques peuplements ligneux d'une savane guinéenne de Côte-d'Ivoire. Thèse de 3° cycle. Fac. Sci. Paris.

———. 1973. Aperçu quantitatif sur les formations ligneuses des savanes de Lamto. *Ann. Univ. Abidjan (Colloque Lamto).* Ser. E. 6(2):19–23.

Monnier, Y. 1968. Les effets des feux de brousse sur une savane préforestière de Côte d'Ivoire. *Etudes éburnéennes* 9, 260 p.

———. 1969. Il était une fois à Ayérémou. . . . un village du Sud-baoulé. *Ann. Univ. Abidjan* 6,1(1) 136 pp.

———, R. Viani, and J. Marchant. 1972. Contribution à l'étude des feux de brousse. Mesure des température. *8th Bienn. Conf. West Afr. Sci. Assoc.* Univ. Ghana-Legon.

Porteres, R. 1964. Le palmier Rônier (*Borassus aethiopum* Mart.) dans la province Baoulé (Côte d'Ivoire). *J. Agr. Trop. Bot. Appl.* 11:499–514, 12:80–107.

Rambelli, A. 1971. Recherches mycologiques préliminaires dans les sols dè forêt et de savane en Côte d'Ivoire. *Rev. Ecol. Biol. Sol* VIII, 2, 210–226.

De Rham, P. 1971. L'azote dans quelques d forêts, savanes et terrains de culture d'Afrique tropicale humide (Côte d'Ivoire). Thèse Fac. Sci. Lausanne.

Roland, J. C. 1967. Recherches écologiques dans la savane de Lamto (Côte d'Ivoire): Données préliminaires sur le cycle annuel de la végétation herbacée. *Terre Vie* 21(3):228–248.

———, and F. Heydacker. 1963. Aspects de la végétation dans la savane de Lamto (Côte d'Ivoire). *Rev. Gen. Bot.* 70:605:620.

Roux-Esteve, R. 1969. Les Serpents de la région de Lamto (Côte d'Ivoire). *Ann. Univ. Abidjan Ser. E* 2(1):81–140.

Thiollay, J. M. 1970. Recherches écologiques dans la savane de Lamto (Côte d'Ivoire): Le peuplement avien. Essai d'étude quantitative. *Terre Vie* 24(1): 108–144.

———. 1971. 1971. L'avifaune de la region de Lamto (Moyenne Côte d'Ivoire). *Ann. Univ. Abidjan E,* 4(1):5–132.

———. 1971. Les Guépiers et Rolliers d'une zone de contact savane-forêt en Côte d'Ivoire. *L'Oiseau et R. F. O.* 41(2–3:148–162.

———. 1971. L'exploitation des feux de brousse par les Oiseaux en Afrique Occidentale. *Alauda* 39(1):54–72.

Viani, R., J. Baudet, and J. Marchant. 1973. Réalisation d'un appareil d'enregistrement magnétique de mesures. Application à l'étude de l'évolution la température lors du passage d'un feu de brousse. *Ann. Univ. Abidjan (Colloque Lamto).* Ser. E. 6(2):295–304.

Villecourt, P. 1973. Premiers éléments du bilan d'azote dans la savane à Rôniers de Lamto-Pacobo (Côte d'Ivoire). *Ann. Univ. Abidjan (Colloque Lamto).* Ser. E. 6(2):33–34.

Villiers, A. 1965. Hémiptères Réduviidés, Phymatidés et Hénicocéphalidés de Côte d'Ivoire. *Bull. IFAN* A, 27:1151–1182.

Vincent, J. P. 1970. Recherches écologiques dans la savane de Lamto (Côte d'Ivoire): Observations préliminaires sur les Oligochètes. *Terre Vie* 24(1):22–39.

Vuattoux, R. 1968. Le peuplement du Palmier Rônier (*Borassus aethiopum*) d'une savane de Côte d'Ivoire. *Ann. Univ. Abidjan Ser. E* 1(1).

———. 1970. Observations sur l'évolution des strates arborée et arbustive dans la savane de Lamto (Côte d'Ivoire). *Ann. Univ. Abidjan Ser. E,* 3(1):285–315.

# CHAPTER 16

# A Critical Consideration of the Environmental Conditions Associated with the Occurrence of Savanna Ecosystems in Tropical America

G. Sarmiento and M. Monasterio

## The Savannas Problem

The savannas cover extensive areas of the American tropics, and, together with rain forests, are the most extended neotropical plant formations. Ecological interpretation of these savannas has been quite divergent due partly to the word "savanna" being applied to completely different ecosystems. Even when limited by a definition like Beard's (1953), it still includes a wide variety of physiognomic types and floristic units under different climatic, topographic, soil conditions, and degrees of human interference. Furthermore, the difficulty in distinguishing secondary herbaceous communities from primary savannas, as well as delimiting savannas and open forests, helps make the interpretation more confused.

It is remarkable that savanna is the only ecosystem in the entire warm tropical region whose origin and permanence have been considered unanimously as a fundamental ecological problem. This is partly due to its peculiar physiognomy, interrupting the continuity of forest formations and suggesting by contrast that it is a secondary or unstable system. A common question is, if rain forests occur under wet tropical conditions and dry deciduous and thorn forests under drier environments, why do grass- and herb-dominated formations appear in the middle of this water gradient?

Almost every savanna region in the American tropics has been the subject of analysis, but each work has been specific in terms of area and viewpoint. Thus most generalizations and their extrapolations to other savanna regions have led to contradictory conclusions. This is even more evident when comparisons are made between savannas on different continents or when the conclusions from the study of African, Asian, or Australian savannas are applied to tropical America.

As a starting point, we will adopt a concept of tropical savanna widely accepted in America, the American savanna being the only region considered here. Following Beard, tropical savanna is defined as a natural and stable ecosystem occurring under a tropical climate, having a relatively continuous layer of xeromorphic grasses and sedges, and often with a discontinuous layer of low trees or shrubs. This definition includes physiognomy (a relatively continuous herb layer, with or without sparse trees or shrubs), floristics (predominance of grasses and sedges in that layer), and ecology (a tropical climatic regimen, a natural, stable ecosystem, and xeromorphic nature of the herbs).

This definition excludes all herbaceous communities where grasses and sedges are not dominant or xeromorphic, as in high mountain vegetation in the Andes (*paramos*), the vegetation of the high Guianan plateau, and the communities of seasonal or permanent swamps. The definition also excludes seral communities as well as obvious man-made systems as pastures and fields. On the other hand, the concept includes pure grasslands, without woody species, as well as plant formations with an almost continuous tree cover, provided that the herb layer is still a continuous, ecologically dominant vegetation stratum.

Other authors have used a broader concept of savanna, as, for example, Dansereau (1957) and Walter (1971), who divide grasslands from mixed communities of herbs, trees, and shrubs. These authors also include in their savanna concept all mixed communities from extratropical regions. Both criteria are fruitless when applied to tropical American savannas. First, because these ecosystems are fundamentally different in rhythmicity from communities of similar physiognomy in subtropical and temperate regions. Second, because in the neotropical region there are closely related ecosystems differing only in the presence or absence of a sparse tree cover, and according to a purely physiognomic definition, these would constitute completely different units. Thus it is neither useful nor convenient to separate pure grasslands from tree grasslands, at least in the American tropics.

Some other authors, like Jaeger (1945) and Lauer (1952), have employed a radically different savanna concept, considering as a "savanna belt" a tropical or subtropical region with a marked seasonal climate. Savanna, then, becomes a geographic unit, defined by its climatic regimen, independent of the actual ecosystems occurring in each area. A substantial part of savanna belts thus defined is in fact covered with various forest types.

The geographic distribution of tropical American savannas, together with a description of the major physiognomic and floristic types, has been reviewed by Sarmiento and Monasterio (1974). Figure 16–1 reproduces the map from that paper showing the main regions of savanna in South and Central America.

Here, the major aim is to discuss the environmental and/or historical factors related to the occurrence of tropical American savannas. We will refer to savanna ecosystems in the plural, because sufficient ecological diversity exists between types to make it impractical to consider all types as one ecosystem. As we will see later, it is necessary to differentiate at least three types of savanna ecosystems.

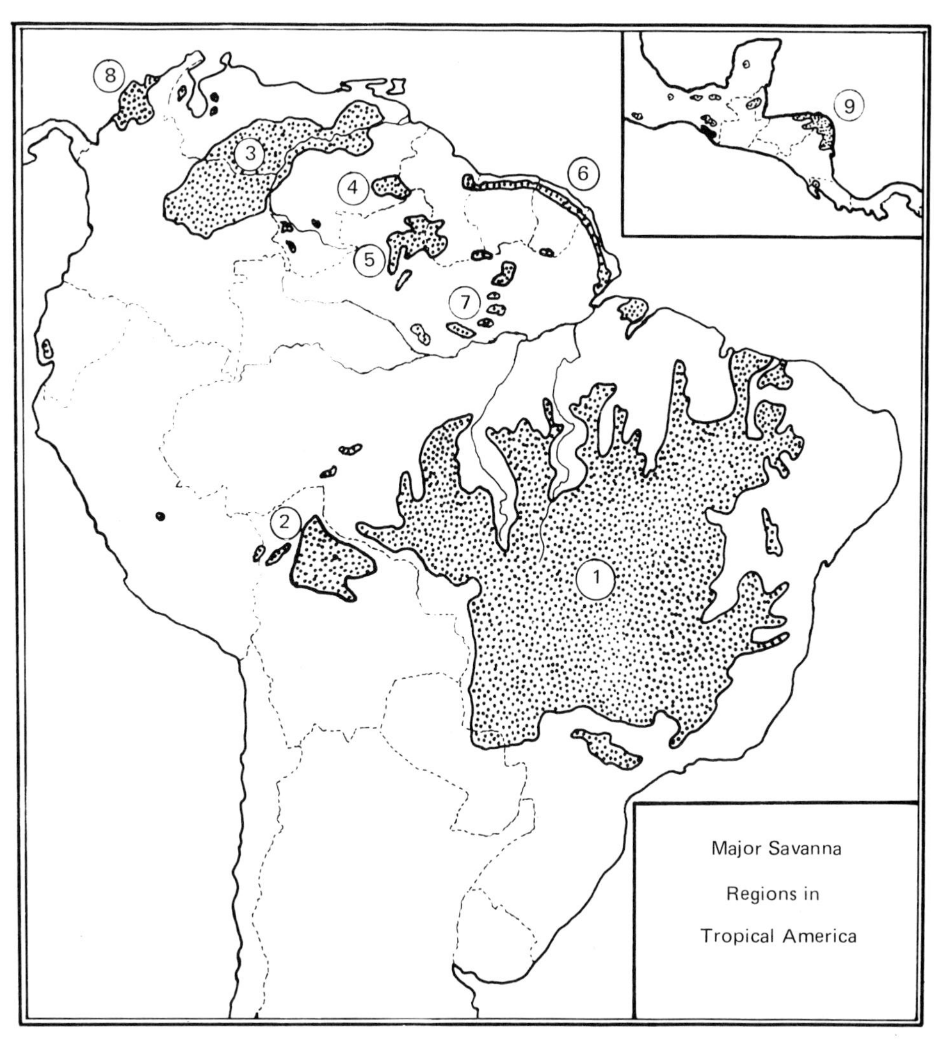


**Fig. 16–1.** Major savanna regions in tropical America. (*1*) Cerrado region; (*2*) Bolivian llanos; (*3*) Orinoco llanos; (*4*) Gran sabana; (*5*) Rupununi-Rio Branco savannas; (*6*) coastal Guianan savanna belt; (*7*) Amazonian campos; (*8*) Magdalena region; (*9*) Miskito region.

A thorough ecological study of the full range of tropical American savannas is still lacking. The first essay on this problem was Beard's (1953). Considering only the savannas extending from Central America and the West Indies to the northern part of the Amazon basin, Beard concluded that these savannas were edaphically determined, appearing in regions of strongly seasonal climate as a consequence of maldrainage problems during the rainy season and acute water deficiency during the dry season.

Hills and Randall (1968) summarize the conclusions of an international symposium held in Caracas in 1964 on the savanna-forest border, emphasizing the points made above. Van Donselaar (1965), considering the savannas only in northern Surinam, critically discussed the origin and evolution of the northern South American savanna flora and vegetation. He speculates about the possible origin of that savanna flora on the ancient plateau of the Guiana Shield. Blydenstein (1967), in Colombia, concluded that several factors acted concurrently, because savannas and forests occurred under identical climatic and soil conditions. Hills (1969) summarizes the studies of a McGill University research team in the Rupununi savannas in Guyana, which considered a hierarchy of operative factors, such as length of dry season, soil fertility, occurrence of laterites, fire, depth of water table, and landscape evolution. Eiten (1972) reviews the major savanna area of South America, the *cerrado* region in central Brazil, concluding that the *cerrado* vegetation is a climatic topographic and edaphic climax, with biotic factors having a much smaller influence. Our team in Venezuela (Monasterio, 1971; Sarmiento and Monasterio, 1969, 1974; Sarmiento et al., 1971; Silva and Sarmiento, 1973) has also considered a variety of factors in the occurrence of the various savanna types, emphasizing the wide diversity of savanna communities and the importance of the evolution of the whole landscape in understanding the distribution of forests and savannas.

In summary, the literature emphasizes the fragmentary knowledge on this subject. The occurrence of savannas in tropical America still remains one of the big biogeographic and ecological problems of the neotropical area. In the following sections we will consider the major ideas encountered in the literature, discussing them in the light of our direct experience.

## The Main Theories on Factors Associated with Savanna Ecosystems

The various explanations of the occurrence of tropical American savannas may be grouped into six categories according to the external factor to which the most emphasis is given. We will consider these theories consecutively without giving a comprehensive list of all supporters, taking into account the fact that many researchers changed their opinions as new evidence became available.

### Climatic Factors

Perhaps the oldest explanation (Grisebach, 1872; Schimper, 1903; Warming, 1908; Bouillene, 1926; Myers, 1936; Jaeger, 1945; Sarmiento, 1968) is that tropical savannas are the result of a particular climate. This climate is defined by Koeppen (1931) as tropical wet and dry, or savanna climate (Aw). The key operative factor in selecting the savanna ecosystem is the regimen of constantly high temperature throughout the year, with alternation of a very rainy season and a prolonged, almost completely rainless season. During the rainy season a large water surplus occurs (Thornthwaithe, 1948), while during the rainless season a strong water deficit prevails, with rainfall much lower than potential evapotranspiration. According to supporters of the theory, savanna is better adapted than any other plant formation to withstand this cycle of alternating soil-water conditions. Rain forests could not resist the extended period of extreme drought, while dry forests could not compete successfully with perennial grasses during the equally extended period with large water surplus.

Extensive areas of tropical America with an Aw climate (Figure 16–2), such as Cuba, the Colombian-Venezuelan llanos, the Bolivian llanos, and the interior Brazilian plateau, have in fact, a plant cover with a predominance of savannas.

## Soil Factors

Several soil features have been postulated as major determinants of savannas. These include soil drainage, soil-water-retention capacity, and mineral supply.

### *Soil Drainage*

According to this theory (Bennett and Allison, 1928; Charter, 1941; Beard, 1944, 1953; Richards, 1952; Miranda, 1952; Walter, 1971), the existence of savanna is related to a soil-water regimen where excessively wet

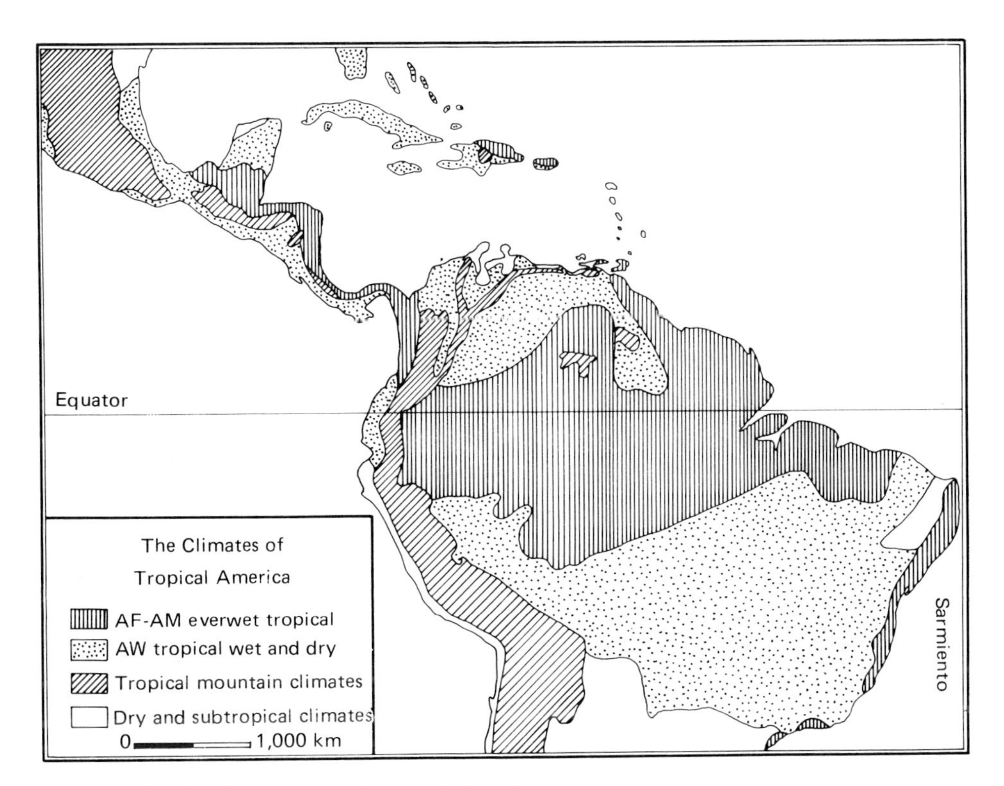


**Fig. 16–2.** The climates of tropical America.

conditions, with anaerobic horizons at or near the surface, alternate with periods where these same soil horizons become dry below the permanent wilting point. This annual cycle can be found under an Aw climate in physiographic and topographic situations with difficult external and internal soil drainage. For instance, seasonally hydromorphic soils frequently occur in alluvial floodplains where a combination of very gentle slopes and superposition of sedimentary layers of different permeability induces a slow drainage during the season of heavy rainfall. As these conditions of impeded drainage are generally associated with impermeable and poorly structured clay horizons, the previously saturated soils not only lose their water rapidly immediately after the end of the rainy season but very soon become physiologically dry, remaining in this state until the beginning of the next rainy season.

Another slightly different situation occurs in well-drained soils with a water table rising near the surface during several months. In both cases the alternation of wet and dry soil phases resembles the annual cycle produced by a savanna climate, and in this sense the explanation is similar to the climatic theory. Nevertheless, the ecological conditions of seasonally saturated or waterlogged soils are more complex, since the biological impact of the climatic cycle is reinforced by the soil-water cycle, resulting in an additional environmental stress on plant life. The potential for various life forms to exploit each soil horizon become significantly different, therefore causing important changes in the structure and composition of the corresponding savanna ecosystems.

### *Dry Soil*

This explanation (Santamaria and Bonnazzi, 1963; Walter, 1969) is completely opposite the previous one, and constitutes a further demonstration of the contradictory interpretations of savanna ecosystems. In this case, the main operative factor is supposed to be an excessively dry soil, determined either by (1) a low water-retention capacity in the upper soil horizons originating in unfavorable texture, as in coarse deposits of gravel or sand; (2) a shallow soil profile as on hard-weather-resistant rocks (quartzites, some sandstones and limestones); or (3) a secondary shallow profile due to the development of a lateritic cuirass near the surface. In each case during a rainless period the soil profile is subjected to a rapid loss of available water.

If these conditions appear under a strongly seasonal climate, the vegetation will suffer from a water stress for almost the entire rainless period. Under an ever-raining climate, a complete soil drought will only occur during short periods between heavy rains, but the level of available soil water will often approach the permanent wilting point.

### *Low Nutrient Supply*

According to this theory (Pulle, 1906; Lanjouw, 1936; Waibel, 1948; Arens, 1958), savanna soils are too poor nutritionally to maintain forest. This deficiency in mineral elements may originate in (1) a very poor parent material such as a pure quartz sand or any long transported alluvial or eolian deposit; and (2) a soil material impoverished through the soil-forming process, such as rapid weathering of ferromagnesian minerals, rapid mineralization of organic matter, the neosynthesis of kaolinitic clays of very low cation exchange capacity, a thorough leaching of soluble elements, increasing desaturation of the cationic complex, and the immobilization of elements in soil complexes.

If both a poor parent material and an extended evolution of the soil profile are concurrent, the resulting soils will be extremely poor, with a critical low level of most essential elements. In less extreme situations, only certain nutrients may be limiting.

Two other factors contributing to soil impoverishment are soil erosion and fire. Erosion with heavy rainfall, open vegetation, and steep slopes may lead to a total loss of the upper soil horizons. Fire results in the accumulation of mineral elements on the soil surface after burning as ash, which can be lost by water or wind erosion.

In any case, be it a senile soil profile, original poverty, or secondary losses, savanna soils in tropical America have in fact an extreme oligotrophic character.

### Fire

Many workers (Rawitscher, 1948; Ferri, 1955; Parsons, 1955; Vareschi, 1960; Tamayo, 1964; Schnell, 1971) maintain that frequent fires in savanna prevent the establishment or persistence of plant formations not resistant to burning. Fire may originate by natural phenomena or by human origin. Some believe that any forest may be degraded to savanna provided that it is burned frequently enough; others think that only some particularly sensitive vegetation types may be transformed into savannas.

Whatever the case, burning will act like a selective filter through which only fire-tolerant or -resistant species pass, and these species are precisely those encountered in present-day savannas.

### The Human Factor

This viewpoint (Aubreville, 1949; Budowski, 1956; Johannessen, 1963; Taylor, 1963; Walter, 1971; Puig, 1972) considers savannas to be relatively recent ecosystems originating from human action upon different forest types. Savannas could have originated and been maintained by aboriginal man through deforestation by burning or with tools to obtain open areas for hunting and later for cattle-raising and crop production. Once the forest has been destroyed and its reestablishment prevented over an extended time, several ecosystem characteristics vital for forest regeneration are irreversibly altered. The normal secondary succession is changed toward savanna, which is maintained as a permanent disclimax.

This explanation, although similar to the fire theory, differs in the fact that fire can be natural and not man induced, while man may resort to other deforestation methods besides burning. According to the anthropic theory, savanna communities would not be older than the first arrival of man in tropical America, that is, they would be postglacial and probably no more than 20,000 years old (Hester, 1966).

### Historicogeographic Changes

Several authors (Aubreville, 1962; Hueck, 1966; Schnell, 1971) have considered the savannas as relics from previous periods when another set of ecological determinants was at work. Similarly, others have considered savannas as the final product of landscape evolution under conditions that are not comparable to the present ones, appearing only in habitats disintegrating under attack by the present morphogenetic agents (Cole, 1960; Egler, 1960). Thus the occurrence of savannas would be additional evidence of dramatic geologic, geomorphologic, and climatic change during the Quaternary Period in tropical American lowlands. These changes have been reported for Brazil, Guyana, Venezuela, and Colombia (see Vuillemier, 1971) and they alter previous ideas about the relative stability of the humid tropical environments.

The savannas problem however remains. What could the environmental conditions be in places where savannas originated and where they actually prevail? If these conditions exist now in any region, savannas would sooner or later become established there, and to explain this establishment we must certainly resort to one of the previous theories. If, on the contrary, these conditions have not been produced again, then the savannas become a paleoecologic puzzle.

An opposite argument is that savannas occur in such a geologically and evolutionarily young environment, that tree species capable of colonizing and forming a closed forest have not yet evolved, or reached these areas. This theory is reinforced by the apparent absence of a fauna of vertebrates from South American savannas. In this way, savannas would belong to the postglacial age, and represent early successional stages in a slow evolution toward a new forest type.

### Holocenotic Theories

Most recent authors (Cole, 1960; Eden, 1964; Hills, 1965, 1969; Goodland, 1970; Monasterio, 1971; Monasterio and Sarmiento, 1971; Eiten, 1972) cite the interaction and concurrence of several factors in the origin and maintenance of savanna ecosystems. Some may be simply conditioning or predisposing factors, others may be a necessary but not sufficient determinant, while others may be important in the genesis of the ecosystem but not in its maintenance. Because these theories always include a set of simultaneous or successive agents, we have called them "holocenotic," to emphasize the interaction of many environmental and evolutionary factors and the holistic nature of their influences.

An example of a holocenotic theory is that, within a region of tropical wet and dry climate, in flat areas with heavy soils of impeded drainage that dry off completely during the long rainless season, the savannas displace any other possible formation not able to withstand the alternation of both water deficiency and water surplus. Another sequence of events leading to a savanna ecosystem would be a tropical rainy climate, with continuous water excess producing intensive soil leaching. Thus well-drained, coarse-textured soils would suffer almost complete loss of critical elements. On these impoverished soils the zonal rain forest is replaced by a lower, more open, and more scleromorphic forest that is also more susceptible to burning. After repeated burnings a savanna becomes established and is maintained by fire.

## Discussion

### Climate

In tropical America a high proportion of savannas occurs under tropical wet and dry climates—Koeppen's Aw climatic type (see Figures 16–1 and 16–2). In this climatic type annual rainfall amounts to 1,000 to 2,000 mm, and 80 to 90 percent of this total falls in a wet season five to eight months long. This climate is produced by the annual cycle of the tradewinds and the seasonal displacement of the Intertropical Convergence Zone between both hemispheres. Most savannas in the Colombian, Venezuelan, and Bolivian llanos, in Cuba, and in the interior plateau of the Guianas and Brazil occur under this climatic type. In contrast, the savannas of the Atlantic coast of Honduras and Nicaragua, Trinidad, the northern coast of Brazil, and the Guianas, as well as the savanna "islands" within the Amazon lowlands, do not occur under seasonal climates but in tropical wet climates (Af-Am types of Koeppen). In these regions annual rainfall generally exceeds 2,000 and even 3,000 mm, and there are at the most three months with rainfall lower than 100 mm.

A first climatic limit evident in the distribution of the tropical American savannas is a drought limit, since there are no savannas in this region with yearly rainfall figures under 800 to 1,000 mm. This is a major difference between America and other continents, where savannas occur under arid and semiarid climates (Walter, 1971). In the neotropical region, with dry climates (B types of Koeppen) and even with Aw climates near the dry limit, dry deciduous and thorn forests occur, and savannas are absent. Only in the subtropical and temperate regions of South and North America do semiarid savannas exist, as in the Chaco region of subtropical South America and the mesquite woodlands of southern United States and northeastern Mexico.

A second climatic limit of savannas in tropical America is altitude. Most savannas appear below 1,000 to 1,200 m above sea level. Along the Andes chain savannas never surpass these elevations. The only American areas where savannas are found at higher elevations are in the isolated ranges and plateau in the interior of the Guianas, where they may reach more than 2,000 m (Myers, 1936). Here they form a shrubby vegetation characterized by its peculiar, highly endemic flora, typical of the "rocky fields" of this region. This altitudinal limitation of tropical savannas could be explained as a thermal limit, not due to low temperatures being unfavorable to growth or establishment of savanna species but constituting a handicap in competition with other growth forms. The latitudinal limit of tropical savannas appears to be a coincident with a frost boundary and with the occurrence of a cold winter season in subtropical regions.

These two climatic limitations reduce the area of tropical savannas to the warm, tropical, rainy climates, the A types of Koeppen classification. According to Koeppen's system the A climates are divided in two main types: the tropical wet climates (Af), where rainfall of the driest month is at least 60 mm; and the tropical wet and dry climates (Aw) with a distinct dry season or at least one month having less than 60 mm. A third type, the monsoon, or Am, climate, is intermediate between Af and Aw, having a short dry season and a total annual rainfall greater than that of Aw climates.

Within the extensive region of Af-Am climates (Figure 16–2), the predominant plant formations are the rain forests (tropical rain forest *sensu stricto*, as well as seasonal evergreen forest and semideciduous forest; see Beard, 1944). Formations other than rain forests appear as islands dispersed in the rain forest matrix; these include savannas, swamps, and open, low forests called "savanna forests," or *caatingas*,[1] in the Amazon region. These three types are undoubtedly controlled by nonclimatic factors, since the wet climate is optimum for rain forest development.

In the tropical American region with wet and dry climates (Aw), the preceding pattern is reversed; here the forests appear as islands surrounded by a more or less continuous savanna matrix. Only in restricted cases do some forest types, mainly semideciduous and deciduous forests, appear over more continuous surfaces. Under these conditions savannas could be considered as "climatic," and the remaining formations could represent soil or topographic climaxes, but we must point out that the disjunction between savannas and dry deciduous forests does not rely primarily on regional climate. For further details about vegetation patterns under different climates in tropical America, see Sarmiento and Monasterio (1974).

## Soils

Most tropical American savannas occur on flat topography, either on Quaternary alluvial plains (as in the llanos of Colombia, Venezuela, and Bolivia) or on ancient, very gently undulating peneplains (as in the Brazilian *cerrados*). A part of these savannas occupies low topographic positions where

[1] It is necessary not to confuse these Amazon *caatingas*, which are sclerophyl, evergreen forests, with the northeastern Brazilian *caatingas*, which are dry, deciduous forests.

rainwater accumulates during the rainy season, leading to the development of hydromorphic soils. However, savannas do not occur on soils that remain water saturated or waterlogged most of the year. In the latter case swamp communities become established, or if the underground water circulates, gallery forests may be maintained.

Savannas occur on ill-drained soils only in conditions of seasonal water saturation, when periods of oxidation and reduction within the soil profile alternate during the year, giving to the soil the characteristic red, yellow and gray mottled colors. These soils are unfavorable for plant growth almost the whole year, remaining anaerobic during the wet season, and then quickly drying shortly after the rains end. On the other hand, even when they are waterlogged, the subsoil at a depth of 0.5 to 1 m may be completely dry, since a clay hardpan near the surface greatly reduces water percolation to deeper horizons. In this way, under a seasonal wet and dry climate, in certain topographic positions, the soil features reinforce the two contrasting climatic extremes, giving rise to a drier habitat in the dry season and a wetter habitat during the rain period. A similar situation occurs in river floodplains when the river water volume changes seasonally and floods its plain almost yearly.

At the other extreme, frequently in savanna regions soils have low water-retaining capacity, due to coarse texture or shallow depth. Many of these soils have developed on coarse alluvial gravels or on alternating depositions of gravels and sands, while in other cases savannas occur on thin soils developed on weather-resistant rocks, such as sandstones, quartzites, and old lateritic cuirasses. Under a seasonal climate these soils are drier during the rainy period than more normal soils, and during the rainless season they remain ecologically dry for a longer time.

Considering nutrients, soils in tropical humid regions (A climates) are usually poor in mineral nutrients, as a direct consequence of the dominant soil genesis processes leading to an increasing elimination of soluble and interchangeable elements from the soil profile. Under particular circumstances the soil may be exceptionally poor in comparison with other tropical soils, as when the parent material is deficient in nutrients, the soil is very permeable, the clay and organic matter content is low, or the soil genesis has been prolonged.

Tropical podzols developed on pure "white sands" (Klinge, 1965; Sioli and Klinge, 1962) are undoubtedly the poorest soils in the American tropics (Stark, 1970). However, the vegetation they maintain is usually an open and low forest and rarely a savanna. In these forests the mineral cycling follows almost entirely biotic pathways through the well-developed organic horizons covering the soil surface where the tree roots are most developed. In this way the nutrient cycles bypass the mineral soil. When this deep humus layer is destroyed, either by fire or human activity, the rupture of the nutrient cycles may be responsible for the lack of regeneration of the forest and the perpetuation of a scrub vegetation or shrub savanna.

Other tropical soils do not show the extreme oligotrophic nature and the labile equilibrium in the nutrient cycles of podzols. However a significant difference exists between forest and savanna soils, not only in turnover rates and total amounts of nutrients in the system but also in the relative importance of the A horizons as nutrient reservoirs. If the soil organic matter content under forests and savannas is roughly equivalent, in a forest ecosystem soil litter plays a first-order role in the annual liberation of the mineral elements accumulated in the aerial parts of vegetation. In contrast, litter production in the savannas is insignificant. The only organic material incorporated directly in the soil originates in underground organs, while the mineral elements of aboveground biomass normally reach the soil as ashes after burning.

In summary, all humid tropical soils seem to be highly oligotrophic, some of them, like podzols, to an exceptional degree. Forests may occupy even the poorest soils. The cycling of nutrients in forests is labile and sensitive to external disturbance.

## Fire

All savannas burn repeatedly at intervals varying from a few months to several years, and the great majority of them burn at least once every one or two years. The main period of burning is during the last half of the dry season or during short rainless intervals. It is difficult to find savannas that have not been burned for long periods, even in small savanna islands surrounded by rain forests. However, in everwet climates without a distinct dry season, fire frequency may be so low as to be unable to hamper woody growth, especially of those fast-growing trees that colonize open sites. Therefore, in this climate, if the savannas are a result of deforestation through fire, it is difficult to understand why the natural regeneration of forest does not proceed. Actually the savanna-forest border remains clean and stable for decennia.

In the case of wet and dry climates, it is obvious that fire does not affect the establishment of woody species in the savannas, and that tree density in these ecosystems is not basically dependent on fires but on certain soil factors (Mooney et al., 1974). If only fire were controlling woody growth and propagation, a forest of woody pyrophytes would become established in a rather short time. Fire in the savannas acts as a filter through which only adapted or tolerant species pass. Thus all woody and herbaceous savanna species are pyrophytes, with different methods of avoiding, resisting, or tolerating recurrent burning. But the grassland or open woody physiognomy of most savannas is not directly related to fire frequency.

In temperate and subtropical regions there are fire-tolerant ecosystems characterized by a closed cover of shrubs or low trees where perennial grasses play, at best, only an accessory role, as for example, the chaparral, *garrigues* and the *Quercus-Pinus* low, evergreen forests of Mediterranean climate of Europe and North America. Why, then, in a tropical savanna, would burning be responsible for an open, grassy physiognomy, while in ecosystems of comparable fire frequency of extratropical regions, closed tree or shrub formations are maintained? Considering also the additional fact that burning woody cover, often of highly resinous wood, produces much higher temperatures for a longer time than savanna fires, which very quickly pass over the herb layer and only reach lower temperatures (Vareschi, 1962).

In present-day conditions, fires are more frequent in tropical savannas than in any other natural system elsewhere in the world. This high frequency is due to the use of savannas for cattle-raising, and is a phenomenon of the last few centuries. However, we have no evidence that these savannas were different 400 to 500 years ago, before European colonizers arrived in tropical America, when they might have been subjected to a much lower fire frequency by the aboriginal population.

In tropical zones natural fires occur mostly in a seasonal wet and dry climate, since commonly in dry climates the vegetation fuel on and near the soil is not continuous enough to propagate the flames, and in everwet climates the standing plant material and the litter are rarely dry enough to become flammable. In everwet climates only the savannas may be easily burned, providing that a long enough dry period has desiccated the standing biomass of the herb layer. Semideciduous forests also are not easily burned, and natural fires are not likely; but if they are intentionally burned, secondary grasslands quickly occupy the former forest land. These herbages differ from natural savannas in both composition and ecology, being dominated by alien grasses like the African species *Panicum maximum, Hyparrhenia rufa,* and *Melinis minutiflora* (Vareschi, 1969; Sarmiento et al., 1971; Parsons, 1972). Through frequent fires these grasslands may be maintained indefinitely without further forest regeneration. Usually the differentiation of these secondary herbaceous communities from primary savannas on the basis of a floristic analysis is not difficult.

## Human Interference and Historic Factors

Human action has and continues to produce vast extensions of herbaceous secondary communities in tropical America. Some of these secondary communities will be maintained by annual burning. On the other hand, many primary savannas have been modified by overgrazing or ploughing, thus changing their original composition and structure, altering the herb-woody equilibrium, and favoring many weeds. Primary savannas, modified savannas, and secondary herbaceous communities may be separated easily by their composition. Without further consideration of secondary systems, we now have much evidence to support the statement that savannas did exist in tropical America before the arrival of the Europeans, before the development of aboriginal cultures, and most probably before the arrival of *Homo sapiens*—even before the birth of this species on the planet.

Palynological evidence, though fragmentary, shows the occurrence of savannas in the Orinoco plains, 5,000 years ago and in the Guianas 14,000 years ago, in both cases comprising the full extent of the available pollen records (van der Hammen, 1963; Wijmstra and van der Hammen, 1966). Palynological records reaching older Quaternary strata show that savannas have occurred in the Amazon basin from the last glacial period, having a probable age of 13,000 to 30,000 years (van der Hammen, 1972). This latter evidence is particularly interesting because the savanna pollen period is located between two forest pollen periods and in an area (near the Brazilian-Bolivian border) where there are no savannas at all.

Indirect evidence of a savanna-like vegetation under a strongly seasonal climate may be obtained by geomorphologic and pedologic data from Quarternary deposits in the Venezuelan plains, since only some soil and erosive paleoprocesses dating from middle Quarternary were possible under this type of climate and vegetation (Zinck and Stagno, 1965; Blanck et al., 1972).

The differentiation of a rich savanna flora in tropical America, distinct from any existing forest flora, is more evidence of the considerable age of the savanna formation. It is very difficult to explain the evolution of hundreds of species with complex and multiple adaptations to wet and dry climates, frequent fires, and low nutrient supply without continuous and prolonged operation of these selective forces. In pollen records from northern South America many savanna species already are present in the middle Eocene (van der Hammen, 1972).

Climatic, orogenic, and geomorphologic Quaternary changes led to dramatic changes in the distribution of lowland tropical ecosystems, with periodic replacement in any given place of savannas, forests, and swamps. These changes are documented in the pollen records from several localities in Colombia, Guyana, and Brazil.

In summary: First, the existence of savannas in tropical America during most of the Quaternary and the very ancient evolution of its flora have been established beyond any reasonable doubt. Second, the replacement of savannas and forests is also well documented. An immediate consequence is that the area of savanna has changed in the past concurrently with major environmental changes. The interpretation of savannas as human-induced ecosystems may be entirely discarded, without ignoring the fact that human action may be partly responsible for the maintenance of savannas.

## Conclusions

We have concluded that the operation of several factors result in and maintain savannas; that is, we have developed a holocenotic interpretation of savanna ecosystems. We now wish to analyze three problems. First, if savannas are sufficiently heterogeneous to be conditioned by a different combination of external factors, the holocenotic interpretation will not be applicable to each type but only to all savannas in a single heterogeneous ecological category. Second, supposing that the holocenotic interpretation is valid for each major savanna type, how will the various factors be arranged in order of importance or in causal sequence? Third, is the impact of the environment on the ecosystem similar in every case, and are the living organisms under similar environmental stresses and identical selective forces?

**Savannas and Rain Forests Under Tropical Rainy Climates**

Let us consider first the situation under tropical rainy climates where long water stress periods do not occur. Under these climates three main types of ecosystems occur (disregarding swamp and other semiaquatic communities): (1) evergreen tropical rain forests; (2) low, open, scleromorphic forests (the Amazon caatingas); and (3) savannas. We have also seen that the pattern of these three types is that of a more or less continuous rain forest, with occasional patches or islands of the other formations.

Most rain forests in this area are climatic climaxes, and the differentiation between various forest formations depends upon the presence or absence of short drier periods (see Beard, 1944, for a classification of tropical American vegetation types). The scleromorphic forests appear in the Amazon region and in the Guianas where they have been called by the unfortunate names of "savanna forests" and "savanna woods" (Heyligers, 1963). They are always associated with white sands and are determined by the highly oligotrophic podzols developed on this parent material (Sioli and Klinge, 1962; Hueck, 1966) or by the constantly dry nature of these soils (Heyligers, 1963).

The savannas appear under two different sets of conditions: (1) on white sands, possibly as a final product of the natural evolution of the *caatinga* forest when the closed nutrient circulation within the ecosystem is no longer possible, or by irreversible rupture of this equilibrium through man-induced fires; and (2) on senile soils of latosolic evolution developed on old sediments of Plio-Pleistocene age. In contrast to podzols, these latosols are neither excessively drained, nor do they have seasonal deficits or surplus of water for long periods, but they have a very low fertility, though not to the advanced degree shown by podzols.

The occurrence of savanna on white sands may be associated with soil mineral deficiency as a determining factor for the open forest and a predisposing factor for savanna, while the natural evolution of the soil profile and/or fire may be the direct cause of savanna establishment and permanence. However, on highly laterized profiles the problem is how is the side-by-side coexistence of rain forest and savanna possible on soils of similar parent material, age, and topographic position, and therefore similar water budget and initial stock of plant nutrients.

Within the wet climate regions of tropical America savanna patches are almost constantly associated with flat alluvial surfaces, slightly elevated over the actual relief, formed by loose and rather coarse material, with frequent occurrences of iron hardpans or disintegrating lateritic cuirasses. These landforms, apparently of late Pliocene or early Pleistocene age, appear as reduced relictual forms in several regions of Central and South America, and are covered to varying degrees by typical savannas. Their soils are, in general, very leached latosols, or, in some cases, podzolized profiles developed on an ancient lateritic soil. Parsons (1955) described this landform in the *misquito* savannas of Nicaragua and Honduras; van Donselaar (1965) recognized this old surface as the main savanna belt in northern Surinam, and Bouillenne (1926) and Egler (1960) report that Amazon savannas appear on these elevated surfaces.

Our hypothesis, to explain the coexistence of rain forests and savannas in the humid regions of tropical America, is as follows: During a climatic period drier than the present one, with a clearly seasonal rainfall regime (an Aw climate), probably during the last glaciation, the Amazon basin and the neighboring regions were covered mostly by savannas, while rain forests were localized along or near the bordering mountains. As the climate became wetter and warmer from 15,000 years ago to the present, the seasonal regime disappeared and everwet climates became generalized in the whole region. Along with this gradual change of climate toward a nonseasonal regimen, rain forests gradually displaced the savannas. This colonization by rain forests proceeded first along river valleys and young terraces that were covering most of the Amazon depression between the old Brazilian and Guianan Shields and were slower on older deposits and landforms with senile soils. This process was certainly not continuous but probably followed the postglacial climatic oscillations. Finally, the rain forest reached the limits of the wet climatic region and only some isolated savanna patches remained where there were the least favorable conditions for forest growth, such as coarse sediments of low water-retaining capacity where lateritic iron pans were frequent and the soils very poor.

Man's arrival in the forest regions of Central and South America stopped forest advance in most areas, since primitive man chose open sites to settle, and savanna areas near the forest border offered him a most suitable environment for protection and hunting. As he employed fire as a hunting tool, savanna burning became more frequent and forest advancement receded. The late intervention of European man completely stopped the process, most markedly during the last century, bordering the former savannas with belts of secondary herbaceous communities.

This hypothetical reconstruction of vegetation displacements in the now wet tropical America is supported by several pieces of evidence. The extreme reduction of the Amazon rain forest area 15,000 to 20,000 years ago seems well documented on the basis of independent biological and biogeographic evidence (Haffer, 1969; Vuillemier, 1971). The corresponding climate was drier and colder, as is suggested by geomorphologic and pedogenetic processes then at work. Consequently, it seems quite probable that a vegetation more or less similar to present-day savannas covered most of this region. As we pointed out earlier, van der Hammen (1972) reported a long savanna phase between two forest periods in the southern Amazonas. Finally, the savanna patches inside this region correspond quite well to the senile soils in old peneplains.

## Savannas, Rain Forests, and Dry Deciduous Forests Under Tropical Seasonal Climates

We have already noted that under a seasonal wet and dry tropical climate three physiognomic units coexist: semideciduous rain forests, deciduous forests, and savannas. These three types form complex vegetation patterns, where as a rule savannas make the dominant continuous matrix and forest appear as more discontinuous patches of varying extension.

**Swamps and Hyperseasonal Savannas**

In flat plains, rain or flood river water accumulates in low topographic situations during the rainy season, depositing the finest sediments carried in suspension. Heavy deposits are thus formed or, in many cases, a sequence of layers of different grain size according to the water-carrying capacity in each period. Heavy textured and poorly structured soils evolve with slow drainage, remaining waterlogged during most of the rainy season. At the end of the rainy season, external drainage and high evapotranspiration begin to desiccate these soils. In a couple of months they become completely dry, very hard, and the surface cracks, resulting in conditions of high water stress for living organisms. The corresponding soil profile is a pseudogley with descending hydromorphism (Zinck and Stagno, 1965; Monasterio and Sarmiento, 1971).

This type of soil-water annual cycle, under the joint action of climate relief and soil, produces a particular kind of savanna, which we have called "hyperseasonal savannas." This type is characterized by an extreme seasonality in vegetative and reproductive activities, a relatively open herb layer even in the period of maximal development (ground cover around 50 percent), a high proportion of xeromorphic sedges besides the grasses, and an almost completely absent tree layer with the noteworthy exception of palms such as species of *Copernicia* in South America and Cuba. Such communities and their habitats have been described in many regions of tropical America (Beard, 1953; Ramia, 1959; Cole, 1960; Blydenstein, 1967; Monasterio and Sarmiento, 1971). In his review, Beard considers that in this habitat the woody species cannot successfully compete with herbs, since, with the possible exception of some palms, most trees are unable to survive alternate periods of soil-water saturation and soil drought.

When the soil remains waterlogged for a longer period or a high water table occurs throughout the year, hyperseasonal savannas are replaced by swamp communities. These later formations vary widely in composition and structure; they may be exclusively herbaceous or have woody elements, in particular some palms such as species of *Mauritia*; but, in any case, they are not seasonal ecosystems. Some intermediate situations arise when the permanent water table remains at a certain depth during the dry season (around 1 m) and in this case some communities appear that are difficult to classify as swamps or savannas.

**Gallery and Semideciduous Forests**

On more normal soils of the seasonal climatic region, where waterlogging and desiccation do not occur, semideciduous or evergreen seasonal forests may form complex patterns with savannas (Monasterio and Sarmiento, 1971; Silva et al., 1971; Monasterio et al., 1971). Forests always appear under the most favorable annual water budgets. In fact, these two forest types only occur in seasonal climate regions where the soil constantly has available water or where it remains below or near the permanent wilting point only for a very short period at the end of the dry season. This shortening of the dry period is not due to climate but to an additional source of soil water available to trees.

The clearest case is gallery forests, which have a permanent water table within the reach of tree roots. When the forests occur further away from rivers, there is evidence that the trees also reach a water table, not only during the rainy season but also during a part of the dry season. In these cases water tables have been shown to oscillate between 1 and 3 to 8 m in depth, and as there is a lag period between the end of rainfall and the downward movement of the water table, the lowest levels are reached not at the end of the dry season (which otherwise could be a critical period for tree water supply) but during the first months of the next rainy season, when again trees have enough water available in the upper soil layers.

In some situations, as in the central Venezuelan plains (Monasterio and Sarmiento, 1971), the source of underground water supply for trees is a perched seasonal water table lying on an impermeable soil layer, or an iron hardpan, or an old lateritic cuirass. This perched water table, although neither permanent throughout the year nor continuous in space, provides an additional water source for certain localized forest patches during the first months of the rainless season, shortening in this way the length of the unfavorable period to no more than two or three months. This discontinuous water table is responsible for the complex vegetation pattern of forest groves in a savanna matrix, leading some authors to believe that this parkland physiognomy is evidence of a formerly continuous forest recently reduced by man's action. But the distribution of forest islands is not at random within the savanna matrix, as we would expect if they were relicts from a former continuous forest, being instead associated with factors favoring the occurrence of perched water tables.

## Seasonal Savannas and Deciduous Forests

When in a lowland region of tropical America under seasonal wet and dry climate, there is neither accumulation of rainwater nor an extra water supply through a high water table, only two primary ecosystems may occur: tropical deciduous forest and savanna. This particular savanna type will be called "seasonal savanna" to distinguish it from the previously considered "hyper-seasonal savanna." We will consider this forest-savanna antinomy and the circumstances favoring the occurrence of one or the other ecosystem. If we analyze the situation in regions where both formations form vegetation mosaics, as in the central Venezuelan plains (Sarmiento and Monasterio, 1969; Monasterio & Sarmiento, 1971) or in the middle San Francisco basin in central eastern Brazil (Azevedo, 1966), we see that each formation appears under a different set of habitat conditions.

In the Venezuelan plains deciduous forests occur on middle Tertiary geological formations formed by clay and shale that give rise to rather heavy soils of moderately low internal drainage and good water-retaining capacity. These medium-textured soils have a high cation exchange capacity, considering the humid tropics, and also a medium base saturation, since the drainage somewhat hampers soil leaching. All these features contribute to maintain soil fertility in medium ranges. The same forest type also occurs in river terraces of similar granulometry and physical and chemical characteristics. In both cases, though internal drainage is slow, the soil does not remain waterlogged or water saturated for long periods, because of adequate external drainage. Otherwise the forest would be replaced by hyperseasonal savannas, as it does in low topographic positions.

In the same region of the Venezuelan plains, seasonal savannas occur over Plio-Pleistocene alluvial sands of coarse texture, often with intercalation of gravel layers, and a less important silt and clay content. These soils are well to moderately well drained, with low water retention, low cation exchange capacity, rather unsaturated, and therefore drier and poorer than the neighboring deciduous forest soils.

In the middle valley of the San Francisco River, in the state of Minas Gerais in Brazil, Azevedo shows a comparable situation in the wide contact area between *cerrados* (seasonal savannas) and *caatingas* (dry, deciduous forests and scrubs). Savannas only occur over sandstones where coarse textured, well-drained, and poor soils develop; deciduous forest appears over limestone, with finer and more fertile soils; and the deciduous scrub, a transition type to more xerophyllous formations, is found over clay-lime alluvium, of intermediate texture and nutrient content.

In both situations of coexistence of deciduous forest and savanna, two main factors seem to control the differential occurrence of each ecosystem. First is the water factor. Deciduous forest has more water available in the upper soil horizons during the wet season, because of higher water-retaining capacity and slower drainage; in contrast, savannas appear on soils with less available water in the upper soil during this same period. Second is the nutrient factor. Forest occurs on richer soils, because a high nutrient content apparently favors deciduous trees in competition with perennial grasses (Mooney et al., 1974).

## Dynamic Relationships Between Different Ecosystems Under Seasonal Climates

Summarizing the preceding discussion, we may consider not only the vegetation but the whole landscape as a dynamic system varying according to long-term climatic fluctuations, geomorphogenesis, and soil evolution. As one example of this landscape evolution, take the case of an alluvial floodplain in a seasonal climate where rivers show great seasonal changes in water volume and almost yearly overflood the adjoining plains.

Figure 16–3*a* is a representation of the initial situation of a portion of a plain between two rivers. Sedimentary processes tend to accumulate coarser sediments near the water courses, forming more or less high sandy banks, while finer materials are transported and deposited further away from the river in the central portion of the plain. Two gradients are then formed outward from

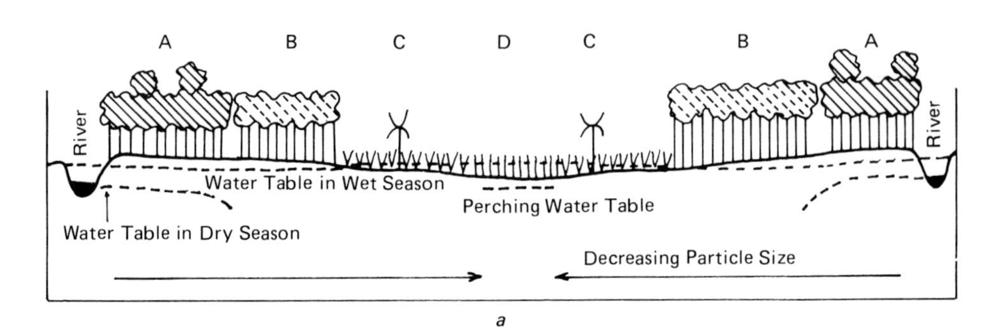


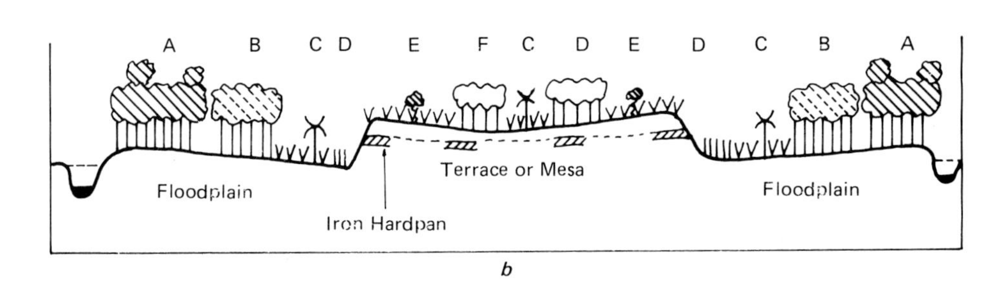


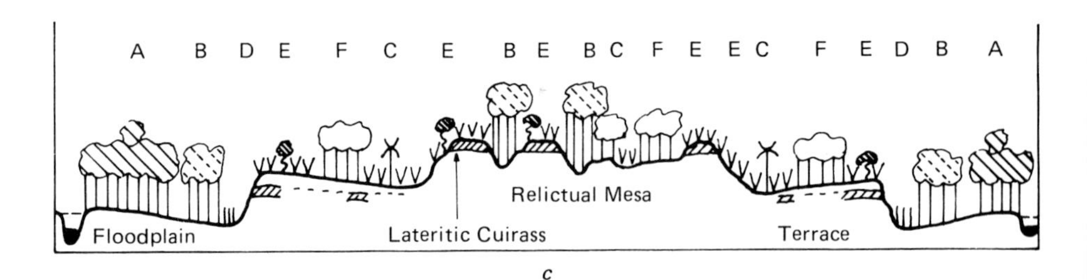


**Fig. 16–3.** Hypothetical reconstruction of landscape and vegetation changes in an alluvial plain under seasonal wet and dry climate along two rejuvenation cycles. (*a*) Wide alluvial plain where vegetation pattern depends on depth of water table in the wet and in the dry seasons, occurrence of perched water tables, and existence of seasonally waterlogged areas. (*b*) After rejuvenation, a terrace or mesa is formed where an indurated iron hardpan occurs at the place of ancient water table level. In the floodplains vegetation pattern is similar to (*a*). In the terrace the pattern depends on drainage, soil-water-retaining capacity, and mineral status of the soil. (*c*) A new rejuvenation cycle has taken place, thus forming two levels of terraces; the oldest has been eroded. The iron hardpan in the old mesa has become a distinct lateritic cuirass acting as a resistant layer to erosion. Vegetation pattern there depends on drainage, occurrence of seasonal or permanent water tables along small channels and rivulets, soil-water-retaining capacity, and mineral status of the soil. $A$ = gallery forest; $B$ = semideciduous forest; $C$ = hyperseasonal savanna; $D$ = swamp; $E$ = seasonal savanna; $F$ = deciduous forest.

the river margin, one of decreasing particle size and another of increasing bad drainage. In such a way the central bottomlands will be more frequently and extensively waterlogged; whereas the banks will be relatively well drained and will have a permanent high water table. The sequence of communities corresponding to these conditions is: A—gallery forest, generally of the evergreen seasonal type; B—semideciduous forest; C—hyperseasonal savanna; D—seasonal semipermanent swamps.

What happens if a new geomophologic cycle of erosion/sedimentation starts because of a climatic change toward wetter conditions? The former floodplain is first desiccated and eroded, forming a new floodplain at lower level and converting the old one into a river terrace (see Figure 16–3*b*). In the newly formed floodplain, the environmental conditions and the vegetation sequence will follow the same previously described pattern, but on the terrace the habitat conditions have been drastically changed. As the external drainage has improved as a result of river dissection, it is possible for the ancient iron- and clay-enriched illuvial soil horizons to change to hardpans or indurated cuirasses. The best drained places in this old but recently raised surface (raised relatively with reference to the new base level) will be occupied by seasonal savannas, while the less well-drained and richer soils could be covered by deciduous forests, restricting the former hyperseasonal savannas and swamps to a few localized habitats with impeded drainage.

If further rejuvenation occurs similar to the previous one, with a like erosion of the former floodplain and consequent formation of a new one, three levels will appear with different water regimens, soil age, soil evolution, intensity of erosion and, therefore, quite divergent vegetation (see Figure 16–3*c*). The first and second levels (the youngest) will repeat the previous process already depicted in the former rejuvenation, while on the oldest and highest level, retrogressive erosion will gradually reduce the primitive flat surface to a hilly landscape, with lateritic capped disorganized hills and many small drainage lines, creeks, and rivulets creeping between. This vestigial mesa will continue to be mostly covered by seasonal savanna and deciduous forests, but semideciduous forests will advance through the drainage network, gradually reducing the surface occupied by other formations. The general picture of community displacement is shown in Figure 16–3*c*. It is interesting to note that a complete "inversion" of the primitive relief has been produced, with resistant hills occupying the former place of bottomlands.

This hypothetical reconstruction of events related to a succession of environmental and vegetational changes would not have much practical value if there were not much evidence to support and document its reality in many lowlands areas of tropical America. During the dramatic climatic changes of the pluvial and interpluvial periods, and also as a result of the spectacular series of uplifts that gave rise to the neighboring mountains in the American *cordilleras,* vegetation changes have undoubtedly occurred. The first palynologic analysis began to disclose some of them while morphogenetic study of landscapes has permitted a reconstruction of ancient soils and ecosystems. In many tropical American lowlands four successive levels of alluvial terraces or rock pediments have been reported (Zinck and Stagno, 1965; Bigarella and Andrade, 1965; Monasterio and Sarmiento, 1971), suggesting that a yet more complex sequence than that depicted in Figure 16–3 took place.

According to this dynamic approach, at the same time that the major landscape features change, ecosystems are displaced and replaced, reflecting changes in drainage, depth of water table, leaching of nutrients, development of soil hardpans, and so on. At any given moment in this dynamic cycle the landscape will be composed not only of the ecosystems already considered but also, and significantly, by numerous more or less transitional forms. By contrast, in areas with an extended evolution or less labile landscape, as in the ancient plateau of Guiana and Brazil, the ecosystem pattern is more stabilized, senile forms become more widespread, and a clear tendency toward predominance of seasonal savannas is evident, the forests being restricted to younger surfaces or richer soils. The Venezuelan plains are a good example of the former situation and the Brazilian *cerrado* area of the latter. In the Amazon basin, which is already under a nonseasonal climate, the previous situation has been entirely reversed; rain forests predominate, and savannas only appear on small relictual landforms.

If to this complex pattern of primary communities we add the evolutionary recent, but impressive, human interference, we get a confused picture in the form of secondary ecosystems, which makes the interpretation of heavily disturbed landscapes quite problematic. Fortunately, the lowland regions of tropical America have not yet changed so much as to be unintelligible, making the interpretation of ecological facts simpler than in other tropical places.

## The Three Major Types of Savanna Ecosystems

After considering the occurrence of several ecosystems under various environmental conditions, we now have the necessary elements to answer the three initial questions asked at the beginning of this section. In the first place, there are at least three main ecological types of savannas, each one associated with a given combination of external factors. One single factor plays a different role in each case, giving rise to quite different global environmental stresses and selective forces on living organisms.

The first type comprises those savannas not subjected to a strong seasonal variation on soil-available water. They are met with only in everwet tropical climates, on well-drained, deep, medium- to coarse-textured soils, with a water table that is far from the rhizosphere at any time of the year. This nonseasonal savanna appears on two different soils: white sands and red sands. The nonseasonal savannas on white sands are less related physiognomically and floristically to all other savanna types, being closely related through their common woody species to the dry Amazon caatingas, or savanna forests and woods. Fragmentary knowledge of these white-sand communities suggests that soil-related factors are considered the main determinants of the series from a high, closed, xeromorphic forest to a low and open savanna (Heyligers, 1963). Their soils have in common two major ecological features: they are the driest and also the poorest soils found in this wet climatic region. On the other hand, the most open communities in the series, some formed by low bushes and an almost bare soil in between, are apparently maintained by a high fire frequency.

The nonseasonal savannas on red soils are intermediate in composition between the white sand and seasonal savannas. Some of them could result from fire action on scleromophic forests, but most have been interpreted as relictual formations in the least favorable environments for forest colonization.

The second type of savanna ecosystem is the most widespread under the wet and dry climates of tropical America, and on the basis of this regional character we have called them *seasonal savannas*. Previously, most authors have referred to them as "dry savannas," in comparison with the "wet savannas" here termed hyperseasonal; but we think that the terminology proposed here is more precise and meaningful from an ecological viewpoint. Seasonal savannas occur on well or moderately well-drained soils, of medium to coarse texture, low availability of nutrients, and deep water table. We hypothesize that this ecosystem represents the biotic response to a seasonal climate with a long rainless period, where the upper part of the soil profile maintains intermediate water levels during the rainy season and dries out slowly after the rains end, reaching a level of ecological drought during the final part of the dry season. The water budget in deeper layers, where most woody species find their water supply, is different (see Mooney et al., 1974). Two factors are necessary for the persistence of this ecosystem: well-drained, deep soils and low soil fertility. Frequent burning contributes to low fertility and is a major selective pressure for savanna species but seasonal savannas could maintain themselves even with a very low fire frequency, as could be the case if only natural fires were considered.

The third type of ecosystem, the *hyperseasonal savanna*, is under the additional environmental stress of an extended period of soil-water saturation. The species in this ecosystem pass from a water-deficient period to another equally long period of water excess. These savannas only occur in regions of seasonal climate, on heavy, ill-drained soils, poor to very poor in nutrients, and are also subjected to periodical fires. As we pointed out earlier, they are mostly grass savannas with no woody cover except the occasional presence of palms. The herb layer has many exclusive species and many others shared in common with seasonal savannas.

The distinction between seasonal, hyperseasonal, and nonseasonal savannas makes the savanna concept more useful and clarifies the ecological context of each type. Many controversies about savannas have been caused by the lack of distinction between these three basically different types of ecosystems.

## Some Essential Factors Acting Upon Different Ecosystems in the Humid Lowland Tropics

In the last part of this chapter we want to summarize in a very concise form the major features associated with the occurrence of various ecosystems in the lowlands of tropical America.

1. With a constantly wet climate, rain forests constitute the zonal climatic formations, disregarding any additional source of plant water supply.
2. In soils with a high water table during the whole year, but with no waterlogging or water saturation in the upper part of the profile, evergreen rain forest is found, independent of the rainfall regimen.
3. In sites with a high water table saturating the upper soil layers for more or less extended periods, swamp-forests develop.
4. On permanently or almost permanently saturated soils, swamp communities occur under any type of rainfall regime.
5. Under an ever-raining climate on very permeable and infertile soils, scleromorphic forests and woods appear, forming a series of plant physiognomies from moderately tall, closed forests to open, patchy, savanna-like vegetation.
6. Under an ever-wet climate in very poor and excessively drained soils a non-seasonal savanna may be maintained, its permanence depending on the burning frequency.
7. In wet and dry climatic regions (Aw climates) on well-drained sites, with poor soils of low water-retaining capacity in the upper horizons, seasonal savannas occur.
8. Under wet and dry climates, in moderately drained soils, with high water-retaining capacity in upper soil layers and medium levels of fertility, dry deciduous forest occurs.
9. On ill-drained, heavy soils, where water saturation and water deficiency periods alternate during a yearly cycle, hyperseasonal savannas are found.

A graphic representation of some of these facts in the particular case of wet and dry climates is shown in Figure 16–4. In that graph the ecological complexity of these environments has been reduced to two major axes of environmental variation, one reflecting soil-water conditions during the most unfavorable period of the year (the end of the dry season) and the other representing soil-water content at the wettest point of the year (the end of the rainy season).

This graphic representation shows the position of eight types of ecosystems in the plane determined by those two axes. Some points of this figure are worth considering: The two axes can discriminate seven out of the eight ecosystems; only the dry deciduous forest overlaps with the seasonal savanna, since their respective occurrence is, as we have already discussed, determined by other factors. Some transitional zones exist, but their nature and precise ecological range could only be established when more quantitative data are available.

In the preceding pages we have postulated the existence of fundamental correspondences between factors and features of the physical environment, mainly those related to annual water budget and nutrient cycling, and the occurrence of different ecosystems in the humid lowland American tropics. Other factors, such as fire, human interference, and paleoecological changes, have been considered under this perspective as agents or processes whose action completes the picture of the present-day distribution of these ecosystems. While additional information is necessary in order to accept or reject our hypotheses, we are convinced that they will help point out precisely the critical data pertinent to the problem, serving as a guideline in establishing priorities for future research.

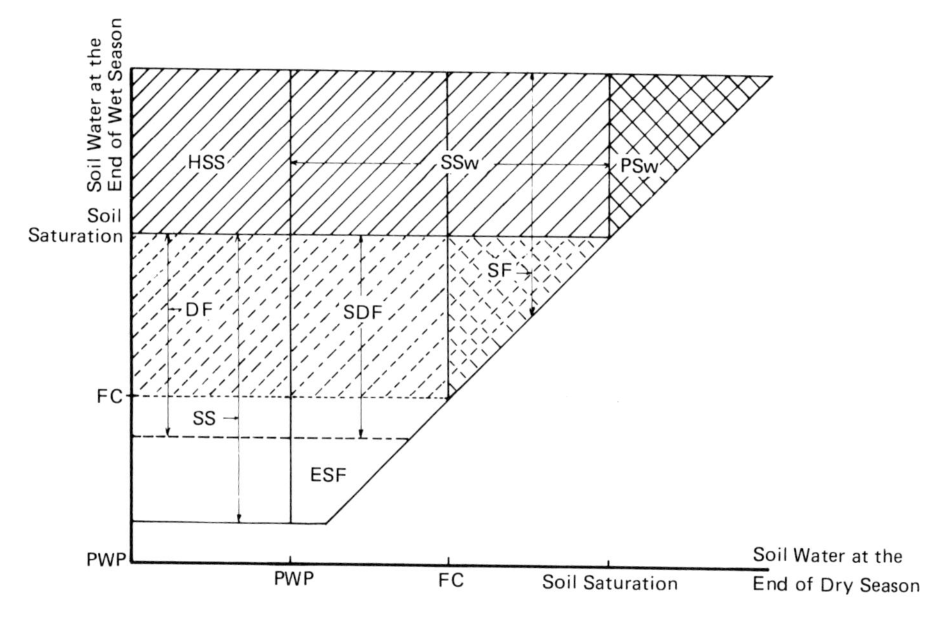


**Fig. 16–4.** Ideal distribution of ecosystems under a tropical wet and dry climate, according to maximal water stresses in the two contrasting rain seasons. Note overlapping of seasonal savanna and deciduous forest, and of seasonal swamp and seasonal forest. *HSS* = hyperseasonal savanna; *SSw* = seasonal swamp; *PSw* = permanent swamp; *DF* = deciduous forest; *SDF* = semideciduous forest; *SS* = seasonal savanna; *ESF* = evergreen sclerophyllous forest. [hatched] = area of seasonal waterlogging; [cross-hatched] = area of permanent waterlogging; [dashed hatched] = area of seasonal soil saturation; [dashed cross-hatched] = area of permanent soil saturation. *PWP* = permanent wilting point; *FC* = field capacity.

## References

Arens, K. 1958. O cerrado como vegetaçao oligotrofica. *Bol. Fac. Fil. Cienc. Letr USP 224 Bot.* 15:58–77.

Aubreville, A. 1949. *Climats, forêts et désertification de l'Afrique tropicale.* Paris.

———. 1962. Savanisation tropicale et glatiations quaternaires. *Adansonia* 2:16–84.

Azevedo, L. G. 1966. Tipos eco-fisionômicos da vegetacao da região de Januária (MG). *Anais Acad. Bras. Cienc.* 38(suppl.):39–57.

Beard, J. S. 1944. Climax vegetation in tropical America. *Ecology* 25:127–158.

———. 1953. The savanna vegetation of northern tropical America. *Ecol. Monographs* 23:149–215.

Bennett, H. H., and R. V. Allison. 1928. *The Soils of Cuba.* Washington: Tropical Plant Research Foundation.

Bigarella, J. J., and G. O. Andrade. Contribution to the study of the Brazilian Quaternary. In *International Studies on the Quaternary*, Wright and Frey, eds. *Special GSA Papers* No. 84, 433–451.

Blanck, J. P., M. Monasterio, and G. Sarmiento. 1972. Las comunidades vegetales y su relación con la evolución cuaternaria del relieve en los llanos Centrales de Venezuela. *Acta Cient. Venez.* 23(suppl. 1):31.

Blydenstein, J. 1967. Tropical savanna vegetation of the llanos of Colombia. *Ecology* 48:1–15.

Bouillenne, R. 1926. Savanes equatoriales en Amerique du Sud. *Bull. Soc. Bot. Belg.* 58:217–223.

Budowski, G. 1956. Tropical savannas, a sequence of forest felling and repeated burnings. *Turrialba* 6:23–33.

Charter, C. G. 1941. *Reconnaissance Survey of the Soils of British Honduras.* Trinidad: Government Printer.

Cole, M. 1960. Cerrado, caatinga and pantanal: The distribution and origin of the savanna vegetation of Brazil. *Geog. J.* 76:168–179.

Dansereau, P. 1957. *Biogeography, an Ecological Perspective.* New York: Ronald Press.

Donselaar, J. van. 1965. An ecological and phytogeographic study of northern Surinam savannas. *Wentia* 14:1–163.

Eden, M. J. 1964. The savanna ecosystem—Northern Rupununi, British Guiana. *McGill Univ. Savanna Res. Ser.* 1:1–216.

Egler, W. A. 1960. Contribuçoes ao conhecimento dos campos da Amazonia. I. Os campos do Ariramba. *Bol. Mus. Paraense E. Goeldi Bot.* 4:1–36.

Eiten, G. 1972. The cerrado vegetation of Brazil. *Bot. Rev.* 38:201–341.

Ferri, M. G. 1955. Contribuiçao ao conhecimento da ecologia do cerrado e da caatinga. *Bol. Fac. Fil. Cienc. Letr. USP 195, Bot.* 12:1–170.

Goodland, R. 1970. The savanna controversy. *McGill Univ. Savanna Res. Ser.* 15: 1–66.

Grisebach, A. H. 1872. *Die Vegetation der Erde nach ihrer klimatischen Anordnung.* Leipzig: Engelmann.

Haffer, J. 1969. Speciation in Amazonian forest birds. *Science* 165:131–137.

Hammen, T. van der. 1963. A palynological study of the Quaternary of British Guiana. *Leidse Geol. Meded.* 29:125–180.

——. 1972. Historia de la vegetación y el medio ambiente del Norte Sudamericano. *Mem. I Congr. Latinoam. Bot. (Mexico)* 119–134.

Hester, J. J. 1966. Late Pleistocene environments and early man in South America. *Am. Naturalist* 914:377–388.

Heyligers, P. C. 1963. Vegetation and soil of a white-sand savanna in Suriname. In *The Vegetation of Suriname*, Lanjouw and Verstugh, eds. Vol. III, pp. 1–148. Amsterdam: Van Cedenfonds.

Hills, T. L. 1965. Savannas: A review of a major research problem in tropical geography. *Can. Geog.* 9:216–228.

———. 1969. The savanna landscapes of the Amazon basin. *McGill Univ. Savanna Res. Ser.* 14:1–41.

———, and R. E. Randall (eds.). 1968. The ecology of the forest-savanna boundary. *McGill Univ. Savanna Res. Ser.* 13:1–128.

Hueck, K. 1966. *Die Walder Sudamerikas.* Stuttgart: Gustav Fischer.

Jaeger, F. 1945. Zur Gliederung und Benennung des tropischen Graslandgurtels. *Verh. Naturf. Ges. Basel.* 56:509–520.

Johannessen, C. L. 1963. Savannas of interior Honduras. *Ibero-Americana* 46:1–160.

Klinge, H. 1965. Podzols soils in the Amazon basin. *J. Soil Sci.* 16:95–103.

Koeppen, W. 1931. *Grundriss der Klimakunde.* Berlin: Walter de Gruyter.

Lanjouw, J. 1936. Studies on the vegetation of the Surinam savannahs and swamps. *Ned. Kruidk. Arch.* 46:823–851.

Lauer, W. 1952. Humide und aride Jahreszeiten in Afrika und Sudamerika und ihre Beziehungen zu den Vegetationsgurteln. *Bonner Geog. Abh.* 9:15–98.

Miranda, F. 1952. *La Vegetación de Chiapas*. Mexico: Departamento de Prensa y Turismo.

Monasterio, M. 1971. Ecologia de las sabanas de América tropical. II. Caracterización ecológica del clima en los llanos de Calabozo, Venezuela. *Rev. Geog.* 21:5–38.

———, and G. Sarmiento. 1971. Ecologia de las sabanas de America tropical. I. Análisis macroecológico de los llanos de Calabozo, Venezuela. *Cuadernos Geog.* 4:1–126.

———, and J. Silva. 1971. Reconocimiento ecológico de los llanos Occidentales. III. El Sur del Estado Barinas. *Acta Cient. Venez.* 22:153–169.

Mooney, H., G. Sarmiento, and M. Monasterio. 1974. Towards an understanding of certain tropical lowland ecosystems. In press.

Myers, J. G. 1936. Savannah and forest vegetation of the interior Guiana plateau. *J. Ecol.* 24:162–184.

Parsons, J. J. 1955. The miskito pine savanna of Nicaragua and Honduras. *Ann. Assoc. Am. Geog.* 45:36–63.

———. 1972. Spread of African pasture grasses to the American tropics. *J. Rge. Mgt.* 25:12–17.

Puig, H. 1972. La sabana de Huimanguillo, Tabasco, Mexico. *Memo. I Congr. Latinoam. Bot. (Mexico)* 389–411.

Pulle, A. A. 1906. *An Enumeration of the Vascular Plants Known from Surinam, Together with Their Distribution and Synonymy*. Leiden: E. J. Brill.

Ramia, M. 1959. *Las sabanas del Apure*. Caracas: Ministerio de Agricultura y Cria.

Rawitscher, F. 1948. The water economy of the vegetation of the "campos cerrados" in southern Brazil. *J. Ecol.* 36:237–268.

Richards, P. 1952. *The Tropical Rain Forest*. Cambridge: Cambridge Univ. Press.

Santamaria, F., and A. Bonnazzi. 1963. Factores edáficos que contribuyen a la creación de un ambiente xerofítico en el Alto Llano de Venezuela: El arrecife. *Bol. Soc. Venez. Cienc. Nat.* 106:9–17.

Sarmiento, G. 1968. Correlación entre los tipos de vegetación de America y dos variables climáticas simples. *Bol. Soc. Venez. Cienc. Nat.* 113–114; 454–476.

———. 1972. Ecological and floristic convergences between seasonal plant formations of tropical and subtropical South America. *J. Ecol.* 60:367–410.

———, and M. Monasterio. 1969. Studies on the savanna vegetation of the Venezuelan llanos. I. The use of association-analysis. *J. Ecol.* 57:579–598.

———. 1974. The distribution and vegetation patterns of the main savanna types in tropical America. In press.

———, and J. Silva. Reconocimiento ecológico de los Llanos Occidentales. I. Las unidades ecológicas regionales. *Acta Cient. Venez.* 22:52–60.

Schnell, R. 1971. *Introduction a la Phytogeographie des Pays Tropicaux*. Paris: Gauthier Villars.

Schimper, A. F. W. 1903. *Plant Geography Upon a Physiological Basis*. Oxford: Groom and Balfour.

Silva, J., M. Monasterio, and G. Sarmiento. 1971. Reconocimiento ecológico de los Llanos Occidentales. II. El Norte del Estado Barinas. *Acta Cient. Venez.* 22: 60–71.

———, and G. Sarmiento. 1974. Un ordenamiento de las sabanas de Barinas (Venezuela) y su relación con las series de suelos. In press.

Sioli, H., and H. Klinge. Solos, tipos de vegetaçao e aguas na Amazonia *Bol. Mus. Paraense Goeldi Avulsa* 1:27–41.

Stark, N. 1970. The nutrient content of plants and soils from Brazil and Surinam. *Biotropica* 2:51–60.

Tamayo, F. 1964. *Ensayo de Clasificación de Sabanas de Venezuela.* Caracas: Universidad Central de Venezuela.

Taylor, B. W. 1963. An outline of the vegetation of Nicaragua. *J. Ecol.* 51:27–54.

Thornthwaithe, C. W. 1948. An approach toward a rational classification of climate. *Geog. Rev.* 38:155–194.

Vareschi, W. 1960. Observaciones sobre la transpiración de árboles llaneros durante la época de sequia. *Est. Biol. Llanos Publ. No. 1* 39–45.

———. 1962. La quema coma factor ecológico en los llanos. *Bol. Soc. Venez. Cienc. Nat.* 101:9–26.

———. 1969. Las sabanas del valle de Caracas. *Acta Bot. Venez.* 4:427–522.

Vuillemier, B. S. 1971. Pleistocene changes in the fauna and flora of South America. *Science* 173:771–780.

Waibel, L. 1948. Vegetation and land use in the Planalto Central of Brazil. *Geog. Rev.* 38:529–554.

Walter, H. 1969. El problema de la sabana. *Bol. Soc. Venez. Cienc. Nat.* 115–116; 121–144.

———. 1971. *Ecology of Tropical and Subtropical Vegetation.* Edinburgh: Oliver & Boyd.

Warming, E. 1908. *Lagoa Santa.* Belo Horizonte.

Wijmstra, T. A., and T. van der Hammen. 1966. Palynological data on the history of tropical savannas in northern South America. *Leidse Geol. Med.* 38:71–90.

Zinck, A., and P. Stagno. 1965. *Estudio Edafológico de la Zona Rio Santo Domingo-Rio Paguey, Estado Barinas.* Caracas: Ministerio de Obras Publicas.

# CHAPTER 17

## Effect of Fire on Organic Matter Production and Water Balance in a Tropical Savanna

J. J. San José and E. Medina

Lowland savannas in northern South America (llanos) occupy a region with a pronounced rainfall seasonality with four to six months dry season and 600 to 1,500 mm rainfall. This seasonality is considered to be determinant in savanna phenology, although species with divergent rhythms have been reported (Monasterio, 1968) due to water availability during the dry season (Vareschi, 1960).

Savanna vegetation growing under high rainfall regimes is a puzzling phytogeographic problem, and hypotheses on its origin have produced an extensive literature (see the chapter by Sarmiento and Monasterio). Savanna vegetation can be formed in response to several limiting factors, such as soil, geomorphology, and fire. The human factor also plays an important role (Walter, 1970; Blydenstein, 1962; Monasterio and Sarmiento, 1968; Bourliére and Hadley, 1970; Daubenmire, 1968).

To explain the savannas in the central llanos of Venezuela, it has been proposed that tree growth is impeded by the presence of a hardpan below the soil surface, built up of fine sediments cemented by ferric oxide (Santa Maria and Bonnazzi, 1963; Monasterio and Sarmiento, 1968). This hardpan prevents penetration of tree roots. The pattern of tree distribution within the grassland, then, is a mosaic where practically no competition exists between trees and grasses and reflects the hardpan structure below soil surface. Grasses use water in the upper soil layers through their intensive root system while roots penetrate to the water table through cracks in the hardpan (Walter, 1964, 1970). Direct evidence for this hypothesis is still lacking.

In addition, annual burning has been proposed as the limiting factor preventing tree growth, because even fire-resistant tree species common in South American savannas, such as *Curatella americana* or *Platycarpum orinocense*, are very sensitive to burning during early phases of growth. On the other hand, fire stimulates grass growth (Blydenstein, 1962; Old, 1969; Daubenmire, 1968) and probably regulates competition between monocots and dicots in the grass cover (Swan, 1970).

This paper reports on the production of assimilatory tissue and water consumption by a protected and burned grass cover in the central llanos of Venezuela.

## Materials and Methods

Two plots (15 by 20 m) of pure grassland were selected in the Estación Biológica de los Llanos (8° 56′ N, 67° 25′ W) at Calabozo, Venezuela. Soils and vegetation of the station have been described in detail (Monasterio and Sarmiento, 1968). The plots studied were located on level terrain with an upper layer of fine sediments, sandy-clay-loam, down to 2 m thick. Plots were covered by *Trachypogon montufari, T. plumosus*, and *Axonopus canescens* in that order of importance. (Differentiation between *T. montufari* and *T. plumosus* was based on leaf hairiness, but identifications cannot be considered as definitive.) Other species present in some areas had less than 2 percent of total aboveground biomass.

At each plot two series of nylon units (Bouyoucos) were inserted at 5-, 10-, 20-, 40-, and 70-cm depth. Only at 70 cm were a few fine grass roots observed, the main root mass being located between 0 and 40 cm (Blydenstein, 1962). After insertion, resistance was measured daily at 8:00 AM for one year. At least once a month soil samples from different depths at two places were taken and water content measured by drying at 105°C. These values were used to adjust resistance measurements to soil-water content in percent dry weight. In homogenized soil samples water-retention capacity and water content at —15 bars were also measured according to Steubing (1965).

Of the two plots, one was burned once on December 28, 1968. From January 1969 on, one to three 1-$m^2$ samples in both plots were harvested for measurement of green biomass. Surface-weight ratios were measured for calculation of total leaf area. Maximal error in sampling was 6.5 percent of mean value, and it was reached in the middle of the rainy season near the peak of aboveground biomass.

Transpiration was measured in detached leaves with a torsion balance according to Stocker (1956) in the three species with intervals of 1 hr during days of different weather in the dry and rainy seasons of 1969. Evaporation with Piche evaporimeter was measured simultaneously. Data of rainfall and evaporation from Tank A were taken from a meteorological station within 400 m of the two plots.

## Results

### Soil Conditions

Soil in both plots is relatively deep and the hardpan is below 1 m. Table 17–1 shows some relevant characteristics for the interpretation of water

**Table 17–1.** Range of soil-water content available for transpiration.

| | Protected | | Burned | |
|---|---|---|---|---|
| | Percent | $1/m^2$ [a] | Percent | $1/m^2$ |
| Content at −15 bars | 7.1 | 79 | 6.5 | 73 |
| Maximum content (laboratory) | 32.8 | 367 | 29.1 | 325 |
| Minimum content (field) | 5.1 | 58 | 5.0 | 56 |
| Maximum content (field) | 31.2 | 350 | 25.8 | 290 |
| Textural class | Sandy-clay-loam | | | |
| Soil bulk density | 1.6 $g/cm^3$ | | | |

[a] In a column of 0.7 $m^3$.

balance. The soil chemistry of both soils is similar; both are acidic (5.1 to 5.6), very low in phosphorus and nitrogen and in exchangeable cations. The profiles do not show marked differences (Table 17–2).

## Seasonal Changes in Soil-Water Content in Relation to Rainfall

Average rainfall in Calabozo is 1,334 mm (Walter and Medina, 1971) but is highly variable: extreme years range from 580 to 1,990 mm (Monasterio, 1968). 1969 was an extremely moist year with 1,839 mm rainfall, and the humid period was extended until November, lasting about eight months. Figure 17–1 shows daily rainfall values and total soil-water content in the two plots. Soil-water content is highly correlated with rainfall in both plots, but maximum values in the burned plot are lower than would be expected from water-retention capacity. Curves of total soil-water content in a column of 0.7 $m^3$ (maximal root depth) were used to calculate real evapotranspiration during dry periods.

The reason for the lower maximum water content in burned plot soil is explained in Figure 17–2. This figure shows the course of real water content during the year at different depths in both plots.

At the beginning during the dry season soil-water content in the burned plot is higher due to the reduced transpiration activity. After the first isolated

**Table 17–2.** Average texture and composition of the soils from protected and burned plots.[a]

| | Percent | | | ppm | | | Percent | | |
|---|---|---|---|---|---|---|---|---|---|
| | Sand | Silt | Clay | P | K | Ca | N | Organic Matter | pH |
| Protected | 55.5 | 19.0 | 25.9 | 4.1 | 27 | 344 | 0.0358 | 1.23 | 5.3 |
| Burned | 57.6 | 16.2 | 26.3 | 3.9 | 26 | 304 | 0.0376 | 1.33 | 5.3 |

[a] Average of 22 samples representing a column of 1-m depth. Analyses were kindly performed by the Soil Service of the Estación Experimental de los Llanos (MAC).

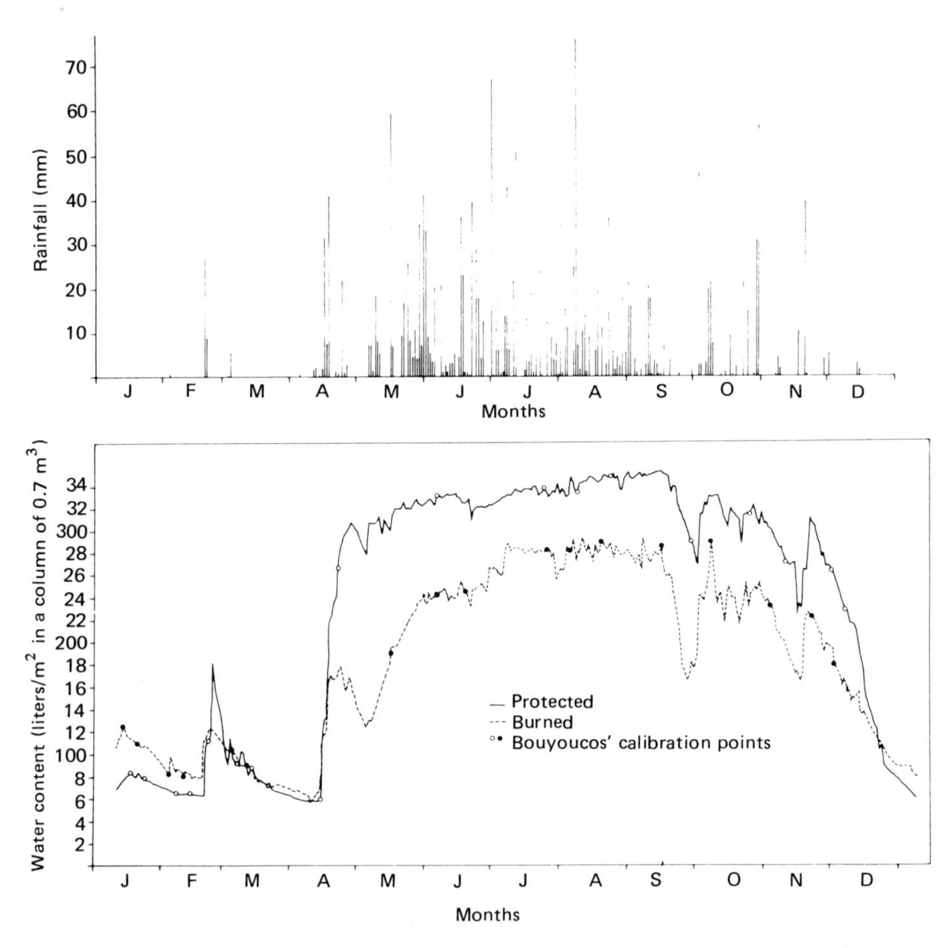


**Fig. 17–1.** Daily records of rainfall (*top*) during 1969 in the meteorological station at Calabozo, and total soil-water content (*bottom*) in a column of 0.7 m³ in the protected and burned plots.

rains in February, this difference disappears and soil-water content is reduced to or below the level present at —15 bars. By this time vegetation on the burned plot has developed significantly.

After the beginning of the rainy season, all soil layers in the protected plot are moistened homogeneously, and the deepest levels, 40 to 70 cm, remain near saturation until the end of the rainy season. During the short dry periods at the end of September and November, the upper soil layer shows strong oscillations in water content. In the burned plot the upper 5 cm never reaches saturation. The water content here changes drastically during rainy and dry days. This is probably due to the absence of complete soil cover, resulting in increased evaporation and runoff. Also, the deepest layers in this plot do not reach saturation either. This is more difficult to explain but may be due to a higher root activity.

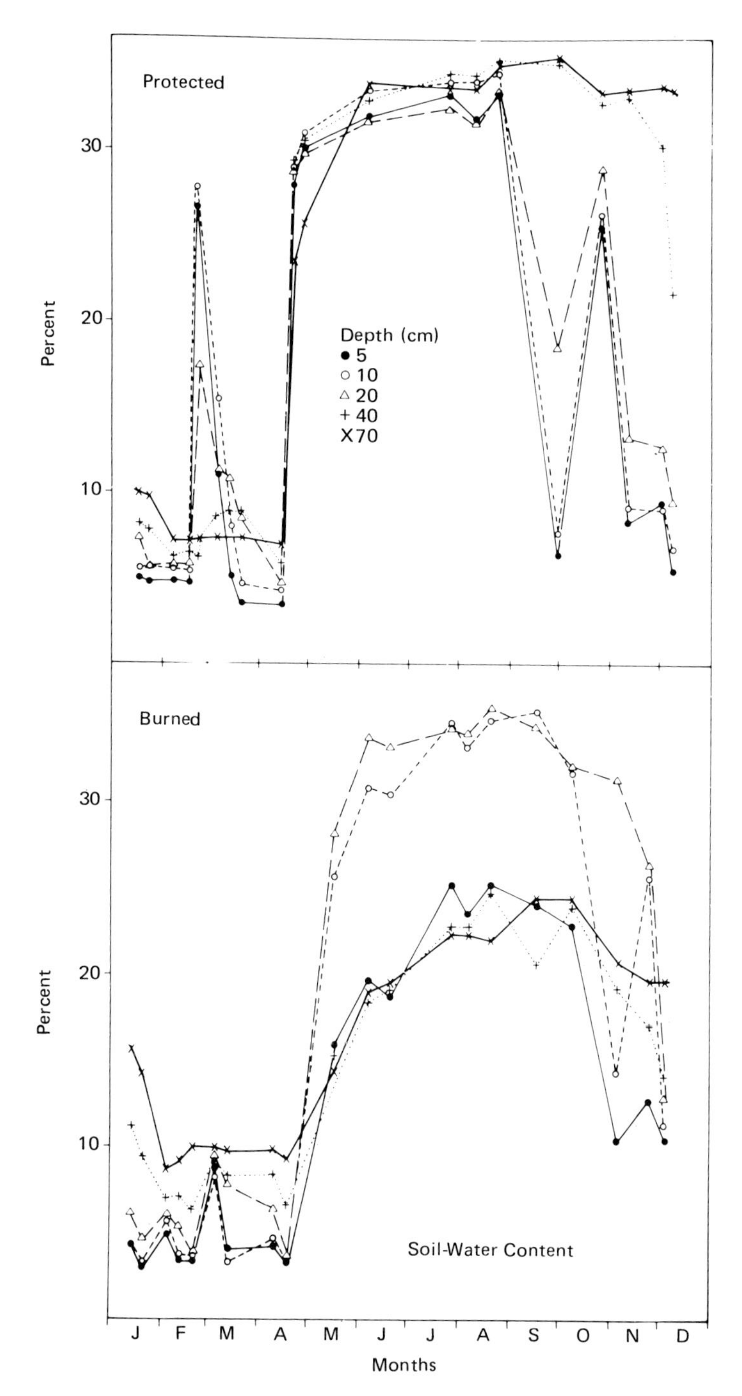


**Figure 17–2.** Soil water content at different depths during the year in protected and burned plots.

## Net Aboveground Production

A comparison between the composition of the grass cover in the plots reveals some differences that are relevant for the interpretation of the production curves. Table 17–3 presents the percentage distribution of living aboveground biomass between the three grass species, both in protected and burned plots and before and after burning. Comparison has been made with biomass values during the same time of the year. The dominance relation between the components, which was similar in both plots before burning, showed an increase in the proportion of the nondominant species after fire.

Figure 17–3 shows the curves of total living aboveground biomass and assimilatory area [leaf area index (LAI)]. It can be seen that growth in the burned plot begins early after fire, but it is restricted until the beginning of the rainy season. Peak values are 415 $g/m^2$ and 4.95 $m^2/m^2$ LAI in the burned plot and 325 $g/m^2$ and 4.19 $m^2/m^2$ LAI in the protected one.

**Table 17–3.** Relative composition of the savanna in percent of aboveground biomass.

| | Burned | | |
|---|---|---|---|
| | Before | After | Protected |
| *Trachypogon montufari* | 57.9 ± 0.2 | 49.1 ± 2.0 | 65.0 ± 0.8 |
| *Trachypogon plumosus* | 28.2 ± 1.1 | 31.2 ± 2.0 | 26.5 ± 0.6 |
| *Axonopus canescens* | 13.8 ± 1.0 | 19.4 ± 0.4 | 8.2 ± 0.8 |

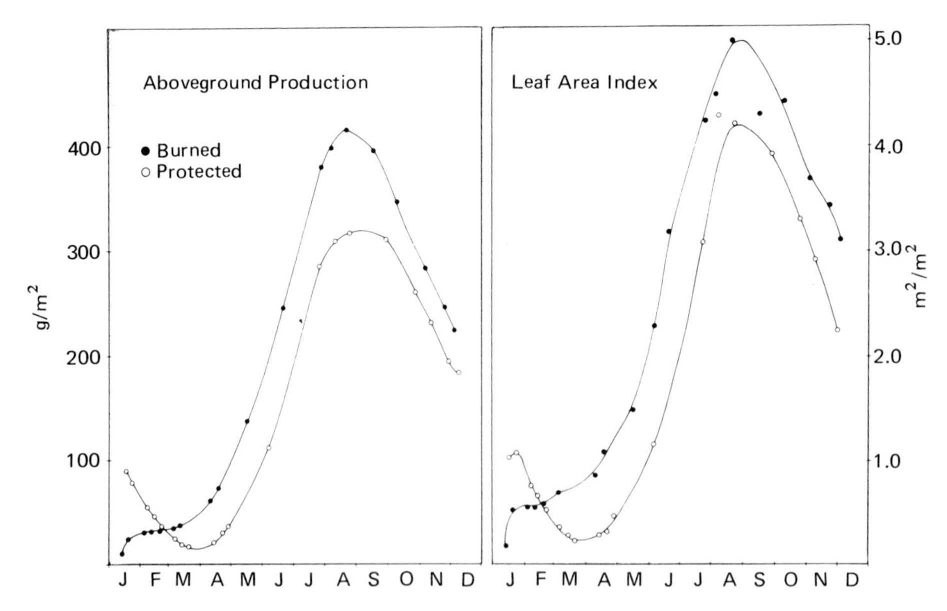


**Fig. 17–3.** Net aboveground production and leaf area index of the protected and burned plots.

Difference in productivity is due mainly to the subordinate components, as can be observed in Figure 17–4. Differences in peak aboveground biomass are negligible in *T. montufari*, while there are increases to 36 g/m$^2$ in *T. plumosus* and to 52 g/m$^2$ in *A. canescens*.

Several authors (Blydenstein, 1962; Daubenmire, 1969; Kucera and Ehrenreich, 1962; Old, 1969) reported that burning grasslands at certain times

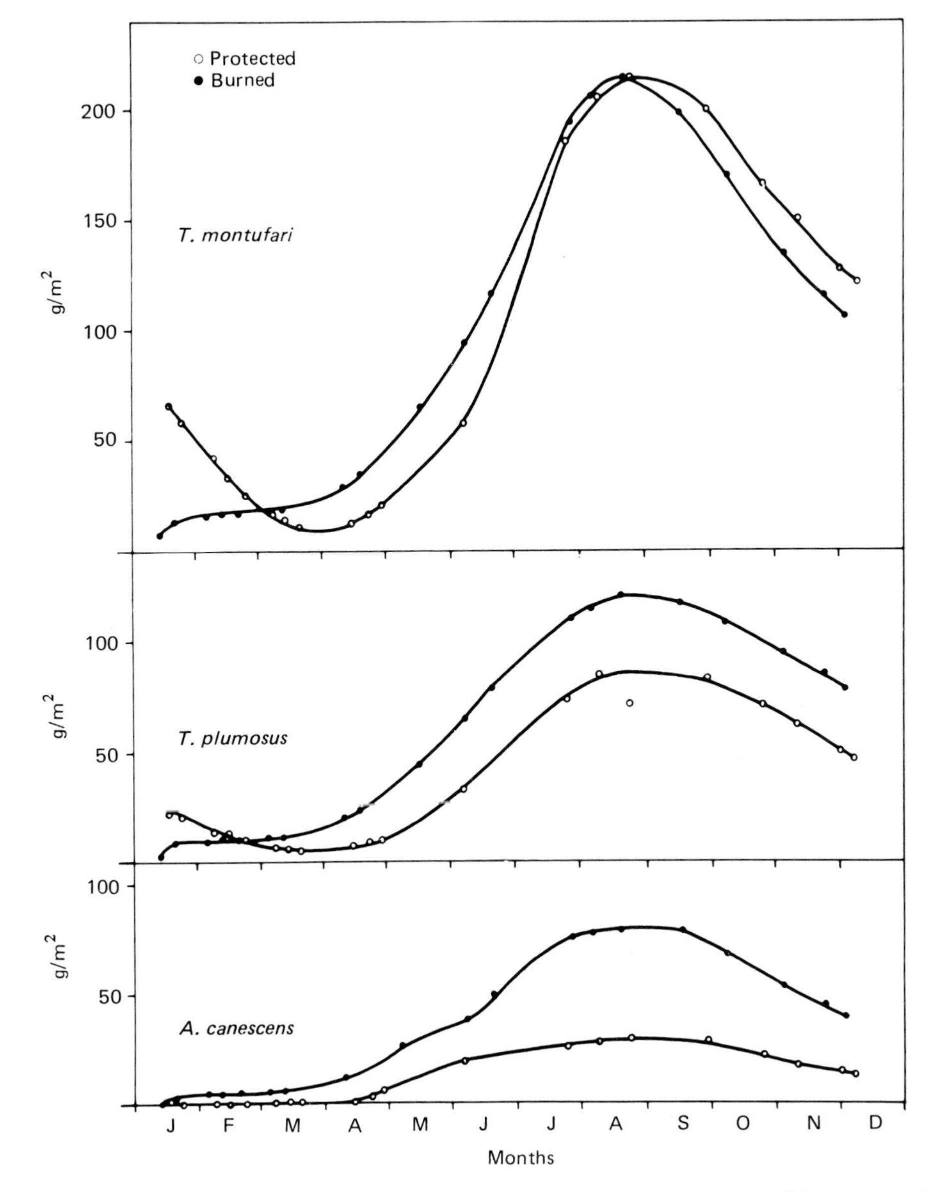


**Fig. 17–4.** Net aboveground production of the three species measured in protected and burned plots.

during the dry season stimulates organic matter production in the following growing season. Results reported here cannot be ascribed to an increase in soil nutrient content in the burned plot, in particular nitrogen and phosphorus, because chemical analyses in both soils do not show significant differences (see Table 17–2). The more likely hypothesis seems to be that accumulated above-ground dead organic matter in protected plots prevents a rapid development of new shoots. Leaching of growth inhibitors from organic matter seems improbable (Old, 1969).

It is noteworthy that in the protected plot *A. canescens* does not produce organic matter in measurable amounts until the end of April after the beginning of the rainy season. In the burned plot this species can develop undisturbed as long as the two *Trachypogon* species have not expanded their maximum leaf area. Increased leaf area development must be accompanied by a higher root growth activity, so that the grass cover after burning also has an increased capacity for water absorption from the soil.

**Transpiration and Water Balance**

Stocker's method for transpiration measurement has been strongly criticized, mainly because it neglects short-term changes in stomatal opening caused by leaf detachment. In dicots, this problem has been clearly demonstrated by Franco and Magalhaes (1965), but in grasses, stomatal control of transpiration seems to be less effective, and they do not restrict water losses under water stress, but transpire further until leaves dry out (Walter, 1964; Stocker, 1956). This methodological uncertainty should be remembered when discussing water consumption by the grass cover.

Transpiration rates are linearly correlated with evaporation rate from a free water surface, provided there is a sufficient water supply and the rates are without stomatal closure.

Daily transpiration of the three species in both plots shows a linear relationship with daily evaporation rates above 4 mm/day. Below these values the slope increases. That would mean that the transpiration rate is relatively higher at low evaporation rates; the experimental points, therefore, are better fitted by a logarithmic linear equation.

As there is no statistical difference in the behavior of the three species, average values were taken to calculate transpiration rate ($T$) from evaporation ($E$) from tank A (Figure 17–5):

$$T\ (\text{ml/dm}^2 \times \text{day}) = 18.53 \log E\ (\text{ml/day}) + 2.59;\ (r = 0.90;\ P = 0.01)$$

This relation indicates that these grasses do not control water losses effectively, and in no case was a decrease in transpiration observed at high evaporation rates. For the purpose of this paper it is assumed that living grass leaves had a sufficient water supply during the period of measurement, and that transpiration continues until the leaves dry out if water reserves are exhausted or limited.

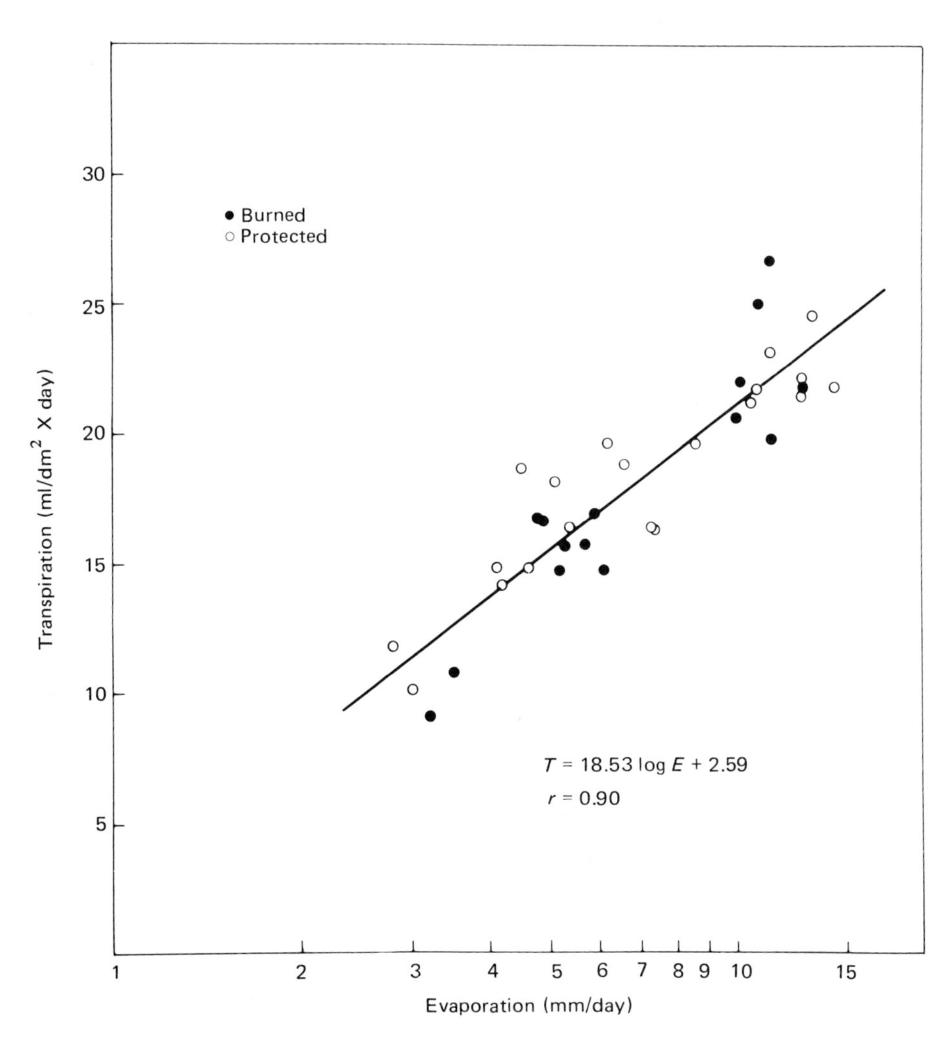


**Fig. 17–5.** Correlation between transpiration per leaf area and evaporation measured in tank A of the meteorological station.

Computing total transpiration per unit area (transpiration rate $\times$ leaf area index), one can obtain the general trend of water loss during the year in the two plots (Figure 17–6). The range of real evapotranspiration measured from the curves of Figure 17–1 (when evapotranspiration $= \triangle$ soil-water content) is also included, and it can be observed that transpiration maxima are always lower than maximal real evapotranspiration.

Given mean monthly evaporation rate and mean leaf area index, monthly water loss rates can be calculated. In Figure 17–7 these figures are compared with available soil-water content (obtained from Figure 17–1 and Table 17–1).

During the dry season the burned plot has a higher water content because there is no transpiration activity, while in the protected plot more than 60 mm

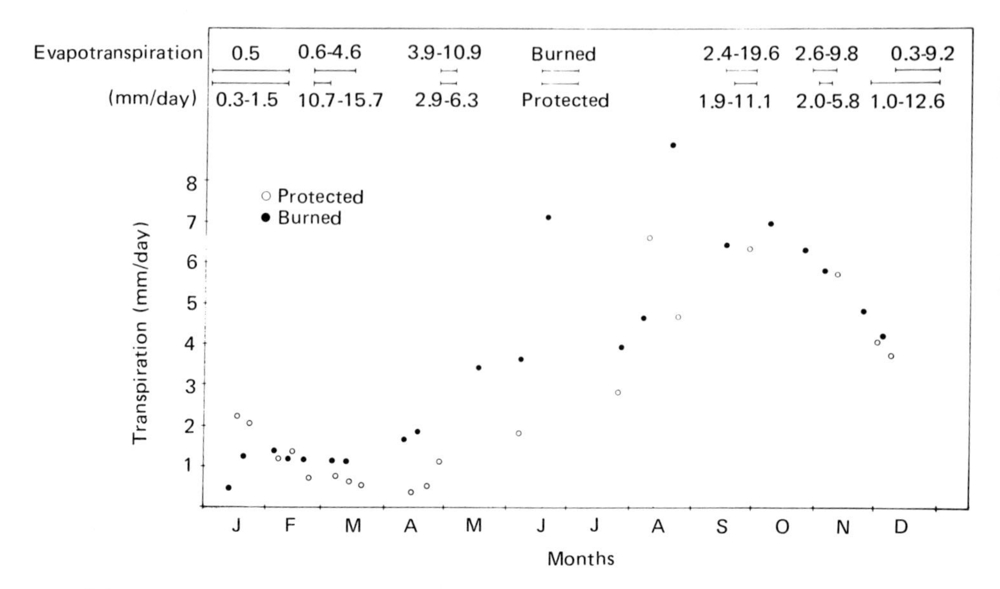


**Fig. 17–6.** Transpiration values measured in the two plots compared with evapotranspiration calculated from Figure 17–1, when evapotranspiration = △ soil-water content. The segments indicate the periods for which calculations were done.

are transpired during January. Soil-water content, then, allows grass growth during the dry season, which consumes the water reserves causing a reduction in growth rate (Figure 17–3). Growth is resumed when the actual rainy season begins.

Totaling rainfall, evaporation, transpiration losses, and total available soil water during the year gives the values shown in Table 17–4. It can be seen that the grass cover uses less water than supplied by rain. The difference must percolate to the water table since runoff was not observed even under heavy rain in the experimental plots. The higher leaf area development in the burned plot is responsible for the high water loss in that plot. Difference in transpiration activity accounts for up to 65 percent of the difference in the total amount of soil water in the two plots available during the year.

**Table 17–4.** Total rainfall, evaporation, available soil water, and transpiration in burned and protected plots (mm).

| | |
|---|---|
| Rainfall | 1,839 |
| Evaporation Tank A | 2,406 |
| Average soil water: | |
| Burned | 1,327 |
| Protected | 1,850 |
| Average total transpiration: | |
| Burned | 1,440 |
| Protected | 1,105 |

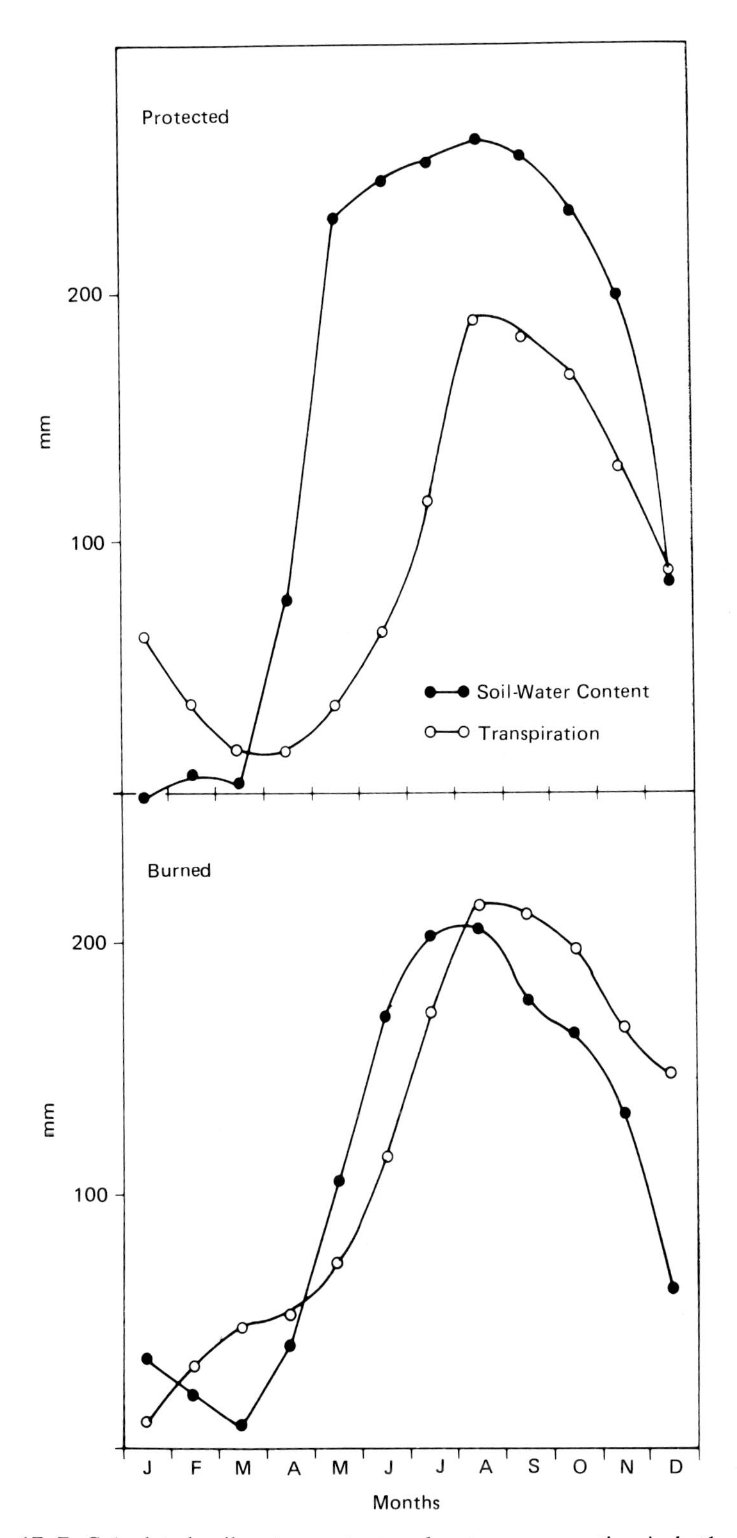


**Fig. 17–7.** Calculated soil-water content and water consumption in both protected and burned plots during the year.

## Discussion and Conclusions

### Grassland Productivity and Changes in Composition Caused by Fire

Burning near the middle of the dry season increases aboveground biomass by almost 30 percent. Similar results for tropical and temperate grasslands have been reported (Daubenmire, 1968; Old, 1969). Moderate fire stimulates growth of perennial grasses, which produce new shoots in a few days after fire; resulting assimilatory tissue is active when the rainy season begins. The trigger factor in this process has not been experimentally established; obviously much more work is needed (Blydenstein, 1962, 1963; Old, 1969).

The amount of organic matter produced after fire and its survival until the beginning of the rainy season depend on time of burning and soil-water availability. Results obtained by Blydenstein (1962) in *Trachypogon* savannas at Calabozo can be explained on the basis of relationships reported here. Too early burning (transition from rain to dry season) induces grass growth, which exhausts soil-water reserves, and the new shoots do not survive or develop very poorly. Too late burning (transition from dry to rainy season) does not permit leaf area development before rains begin. Therefore no difference between burned and protected plots will be observed. In the latter case, even a reduction in organic matter production in the burned plot could be observed, since at this time soil-water reserves are practically exhausted. Growth is extremely restricted by water stress, and production of new shoots by fire can result in damage to the growth points in grasses.

It seems that protection against fire also causes changes in competition relations between grass cover components. In temperate grassland it has been found that fire can increase the proportion of dicots against monocots (Swan, 1970). In tropical highland grasslands it has been reported that it causes drastic changes in proportion of grassland components. Fire induces a stronger development of dominated species of unburned or protected savannas, either grasses or herbs. Similar behavior is observed by *Trachypogon* savannas in Calabozo. Protection against fire increases the proportion of *T. montufari* against *T. plumosus* and *A. canescens*. The increase in aboveground production in the burned plot is due to higher productivity of subordinate, not of the dominant species. On the other hand, long protection increases the amount of litter accumulated (Blydenstein, 1963). Subsequent fires are much more destructive and perennial grasses die, opening a rapid successional change in the savanna initiated by a strong invasion of dicots (Aristeguieta and Medina, 1965).

### Productivity and Water Consumption

Transpiration is a function of evaporative force of the surrounding air and the leaf area, provided enough soil water is available. In the experiments reported here, the amount of leaf area developed is controlled by water availability from soil, so that the leaf area present at each time has sufficient water supply. Thus it shows unrestricted transpiration. Restriction of leaf area development can be observed during the dry season after fire when the growth rate of new shoots is reduced until the rainy season properly begins.

Although total transpiration losses are distinctly lower than total rainfall, transpiration values seem very high in comparison with the few data available in the literature. In fact, transpiration losses in the two plots are twice as high if Walter's relation of 100 g organic matter/m$^2$/100 mm rain is used as the basis (Walter, 1964). That would mean a transpiration coefficient of 1,000 liters/kg, while in the two plots studied here, considering root production as 20 percent of aboveground production (Medina, 1970), those coefficients are almost three times higher. In addition, grasses in experimental plots are $C_4$ plants, which are known for their high water use efficiency (Black, 1971). The explanation is that Walter's relation was calculated roughly for climatically determined savannas, that is, those growing with 600 or less mm rain/y. There, water is the limiting factor, while in Calabozo nutrients rather than water are the limiting factors for grass growth. Unrestricted transpiration also means continued photosynthesis, but photosynthetic rate can be strongly reduced by nutrient deficiency as has been demonstrated in many plants (Medina, 1971). Moreover, evaporation in the central llanos is extremely high, causing a decrease in water use efficiency (Vareschi, 1960; Monasterio, 1968).

It can be expected that in a less humid year, transpiration will be proportionately lower, not because of a lower relative transpiration rate but because life time and amount of assimilatory area will be reduced. In this sense, for productivity of savanna under strong rain, seasonal climates, the extension of the rain period is more important than total amount of rainfall.

## References

Aristeguieta, L., and E. Medina. 1965. Protección y quema de la sabana llanera. *Bol. Soc. Ven. Cienc. Nat. No. 109* 129–139.

Black, C. C. 1971. Ecological implications of dividing plants into groups with distinct photosynthetic production capacities. *Advan. Ecol. Res.* 7:87–114.

Blydenstein, J. 1962. La sabana de Trachypogon del Alto Llano. *Bol. Soc. Ven. Cienc. Nat. No. 102* 139–206.

——— .1963. Cambios en la vegetación después de la protección contra el fuego. I. Aumento anual en material vegetal en varios sitios quemados y no quemados en la Estación Biológica. *Bol. Soc. Ven. Cienc. Nat. No. 103* 233–238.

Bourliere, F., and M. Hadley. 1970. The ecology of tropical savannas. *Ann. Rev. Ecol. Syst.* 1:125–152.

Daubenmire, R. 1968. Ecology of fire in grasslands. *Advan. Ecol. Res.* 5:209–266.

Franco, M., and A. C. Magalhaes. 1965. Techniques for the measurement of transpiration of individual plants. In *Methodology of Plant Eco-physiology*, Eckardt, F. E., ed. pp. 211–224. Paris: UNESCO.

Kucera, C. L., and J. H. Ehrenreich. 1962. Some effects of burning in central Missouri prairie. *Ecology* 43:334–336.

Medina, E. 1970. Estudios eco-fisiologicos de la vegetacion tropical. In *Ciencia en Venezuela 1970*. pp. 63–88. Univ. Carabobo, Venezuela.

———. 1971. Relationships between nitrogen level, photosynthetic capacity and carboxydismutase activity in *Atriplex patula* leaves. *Carnegie Year Book* 69: 655–662.

Monasterio, Maximina. 1968. Observations sur les rythmes annuels de la savanne des "Llanos" du Venezuela, thesis, Univ. Montpellier, France.

———, and G. Sarmiento. 1968. Análisis ecológico y fitosociológico de la sabana de la Estación Biológica de Los Llanos. *Bol. Soc. Ven. Cienc. Nat. No. 113-114* 477–524.

Old, Sylvia M. 1969. Microclimate, fire and plant production in an Illinois prairie. *Ecol. Monographs* 39:355–384.

Santa Maria, F., and A. Bonazzi. 1963. Factores edáficos que contribuyen a la creación de un ambiente xerofítico en el Alto Llano de Venezuela: el arrecife. *Bol. Soc. Cienc. Nat. No. 106* 9–17.

Stocker, O. 1956. Die Abhängigkeit der Transpiration von den Umweltfaktoren. In *Hb. Pfl. Physiol.* Bd. III, pp. 436–488. Berlin: Springer-Verlag.

Swan, F. R. 1970. Post-fire response of four plant communities in south-central New York State. *Ecology* 51:1074–1082.

Vareschi, V. 1960. Observaciones sobre la transpiración de árboles llaneros durante la época de sequía. *Est. Biol. Llanos. Publ. No. 1* 39–45.

Walter, H. 1964. *Die Vegetation der Erde*. Vol. I. Jena: G. Fisher. Verlag.

———. 1970. *Vegetationszonen und Klima*. Stuttgart: Ulmer Verlag.

———, and E. Medina. 1971. Características climáticas de Venezuela sobre la base de climadiagramas de estaciones particulares. *Bol. Soc. Ven. Cienc. Nat. No. 119-120* 211–240.

# CHAPTER 18

# Net Primary Production in Grasslands at Udaipur, India

R. L. Shrimal and L. N. Vyas

## Introduction

Although they are not the climatic climax of the region (Bor, 1942), grasslands are found in extensive patches all over India, forming important aspects of terrestrial vegetation. Information on primary production and turnover is available for grassland ecosystems of temperate regions (Odum, 1963; Ovington and Lawrence, 1967; Westlake, 1963), and for grassland communities from the Indo-Gangetic plain (Singh, 1968; Singh and Misra, 1969), but the grasslands of the semiarid regions of Rajasthan have not received much attention (Vyas et al., 1972).

The present report deals with the changes in net green biomass, dead organic matter, litter, and underground biomass in protected and unprotected grassland ecosystems near Udaipur.

## Study Area

The study was conducted in Mota Magra forest block 14 km east of Udaipur (24° 35′ N latitude and 75° 49′ E longitude) situated at an altitude of 587 m above sea level. The site selected for study has a plain topography and sandy loam soil. The soil profile does not show clear differentiation of horizons.

The climate of the area is monsoonal (Figure 18–1). The area receives a mean annual rainfall of 660 mm distributed over 41 rainy days. Most of the rainfall (92 percent) is received during the rainy season, which extends from the middle of June to September. Mean maximum temperature varies from 24.3° (January) to 38.6°C (May), while the mean minimum temperature varies from 7.7° (January) to 25.4°C (June).

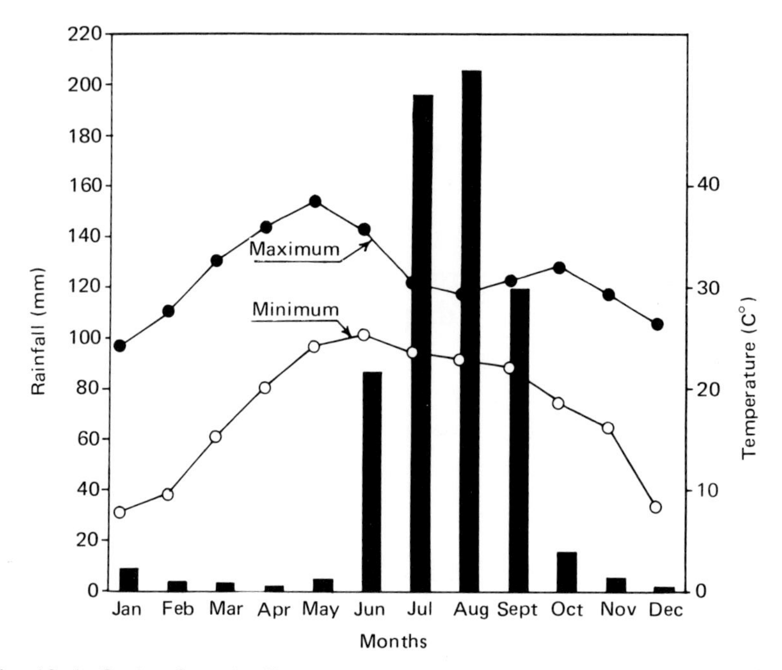


**Fig. 18–1.** Ombrothermic diagram for Udaipur based on the average of five-year data. Bars represent rainfall.

The vegetation is composed of 17 species of flowering plants, including eight species of grasses, four species of legumes, and five species of other herbs. On the basis of cover and density, the grassland community is dominated by *Heteropogon contortus* and *Dichanthium annulatum.*

The site of investigation is divided into two stands of different histories, that is, unprotected and protected for three years. The protected area has a permanent wall constructed by the forest department; hence it is free from biotic disturbances. The unprotected area is under biotic pressure of grazing and scraping.

## Methods

The grassland vegetation of the area was studied during the period June 1971 to May 1972. A sample plot of 50 by 50 cm was used for estimating aboveground standing crop. For the estimation of the standing crop of underground parts, a soil monolith of 25 by 25 cm was dug to a depth of 30 cm. The plant material was harvested in replicates of five plots (chosen at random) in the last week of each month. Care was taken to sample a quadrat only once.

The harvested material was sorted in the field according to species and brought to the laboratory in polythene bags. The material was further separated into green and dead tissue in the laboratory. All plant material was dried at 85° C for 48 hr and weighed.

## Observations

### Standing Crop Biomass

In both the stands the aboveground biomass increases from June to September (Figure 18–2 and 18–3). In September most of the species reach maturity and start dying, causing a decline in the aboveground biomass from the month of October. The total biomass was found to vary from 103 in May to 240 g/m$^2$ in September in the unprotected stand and from 102 in May to 204 g/m$^2$ in September in the protected one. The dominant species on each site showed the same trends in standing crop. In all the months the total net biomass was slightly higher in the unprotected stand.

The trends in the standing dead material and litter follow the trend in aboveground biomass. There is a steady accumulation of litter from the month of September, and in November and December the amount of litter exceeds that of the green and nongreen biomass combined.

The underground biomass (Figures 18–2 and 18–3) increases with the aboveground material during June and July. The underground biomass then continues to increase to the next May and the aboveground biomass declines after the September peak. In contrast to the aboveground biomass, underground biomass was slightly higher in most months in the protected stand.

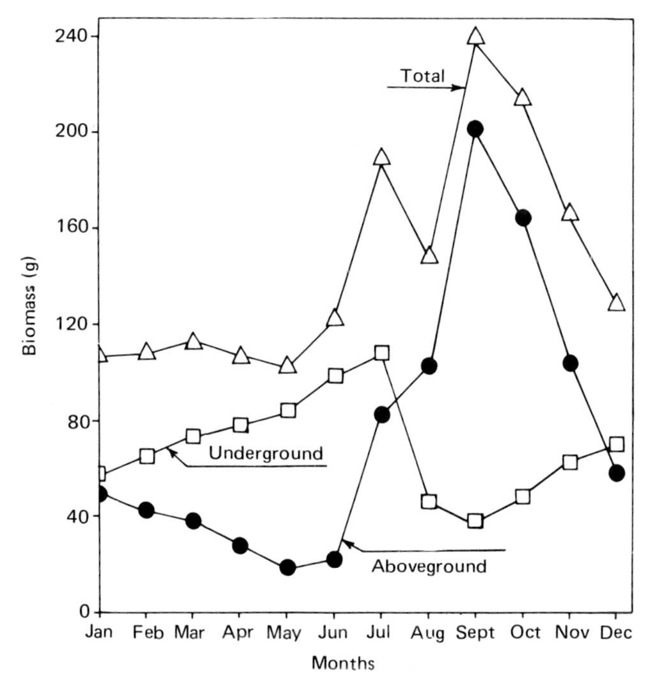


**Fig. 18–2.** Monthly variation in biomass (g/m$^2$) in an unprotected grassland near Udaipur.

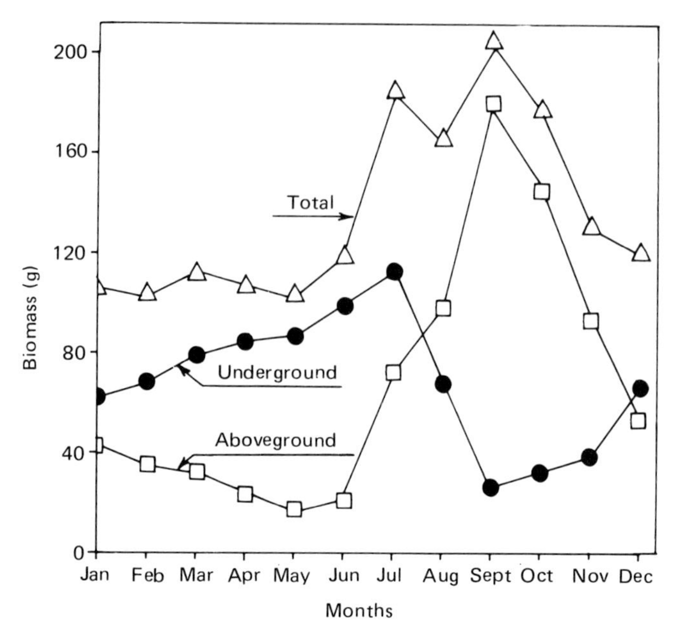


**Fig. 18–3.** Monthly variation in biomass (g/m$^2$) in a protected grassland near Udaipur.

## Net Primary Production

Net primary production for each month from June to October was calculated as the difference between the biomass values on successive sampling dates (Table 18–1). At both the sites aboveground net production was maximum during the month of September. In contrast, the underground biomass production was maximum in June.

Production was also estimated by adding the peak standing crop of each species (Table 18–2). *Heteropogon* and *Dichanthium* contribute most to production in both plots. By this method of estimation the net primary productivity on the unprotected site (202.7 g/m$^2$/yr) is significantly higher than that on protected site (163.2 g/m$^2$/yr).

# Discussion

In Udaipur grasslands, the standing crop of green material was found to be relatively high during monsoon and postmonsoon months. The lowest value of 7 g/m$^2$ was recorded in May and the highest value, 182 g/m$^2$, was recorded in September. A comparison of Figures 18–1, 18–2, and 18–3 suggests a correlation between the pattern of rainfall and biomass. Negligible rainfall during winter and summer months limits production through soil moisture. A decrease

**Table 18–1.** Net primary production in the grassland near Udaipur during the months June to October, 1971.

| Month | Aboveground net production ($g/m^2$) | | Underground net production ($g/m^2$) | | Total net production ($g/m^2$) | | Productivity ($g/m^2/day$) | |
|---|---|---|---|---|---|---|---|---|
| | Unprotected | Protected | Unprotected | Protected | Unprotected | Protected | Unprotected | Protected |
| June | 5 | 3 | 15 | 14 | 20 | 17 | 0.66 | 0.53 |
| July | 59 | 52 | 9 | 13 | 68 | 65 | 2.26 | 2.2 |
| August | 21 | 25 | −63 | −45 | 21 | 25 | 0.7 | 0.83 |
| September | 101 | 82 | −7 | −42 | 101 | 82 | 3.36 | 2.73 |
| October | −38 | −35 | 11 | 7 | 7 | 11 | 0.36 | 0.23 |

**Table 18–2.** Net aboveground production ($g/m^2/yr$) on the two stands with different histories; stand 1 (unprotected) and stand 2 (protected for 3 years).

| Species | Maximum Green Standing Crop ($g/m^2$) | |
|---|---|---|
| | Stand 1 | Stand 2 |
| Grasses: | | |
| *Apluda aristata* | 9.2 | 14.1 |
| *Aristida funiculata* | 5.4 | 3.2 |
| *Chloris virgata* | 6.7 | 4.0 |
| *Dichanthium annulatum* | 35.1 | 29.3 |
| *Eragrostis tenella* | 2.3 | 4.3 |
| *Heteropogon contortus* | 45.8 | 39.7 |
| *Themeda quadrivalis* | 7.4 | 11.3 |
| *Tragus biflorus* | 3.0 | — |
| Legumes: | | |
| *Alysicarpus hamosus* | 6.3 | 8.4 |
| *Desmodium triflorum* | 8.6 | 6.1 |
| *Indigofera cordifolia* | 24.5 | 19.6 |
| *I. linnaei* | 13.0 | 9.2 |
| Other species | | |
| *Boerhaavia diffusa* | 11.4 | — |
| *Evolvulus alsinoides* | 4.8 | 2.8 |
| *Glossocardia bosvallea* | 9.2 | — |
| *Justicia procumbens* | 4.7 | 11.2 |
| *Lepidagathis cristata* | 5.3 | — |
| Net aboveground production ($g/m^2/yr$) | 202.7 | 163.2 |
| Rate of production ($g/m^2/day$) | 0.55 | 0.44 |
| Number of dominant species | 17 | 13 |

in biomass after September is the result of the normal decline following maturity with the approach of dry season. The negative values of net production obtained in the dry months show that in these months growth is very little as compared to the losses from respiration, drying, and withering owing to dry hot conditions.

In these grasslands about 40 percent of the community biomass is shared by *Heteropogon contortus* and *Dichanthium annulatum*. By the peak standing crop method, the community net primary productivity is between 160 and 200 $g/m^2/yr$. At Varanasi the productivity figures for *Dichanthium* grassland has been reported as 400 to 500 $g/m^2/yr$ (Singh, 1968), and at New Delhi *Heteropogon* produces 800 $g/m^2/yr$ (Varshney, 1972). These differences illustrate the unfavorable effects of semiarid climate of the area.

The inverse relationship between aboveground and underground biomass (Figures 18–2 and 18–3) suggests that there is a greater accumulation of organic matter in underground parts during hot or cold dry periods. This observation agrees with the findings of Choudhary (1972). The lower values of underground biomass observed during the wet months may be due to the upward translocation of food material or a high rate of death and decay of the root system (Singh and Yadav, 1972).

In the present study both number of species and aboveground production showed a decline as a result of protection. These findings are in line with the views of Singh and Misra (1969), that species diversity increases production while dominance increases community stability but reduces production.

## Acknowledgment

Our sincere thanks are due to the authorities of forest department of Rajasthan, especially Shri A. L. Sankhla, Divisional Forest Officer, Udaipur, for facilities and permission to work in the area. We also thank Professor R. D. Kumar for laboratory facilities.

## References

Bor, N. L. 1942. Ecology, theory and practice. *Proc. Ind. Sci. Congr.* Presidential address (Bot.).

Choudhary, V. B. 1972. Seasonal variation in standing crop and net above ground production in *Dichanthium annulatum* grassland at Varanasi. In *Tropical Ecology, with an Emphasis on Organic Production*, P. and F. Golley, eds. pp. 51–58. Athens.

Odum, E. P. 1963. *Ecology*. New York: Holt, Reinhart and Winston.

Ovington, J. D., and D. B. Lawrence. 1967. Comparative chlorophyll and energy studies of prairie, savanna, oakwood and maize field ecosystems. *Ecology* 48: 415–524.

Singh, J. S. 1968. Net above ground community productivity in the grasslands at Varanasi. *Proc. Symp. Rec. Advan. Trop. Ecol.* 2:631–653.

———, and R. Misra. 1969. Diversity, dominance, stability and net production in the grasslands at Varanasi, India. *Can. J. Bot.* 47:427–525.

Singh, S. S., and P. S. Yadav. 1972. Biomass structure and net primary productivity in the grassland ecosystem at Kurukshetra. *Symp. Trop. Ecol.* 59–74.

Westlake, D. E. 1963. Comparison of plant productivity. *Biol. Rev.* 38:385–425.

Varshney, C. K. 1972. Productivity of Delhi grasslands. In *Tropical Ecology, with an Emphasis on Organic Production*, P. and F. Golley, eds. pp. 27–42. Athens.

Vyas, L. N., R. K. Garg, and S. K. Agarwal. 1972. Net aboveground production in the monsoon vegetation at Udaipur. *Symp. Trop. Ecol.* 95–100.

# Tropical Water Bodies

Aquatic systems can be considered as separate subjects of inquiry or as subsystems that, with the terrestrial systems, form landscape units. This latter orientation is the common theme of the following papers, as well as an earlier volume of *Ecological Studies* (Hasler, 1974).

In this and the earlier volume Sioli stresses the point that rivers reflect the terrestrial systems that make up their watersheds, and discusses the Amazon, Congo, and Nile rivers from this point of view. His conclusions are convincing, especially for the Amazon and its tributaries. The paper by Fittkau and associates provides detailed data and further conclusions on this massive river system. These two papers reflect the long experience of these workers in Amazonia.

Rodriguez examines another type of aquatic system—the tropical estuary—and describes the interaction between the oceans and lands in this unique habitat. Rodriguez also takes a broad view of the estuary and reviews the extensive literature from the point of view of his own detailed studies on the the Venezuelan estuarine environments.

Finally, Lugo and colleagues present data on yet another aquatic system, the mangrove forest. This vegetation is characteristic of the tropics and has received considerable study, especially from a structural point of view. Lugo and co-workers focus on the dynamics of the mangrove and their contribution illustrates the type of study needed to understand not only the structure but also the management and preservation of this unique vegetation.

## Refcrence

Hasler, A. D. 1974. *Coupling of Land and Water Systems.* New York: Springer-Verlag.

# CHAPTER 19

## Tropical Rivers as Expressions of Their Terrestrial Environments

HARALD SIOLI

### Introduction

When limnology was founded about 90 years ago, the only object of that new science was the study of inland lakes. A lake was then considered to be a microcosm—"a little world for itself" (Forbes, 1887). Some years later, Forel (1901) discovered that a lake is in direct or indirect connection with the atmosphere, with its surrounding firm land, with its catchment area, and by its outflow with the ocean. He concluded "every lake is an organ of the earth." Later on, Naumann (1932) developed the concept of regional limnology, basing it on the interrelations of the lakes with their environment. Even so, present limnological studies in relation to lakes concentrate mostly on the network of intralacustrine interactions, and on that structure of causes and effects that describes a lake as an ecosystem, and not on the connections of the lake to its terrestrial environment.

Much later than lakes, rivers were of interest to limnologists. But the principles and ideas developed from the study of lakes could not be transferred and applied to rivers. Lakes are systems in which circulation processes of matter predominate, making them, to a certain degree, independent of outside contributions and guaranteeing their integrity.

### Characteristics of Rivers

Rivers cannot in the slightest degree be considered as microcosms. Instead, rivers are throughflow systems that receive all matter passing through them from the surrounding landscape and conduct it to the ocean or lakes. For their existence, rivers depend on continuous input as well as continuous output. Both are factors, however, that are not part of the rivers themselves. Thus, when one of these factors stops, rivers cease to exist. They may either dry up or turn into swamps or lakes. In any case, they are rivers no more.

Thus rivers are nothing else than functional parts of higher units: of landscapes, of geosynergies (Schmithüsen, 1963), or biogeocoenoses (Sukachev, 1964), on whose existence they depend. They receive their waters from surplus rain falling on the landscapes of their catchment areas that does not return directly into the atmosphere by evaporation. This surplus partly runs off superficially and partly percolates through the soils, collecting itself in the groundwater and then is drained from the landscapes through river systems.

In this way the rivers are not only fed with the water itself as *conditio sine qua non* for their existence, but the rainwater on its way over and through the soil and rock takes with it all that can be mobilized by its physical and chemical action. These include the particulate and soluble products of decomposition and the remineralization processes occurring with living matter. These products may differ from region to region according to the geological-lithological, geomorphologic, climatic, edaphic, and vegetation characteristics.

The river thus receives all its individual characteristics from the landscapes of its drainage basin. From it the river is provided with its bottom load of different kinds and sizes of particles, with its greater or smaller quantities of suspended matter of silt and clay, with the inorganic ions in its water, and with the organic matter, as humic substances, originating in the biota of the landscape. Also, the form of the river bed is at least partially determined by the geological history of the land area around it; on the other hand, feedback effects from the activity of the running water of the river itself also affect its surroundings.

In all its characteristics we see a river being shaped and conditioned by the parameters of the biogeocenosis of its catchment area. If we tend toward a physiological comparison, we must say that its function and significance are that of an excretory system of the landscape. It is through the rivers that the end products of the landscapes metabolism, abiotic as well as biotic, are eliminated. And in the rivers the remineralization of those end products is further processed more or less to completion, if not yet achieved on or in the landscapes themselves.

In this regard there is no principal difference between tropical rivers and those of other geographic or climatic zones. A practical difference may be only that most river systems in the temperate zone are already so polluted that pathological conditions prevail, and those rivers cannot be used as indicators for baseline conditions in their natural environment. In the tropics, at least the Amazon is practically unaltered by human interactions and not contaminated by man-made pollution. The Congo is not yet much influenced, but other great African rivers such as the Nile, Sambesi, Volta, while not polluted, are dammed and partly transformed into man-made lakes. Even in those sections of their courses where the character of a river has not yet disappeared, the regime of their waters has completely changed and has little relationship to the conditions of their environments.

The regime of one of these rivers, the Nile, was studied in connection with environmental conditions in the last century when modern research on that river started and when dams had not yet been built. The regime of the Nile had been known in the time of the ancient Egyptians who installed their famous "Nilometer" to observe and follow the changes of water level. The high summer floods remained a puzzle to the Egyptians, although it was guessed correctly that the summer floods must indicate climatic conditions in the drainage areas of some tributaries different from those in the region where the high floods occurred. However, this dependency could only be explained after the specific conditions of rainfall, discharge, geomorphology, evapotranspiration, and other factors over the course of several years were known for the main tributaries, and these facts brought together and compared with what happened along the Nile itself.

This was a complicated task in such a vast river system with extremely unequal climatic and geomorphologic conditions in the landscape of the most important tributaries. For example, one of these, the White Nile, has its headwaters in the central African lake region, which receives only small amounts of rain. On the way through the enormous swamp area of the Sudd the river loses water by evapotranspiration. Finally, the river collects again in the White Nile, which further on flows mostly through very flat land. Other tributaries, such as the Blue Nile, have their sources in the heights of the Abyssinian highlands, with reasonably heavy rainfall. When the Blue Nile carries high water and connects with the White Nile, it dams up the water of the White Nile, flooding thousands of square kilometers of flat land around it. Lower down, it then causes the high floods of the Nile itself (cf. Lockermann, 1958). In this example, climate and geomorphologic factors of the terrestrial environment act together as an intricate complex difficult to analyze.

Fortunately, there are much simpler and more clear-cut dependencies between rivers and their environments. On the Congo, when Berg (1961) studied the distribution of the water hyacinth *Eichhornia crassipes*, introduced into the Congo system around 1950, he found the occurrence of this plant bound to certain chemical conditions of the waters of different tributaries. *Eichhornia* thrived only in relatively neutral waters with a pH above 5.2 (pH values from 4.5 to 4.2 were inhibitory for its growth, and a pH $\leqq$ 4.2 was so toxic that the plants died). Furthermore it was found that the pH values of the tributaries were characteristic for the region from which the river came. All upper tributaries of the Congo and those of it right side conduct more or less neutral or only slightly acid water, while the tributaries from the left come from the forest-covered Central Congolese basin and have very acid brown water. Based on these relationships, Berg could predict where *Eichhornia* could be expected to establish itself and also where no danger of an eventual invasion existed. In this example the chain of dependencies was followed upward from a biological event over the chemical peculiarities of the rivers to the regions and landscapes of the headwaters and then back again from the region to the river for further interpretation.

## The Amazon River System

More detailed examples of what tropical rivers may eventually tell about their terrestrial environments are cited using the Amazon system as an example (cf. Sioli, 1974). The tributaries of the lower and middle Amazon—Rio Xingú, Rio Tapajós, and all the others from there to far above the mouth of the Rio Negro—exhibit a very strange and peculiar topography in their lower courses. Figure 19–1 shows that these rivers possess, in their upper course, a completely "normal" river bed, but that they suddenly enlarge disproportionately at a certain point; from there on, in their lower course, they represent an elongated lake. In those wide "mouthbays" the current has slowed down to a degree that no depth erosion takes place. Nevertheless, these mouthbays can be very deep, reaching, in the case of the Rio Negro, 100 m. The sediment accumulating on the bottom of these mouthbays is a fine loose ooze, as in a lake, and would be taken away by the slightest true river current.

An aerial view of the mouthbays (Figure 19–2) suggests an explanation of their form: as dammed up water bodies that have filled and drowned their original river valleys. True rivers that carved the wide and deep valleys into the soft sedimentary substratum of the Amazon basin when the water level

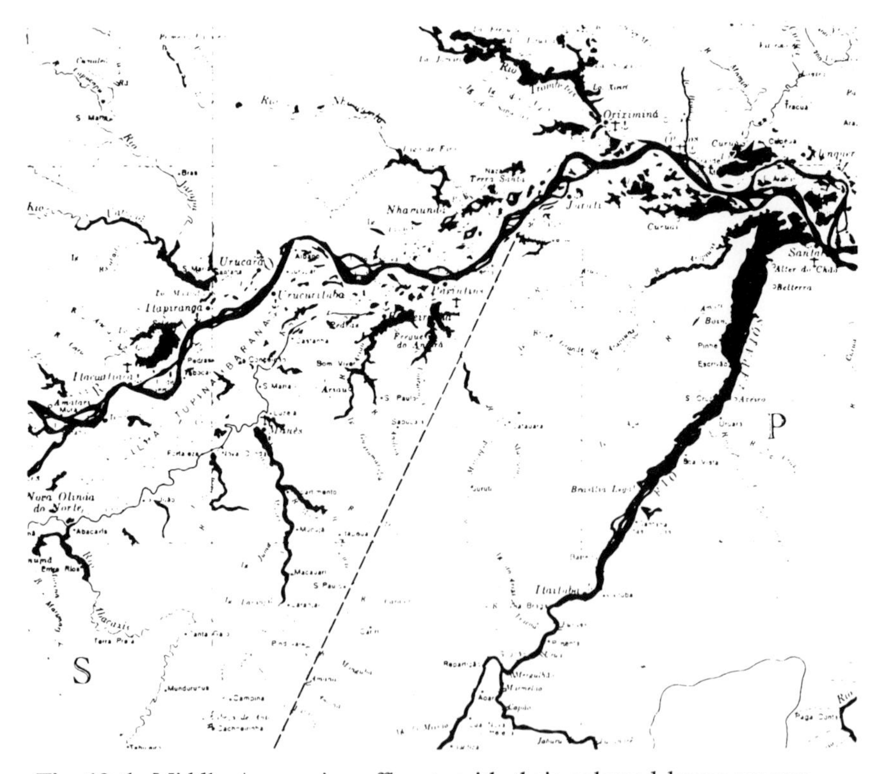


**Fig. 19–1.** Middle Amazonian affluents with their enlarged lower courses.

**Fig. 19–2.** Lower course (mouthbay) of Rio Arapiuns, a drowned valley. (Photo by H. Sioli.)

was lower and the rivers flowed on the thalweg of the valley with steeper gradient and stronger current. One reason for a change in the height of the water level must have been a rise of the ocean level that happened, as we know, some 12 to 20 thousand years ago with the end of the last glacial period. But besides this, there has also been regional vertical movement of the land.

River water chemistry permits conclusions not only about the lithology and pedology of its catchment area but also about parameters of the terrestrial ecosystems important for the eventual utilization of the area. In middle and lower Amazonia these are clearly related to the geology of the region (Figure 19–3). Most river waters are chemically extremely poor and pure (Table 19–1). With the exception of the creeks of the Carboniferous strips and the floodland of the Amazon itself, all rivers, clear as well as the black ones, are best described as slightly contaminated distilled water (Katzer, 1903; Sioli, 1967, 1968).

Practically the whole region is covered by forest that approximates a climax state. Its biomass is constant over a very long time, and nothing that is liberated by weathering processes in the soil will be additionally absorbed and accumulated in the vegetation as would be the case in forests with increasing biomass or in cultivated fields with the harvests being exported from the plantation region. Also, under the Amazonian wet climate, with ± 1,800 to

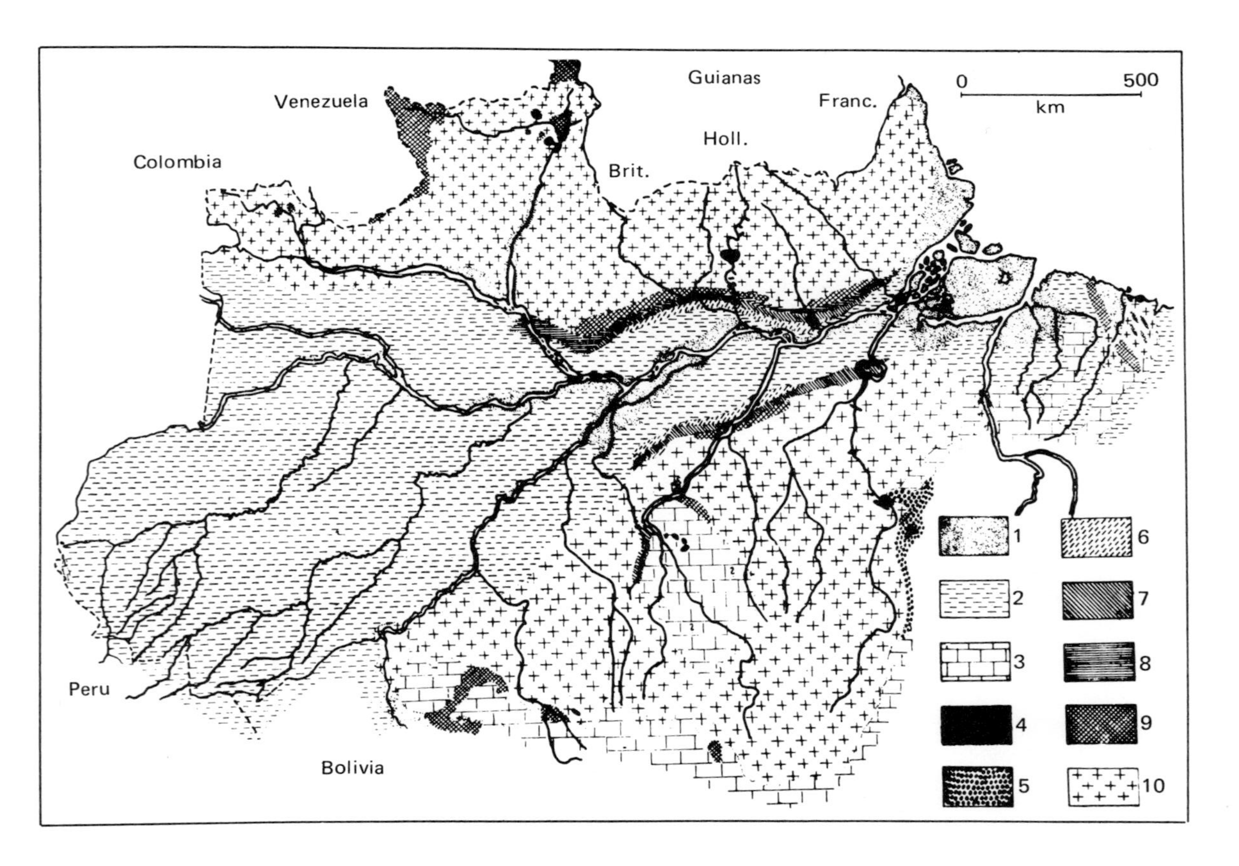


**Fig. 19–3.** Geological map of the Brazilian Amazon region. (After Avelino Ignacio de Oliveira, 1938). *1*, Quaternary; *2*, Tertiary; *3*, Cretaceous; *4*, Jurassic diabase eruptions; *5*, Permian; *6*, Carboniferous; *7*, Devonian; *8*, Silurian; *9*, Presilurian; *10*, Archean.

**Table 19–1.** Hydrochemistry of running waters from the geological regions of Amazonia.

| | Tertiary Series of the Barreiras | Archean | Carboniferous |
|---|---|---|---|
| pH | 4.2–5.5 | 4.0–6.6 | 5.2–7.8 |
| $HCO_3'$ mval/liter | 0.00–0.04 | 0–0.174 | 0.026–6.311 |
| $Ca^{\cdot\cdot}$ mg/liter | 0– ± 5 | 0–18.4 | 2.6–204 |
| $Mg^{\cdot\cdot}$ mg/liter | 0–0.38 | 0–5.6 | — |
| $Na^{\cdot}$ mg/liter | 0.847–2.530 | 0.245–2.060 | — |
| $K^{\cdot}$ mg/liter | 0.534–1.52 | 0.143–1.000 | — |
| $Li^{\cdot}$ mg/liter | — | 0–0.160 | — |
| $Fe^{\cdot\cdot} + Fe^{\cdot\cdot\cdot}$ γ/liter | 0–143 | 0–250 | 0–1200 |
| $Mn^{\cdot\cdot}$ γ/liter | 0–82 | 0–212 | 0–160 |
| $Al^{\cdot\cdot\cdot}$ γ/liter | 0–488 | 0–314 | 0 |
| $Cl'$ mg/liter | 0–3.5 | 0–2.5 | 0–16.5 |
| $SO_4''$ mg/liter | 0.000–0.480 | 0–2.690 | 0–556.7 |
| P ($PO_4'''$) γ/liter | 0–50.2 | 0–110 | 0–42 |
| N ($NO_3'$) γ/liter | 0– ± 200 | 0– ± 150 | 0– ± 550 |
| N (Kjeldahl) γ/liter | 138–724 | 0–2620 | — |
| Si diss. mg/liter | ± 0.5–4.5 | 0.502–6.650 | 1.5–22.4 |

3,500 mm of rain per year, no weathering products of the lithosphere or the soils are stored in the soils but rather are washed out by the percolating water and appear in the springs, creeks, and rivers. Despite all that strong weathering occurring in a humid warm climate, almost no soluble inorganic substances, including nutrients for plant growth, are being liberated. Although the soils do not dispose of nutrient reserves for the growth of the vegetation, luxuriant high forest grow on these nutrient-poor soils. This would not be possible if the forest lived in an open throughflow system; instead, much of its nutrient metabolism must be that of a closed circulation system in which the same nutrients continuously circulate through the generations of forest vegetation, provided with effective adaptations against leakage (Sioli, 1954).

## Effect of Land Utilization on the River

Serious consequences are implied for a rational utilization of the Amazonian soils occurring in most parts of the region. It is absolutely contraindicated to cut down the forest, burn it, and plant in the clearings, according to the famous slash-and-burn system ("roça") so widely distributed in the tropics. The crop gets its nutrients out of the ashes of the burnt vegetation that accumulated them during hundreds, if not thousands, of years. After an average of two years the ashes are gone, partly taken up by the crop and taken away with the harvest, but with the greatest part washed out by the rains and drained into the creeks and rivers. After exhaustion of the ashes, the impoverished roça is abandoned, and at another place the same method is repeated, with the same consequences.

This process is reflected in the chemistry of the creek water. In the catchment area of Rio Arapiuns, of absolute uniformity within the tertiary "Barreiras" sediments of lower Amazonia, all creeks had almost equal pH values of 4.5 to 4.6. Only one single creek once showed surprisingly a somewhat higher pH of 4.8. Looking for an explanation of this strange and unexpected finding, a fresh roça was discovered some hundred meters upstream on the bank of that creek. It had been burnt a few days ago and the night before the examination of the creek it had rained, whereby part of the fresh ashes had been washed into the creek raising the pH of its extremely chemical poor and unbuffered water.

From the results of the chemical river studies, we must conclude that the Amazonian rain forest area cannot easily be transformed into extensive agricultural fields. As long as the population density remains as low as it is in the interior of the vast region, the traditional shifting cultivation does not cause great harm. The small roças, distant from each other, are only needlepricks in the vast continuous forest area that heal quickly. Protected against erosion by the forest, out of which some nutrients may also flow into the abandoned little plantation, the forest quickly starts to grow again, and after 30 to 40 years only a botanist can distinguish such a place of secondary growth, called "capoeira," from the true virgin forest.

But the situation changes immediately when the shifting cultivation lays a dense and coherent net of clearings in the former forested area. Previously, all such large-scale agricultural experiments have failed in Amazonia. Also, the application of mineral fertilizers is no solution, since most Amazonian soils do not possess sufficient sorption-capacity to fix the fertilizer salts; the first rain will wash them out. For any development scheme for the great region, as presently being carried out, new and special techniques must be invented and applied to insure adequate and long-lasting fertilization (cf. Sioli, 1969, 1973)—without considering other and still more serious problems which will unavoidably arise with a deforestation of a great scale.

On the other hand, we have already mentioned two strips of marine sediments of Carboniferous age in lower Amazonia. At these locations limestone and gypsite deposits occur, and Jurassic diabase eruptions reach the earth's surface. Most creek waters of this zone have a pH around the neutral point and are rich in dissolved ions, including nutrients for plant growth. In this zone there must be richer, more fertile soils that release more nutrient reserves than the other terra firme soils of the Amazonian basin from which the pure acidic waters come. Indeed, some agricultural experiences in the Carboniferous strips support that conclusion. Also in the present development scheme of the Brazilian Amazonas region, starting with the famous construction of the Transamazonica, the southern Carboniferous strip is especially planned for agricultural purposes.

At the same time, however, these neutral waters, particularly the forest creeks when exposed to solar light after the shadowing trees with their closed canopy along the banks are removed, are predisposed as excellent habitats for planorbid snails, which are vectors of human schistosomiasis (*Sch. mansoni*). Such a situation was shown by the occurrence of endemic schistosomiasis at Fordlandia, in the Carboniferous area at the Rio Tapajós, some 20 years ago (Sioli, 1953). Schistosomiasis became endemic just at that single place in the Amazonian interior due to the chemical peculiarity of the rivers of that area.

Fertile soils in Amazonia occur on the vast floodplain of the middle and lower Amazon itself, the so-called "várzea" of 20 to 100 km in width. It is a foreign object in the landscape, built up by the sedimentation of the Amazon's turbid "white" water around 100 ppm of suspended particles. The wide valley of the middle and lower Amazon is understood as having drowned with the rise of the ocean level after the last glacial period. The Amazon's ochre-yellow water has long since filled the drowned valley and also has laid a wide sedimentation area along the ocean coast up to French Guiana. From the headwaters in the Andes, with their variegated lithology, their strong weathering processes, and steep slopes, fresh erosion products are transported downstream and first deposited along the foot of the Andes, carried on, eroded, and sedimented again probably many times, until they reach the lower Amazon deposition and erosion várzea, which as a whole is in a steady state. It thus forms a fresh soil, the constant fertility of which is guaranteed by new annual layers of fresh sediments on its flooded várzea foreland (Fig. 19–4).

Each step indicates something about the former. The white water of the Amazon is an expression of the strong erosion in its zone of origin; the várzea is due to the sedimentation capacity of that white water with its sediment load. The fertility of the várzea soil indicates the recent weathering and erosion processes that effectively attack the different rocks in the headwater areas,

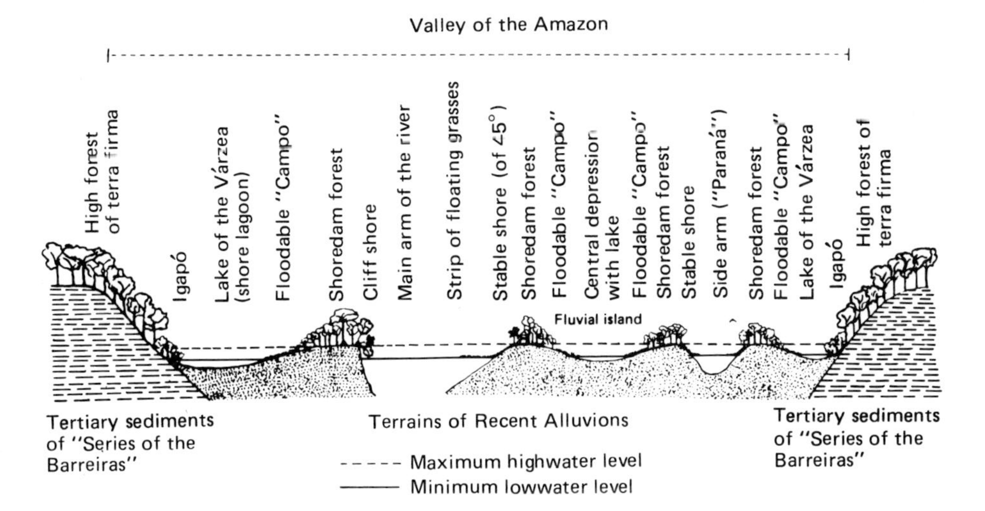


**Fig. 19–4.** Schematic cross section through the lower Amazon Valley; height exaggerated.

furnishing products richer in nutrients than do the granites, gneisses, and sandstones of the Archean shields of the Guianas and of Central Brazil that gave rise to the poor soils of most Amazonian terra firme. Camargo (1958) based his ideas of a rational utilization of the Amazonian várzeas by short period crops on this insight, and Fittkau (1971) has used it also to develop an ecological subdivision of Amazonia.

## Black Water Rivers

The Amazon region is famous for tropical black waters, which, however, also occur in other tropical areas, such as the Central Congo basin and Southeast Asia (Borneo). These are the other extreme of natural waters in Amazonia as opposed to the white water. They are free of any inorganic suspended particles, and thereby are transparent. They are brown to even red-brown in color, caused by dissolved organic humic matter, and they are extremely poor in inorganic ions, and acid down to pH values below 4.

The origin of these black waters in Amazonia was an enigma for a long time, but the problem was finally solved at the upper Rio Negro region, which is the classical black water river. There it had been noticed by scientific travelers, for example Koch-Grünberg (1909), that many of the forest creeks conduct crystalline clear, colorless water while others are deep brown. Examination of the soils around their origin areas showed that the clear waters came out of ochre to brownish clayey soils, covered by high forest, while the black water creeks originated from soils of pure white bleached sands on which occurs a shorter and more open forest called caatinga. Comparative chemical analyses of the waters indicated, mostly by absence or presence of aluminum in the waters, that the ochre soils of the clear water area are latosols, while in the bleached sands podsolization processes occur (Sioli, 1954) (Table 19–2). That hint of the existence of true tropical lowland podsols was later verified and confirmed by finding a compact black illuvial B horizon in many of those bleached sands and by discovering classical podsol profiles (Klinge, 1967). The absence of suspended matter in those black waters is also explained by the pure sands in their catchment areas: they have no fine clay particles among them that could make the water turbid.

Thus, as in the case of the Amazonian black waters, the water expressed very specific conditions in their source regions.

**Table 19–2.** Hydrochemistry of clear and black waters of Upper Rio Negro region.

| Water | pH | Aluminum | Total Iron (mg/liter) | Dissolved $SiO_2$ (mg/liter) | $N_2O_5$ (mg/liter) | $NH_3$ | $KMnO_4$ Consumption (mg/liter) |
|---|---|---|---|---|---|---|---|
| Clear water | 5.2 | 0 | 0.03 | 3.5 | 0.2 | 0 | 9.9 |
| Black water | $\leq$4.1 | + | 0.15 | 3.0 | 0 | ++ | 92.3 |

**Clear Water Rivers**

The third category of Amazonian river types is that of clear water rivers and creeks. Their acidity ranges from pH ~ 4.5, in the zones of Tertiary "barreiras" sediments, to more than 7 in the Carboniferous strips. Accordingly, their inorganic ion content also is not uniform and corresponds to the lithological substratum. Also some big rivers have clear water, as the Tapajós and the Xingú, with chemically poor waters and pH values around 6.5.What all the clear waters have in common, however, is that they are colorless due to lack of coloring humic matter, and have low transparency due to a minute amount of suspension load. In summary, we can compare the three main Amazonian water types—white (turbid), clear, and black—in Table 19–3.

The clear waters come from more or less flat, forest-covered zones. This is important since the dense forest cover inhibits soil erosion by the heavy tropical rains. The impact of the raindrops falling from great heights with high velocity is absorbed by the closed canopy of the forest trees and most of the rainwater reaches the soil by stem runoff or by the droplets falling from the tree's leaves. Thus practically all rainwater penetrates the soil and no surface runoff occurs. Also, the erosive action of the droplets that do not strike the soil but only gently drop on it from little height is in no sense comparable to what a heavy rain would do to a naked soil, washing from its surface layers all finer particles until only bare sand is left behind and even this goes into the river bed.

This is what will happen on a tremendous scale to the soils of Amazonia by deforestation of the "hylaea," as Alexander von Humboldt has characterized the region. The creeks and rivers will receive at once the entire rain and all the erosion products by quick surface runoff, with almost no penetration of the water into the soils. These waters will turn into turbid floods that run down the beds with higher water levels and with faster currents than ever before. Their erosion attack on the ground and banks of their beds will also increase correspondingly and enhance still more the whole erosion process of the land. The terrestrial matter and the soil ultimately will go into the ocean.

## Conclusions

Thus rivers are, in their basic character, products of the water economy of their drainage area. The quantity of water they drain calculated per area and time unit (not their discharge in $m^3$/sec), as well as the distribution of that amount over the seasons of the year, is determined by the water economy of the surrounding landscape. The water economy depends not only on the geographic-geomorphologic situation of the drainage area as external climatic factors but also very strongly on the vegetation cover of the drainage basin.

The Amazonian lowland is the largest and most typical zone of the humid

**Table 19–3.** Types of Amazonian waters as expressions of the environmental conditions in their headwater areas.

| Water | Water Color | Headwater Area and Its Relief | Soils | Vegetation | Examples |
|---|---|---|---|---|---|
| Black water (água preta) | Olive to coffee-brown, transparent | Plain lands | Podsols (bleached sands) | Caatinga, sandy campo, campina | Rio Negro, Rio Cururú, creeks from campinas |
| Clear water | Yellow to olive-green, clear, transparent | ± Plain lands as shields of central Brazil and the Guianas, Tertiary terra firme of Amazonia | Brown clays | Dense Amazonian high forest, (at least gallery forests in southern and northern border zones of Amazonia) | Rio Tapajós, Rio Xingú, most creeks of tertiary terra firme |
| White water (água branca) | Clayey yellow, turbid | Mountains and mountain slopes (as primary suppliers of suspended matter) | Brown clays, end moraines in high altitudes | Andean high forest (with solifluction), nonforest vegetations | Rio Amazonas, Rio Madeira |

equatorial tropics. The rains are not equally distributed everywhere in that region over the course of the year but are well enough distributed so that the area is practically completely covered by high rain forest, and the rivers conduct water all year around. Rain forest and the regime of running waters thus seem to be, and indeed are, consequences of the climate via the water economy of the landscape.

However as many times happens in nature, the dependency from climate (quantity and distribution of rains) to water economy of the landscape to regime of the rivers as well as to the forest vegetation are not a one-way relationship. Feedback is also built into that system, namely, from the forest vegetation back to climate and from there again to the water economy and to the regime of the rivers and vegetation: rainwater falling on densely forested land penetrates into the soil; there it is taken up, to a great part, by the dense root system of the forest, which, as Klinge stated (personal communication), is three times as dense as that of a temperate forest; by evapotranspiration this water returns to the atmosphere where it humidifies the air, condenses anew to clouds, and, as local rains, comes back to the forested earth, and so on.

This process means that the great seasonal cycle caused by the external climatic factors is superposed by many small local circulations of water. Indeed, when flying over the Amazonian lowlands one always observes local rainfalls, even in the so-called dry season. These continuous small circulations of water maintain the humidity that the forest constantly needs, even in the dry months of the year. The forest, through its effect on the impact of rainfall, its dense root system, its high evapotranspiration, and by recycling of a high percentage of the rainwater, acts as a buffer mechanism that bridges the dry season.

It is now easy to imagine what will happen when the forest over large stretches of terrain are cleared and open and stunted shrub (secondary growth) and short period crops or pasture land takes its place. Instead of penetrating the soil, most of the rainwater will run off superficially and go directly into the rivers. The water will no longer be absorbed by the dense root system of the former forest. Since the small bushes and the pasture grasses offer much less evapotranspiration than the forest (perhaps one-third), a relatively insignificant percentage of the rainwater will return into the atmosphere, and the local circulation of water will be interrupted. The dry season will become much more accentuated, and the regime of the running waters will be altered. The first heavy rains will cause immediate high floods, and in the dry season, the rivers will have little water and may eventually even dry up completely. The Zona Bragantine, east of Belém, is deforested in vast areas and is already a warning against such changes in the water economy of a formerly densely wooded landscape (cf. Sioli, 1957). Thus the changed regime of the rivers will become an expression not of natural conditions in the environment but of what man is able to inflict on a tropical landscape when he does not consider the ecological structure of that system to which the rivers belong as functional parts.

Let me conclude with the hope that this last example will be a purely theoretical one, and not be realized in what is going on throughout the tropics under the slogan of "development."

## References


Berg, A. 1961. Rôle écologique des eaux de la Cuvette Congolaise sur la croissance de la jancinthe d'eau (*Eichhornia crassipes* Mart. Solms). *Mem. ARSOM Bruxelles Nouv. Ser. 12* 3:1–120.

Camargo, F. C. de. 1958. Report on the Amazon region. Problems of humid tropical regions. In *Humid Tropics Research.* pp. 11–24. Paris: UNESCO.

Fittkau, E. J. 1971. Ökologische Gliederung des Amazonasgebietes auf geochemischer Grundlage. *Münster. Forsch. Geol. Paläont.* 20/21:35–50.

Forbes, St. A. 1887. The lake as a microcosm. *Bull. Peoria Sci. Assoc.* 1887:1–15.

Forel, F. A. 1901. *Handbuch der Seenkunde.* Stuttgart: Engelhorn.

Katzer, F. 1903. *Grundzüge der Geologie des unteren Amazonasgebietes (des Staates Pará in Brasilien).* Leipzig: Max Weg.

Klinge, H. 1967. Podzol soils: A source of black water rivers in Amazonia. *Atas do Simpósio sôbre a Biota Amazônica, Rio de Janeiro (Limnologia)* 3:117–125.

Koch-Grünberg, Th. 1909. *Zwei Jahre unter den Indianern.* Berlin: Ernst Wasmuth.

Lockermann, F. W. 1958. Zur Flußhydrologie der Tropen und Monsunasiens. Inaug. Diss., Bonn.

Naumann, E. 1932. *Grundzüge der Regionalen Limnologie.* Stuttgart: Schweizerbart.

Schmithüsen, F. 1963. Der wissenschaftliche Landschaftsbegriff. *Mitt. Flor.-Soz. Arbeitsgem. Stolzenau, N.F.* 10:9–19.

Sioli, H. 1953. Schistosomiasis and limnology in the Amazon region. *Am. J. Trop. Med. Hyg.* 2(4):700–707.

———. 1954. Betrachtungen über den Begriff der "Fruchtbarkeit" eines Gebietes anhand der Verhältnisse in Böden und Gewässern Amazoniens. *Forsch. Fortschr. Berlin* 28(3):65–72.

———. 1955. Beiträge zur regionalen Limnologie des Amazonasgebietes. III. Über einige Gewässer des oberen Rio Negro-Gebietes. *Arch. Hydrobiol.* 50(1):1–32.

———. 1957. Beiträge zur regionalen Limnologie des Amazonasgebietes. IV. Limnologische Untersuchungen in der Region der Eisenbahnlinie Belém-Bragança ("Zona Bragantina") im Staate Pará, Brasilien. *Arch. Hydrobiol.* 53(2):161–222.

———. 1967. Studies in Amazonian waters. *Atas do Simpósio sôbre a Biota Amazônica, Rio de Janeiro (Limnologia)* 3:9–50.

———. 1968. Hydrochemistry and geology in the Brazilian Amazon region. *Amazoniana, Kiel* 1(3):267–277.

———. 1969. Entwicklung und Aussichten der Landwirtschaft im brasilianischen Amazonasgebiet. *Die Erde, Berlin* 100(2–4):307–326.

———. 1973. Recent human activities in the Brazilian Amazon region and their ecological effects. In *Tropical Forest Ecosystems in Africa and South America: A Comparative Review,* B. Meggers, E. Ayensu, and W. D. Duckworth, eds. pp. 321–334. Washington, D.C.: Smithsonian Press.

Sukachev, V. N. 1964. *Fundamentals of Forest Biogeocoenology.* London: Oliver and Boyd.

# CHAPTER 20

## Productivity, Biomass, and Population Dynamics in Amazonian Water Bodies

E. J. FITTKAU, U. IRMLER, W. J. JUNK, F. REISS, AND G. W. SCHMIDT

### Ecological Division of the Amazon Region

The rain forest area of Amazonia, although topographically and climatically only slightly differentiated, is inhomogeneous in its ecological structure. According to its geological makeup it is divided into several regions that are defined geochemically and can be characterized ecologically (Fittkau, 1969, 1970, 1971). The running waters are useful indicators of the geochemical character of the Amazonian forest landscapes because their water chemistry reflects the availability of nutrients in the soils (Sioli, 1954a, 1954b, 1957).

The division of the "hylaea" (shown in Figure 20–1) into three regions of differing nutrient supply is based on numerous comparative limnological investigations of creeks and small rivers (Fittkau, 1964, 1967, 1974).

Central Amazonia is made up of deposits of weathered, redeposited Tertiary and Pleistocene sediments of fluvial and lacustrine origin on which only extremely nutrient-poor soils can be formed (Anon., 1969). The creeks and rivers of this area belong to the most electrolyte-poor natural waters on earth. Frequently, they are discolored to black waters due to a high content of humic substances (Santos et al., 1971; Schmidt, 1972a). In the north and in the south the central Amazonian sedimentation zones connect with the crystalline pre-Cambrian continental shield, which is rising but slowly and in places is interspersed with diabase or covered by Paleozoic and younger, predominantly continental sediments. Corresponding to the heterogeneous origin of the surface stratum in these northern and southern peripheral zones of Amazonia, the nutrient supply is variable and in general better than in central Amazonia. The rivers of these areas generally belong to the clear water type; however, according to local conditions and season of the year they may assume black or white water characteristics, or a mixture of the two (Sioli, 1963, 1967). Toward the west the extremely nutrient-poor sediment formation of central Amazonia is replaced by a broad pre-Andean zone, which becomes broader toward the

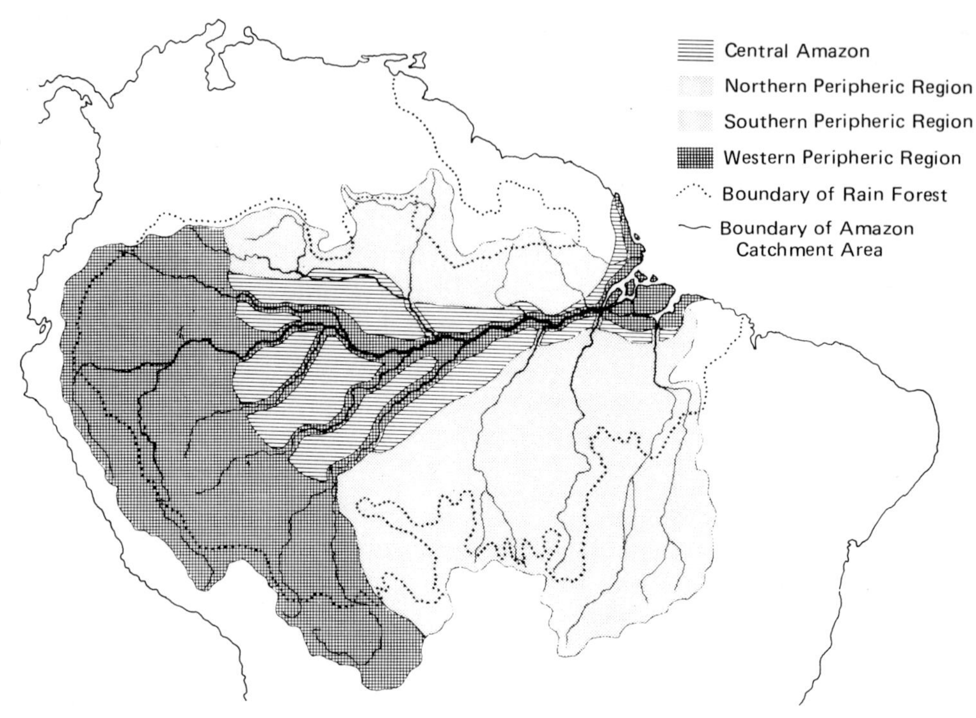


**Fig. 20–1.** Ecological divisions of the Amazon region, based on limnological investigations of creeks and rivers.

south and consists mainly of less weathered sediments, originating to a great extent from basic rocks of the Andean region. The waters originating in the Andes are loaded with suspended inorganic matter when passing through the pre-Andean area and become white water rivers with a relatively high electrolyte content (Gibbs, 1967). Corresponding with this geological division the geochemical conditions in the Amazonian catchment area deteriorate from the periphery toward the center.

The broad valleys of the rivers coming from the border areas cut through the central Amazonian lowland and correspond geochemically, in their relatively high nutrient content, with the alluvia and waters of their upper catchment areas.

The extreme electrolyte-poor conditions existing in the aquatic biotopes of central Amazonia directly, or indirectly as a result of the absence of primary production, exclude a number of organisms (Fittkau, 1974). The following biological production data of the aquatic ecosystems of Amazonia relate to central Amazonia and therefore include the waters of this geochemically poor region as well as the richer Amazon valley that cuts through the central area.

Besides the division caused by the nutrient supply, another ecological differentiation of the landscape results due to the annual water level fluctuations: the "terra firme"—the land not reached by high water and the regularly flooded land. The average annual difference between the low and high water level of the main stream system is about 10 m in the center of central Amazonia.

These annual water level fluctuations determine the structure of all aquatic and semiaquatic biotopes of the main river system, as for example, those of the Amazon valley. These annually flooded areas are called "várzea" (Sioli, 1964). The higher lying and therefore only shallowly flooded várzea areas were originally covered by forest but today they are often used for agriculture. The lower lying areas are often covered by a layer of grasses and weeds that grow after the water recedes.

The flooded areas of the central Amazonian black water rivers are covered to the low water level with "igapó," a forest type that is adapted to high water for many months and even to partially to complete inundation (Takeuchi, 1962). The waters of these rivers are poor in nutrients and suspended solids.

Wherever the central Amazonian waters reach the várzea region of the Amazon valley, there results, according to the topography of the valley, a mixed zone of different types of standing and running waters, the ecological aspects of which change in the course of the year. Thus the "lakes" (Fig. 20–2) of the várzea can be exposed in the course of one year to the influence of the two basically different water types of the central Amazonian region. On the one hand, the terra firme and its drainage system supply the várzea with black and rain water, which is extremely electrolyte poor, very acid, and enriched with humic colloids but without significant amounts of suspended inorganic material. On the other hand, at high water the main stream supplies the várzea and its lakes with white water with an electrolyte content 10 times higher, of almost neutral reaction, and low humic colloid concentration, but highly enriched with inorganic suspensoids (Schmidt, 1972b).

How strong the influence of the two components, black and white water, is on the várzea lakes depends on the position relative to the terra firme and the main stream. Lakes situated on river islands are almost exclusively determined by white water. Lakes situated close to the terra firme are governed at low water only by the black water, whereas at high water phase white water enters from the main stream and a mixed water body results. The influx of white water can take place directly through the connecting channels or indirectly by overflowing of the river's levees, thus supplying water in a broad front to the várzea whereby with increasing distance from the river the suspensoid concentration decreases due to sedimentation. Between the pure white water lakes on islands and the lakes strongly influenced by the black water running off the terra firme, a continuous spectrum of mixed waters exists in the várzea. Similarly, the várzea forest under the influence of black water changes with increasing distance from the white water stream and increasingly acquires the character of the Igapó the closer it comes to the terra firme. The Igapó is, however, only fully developed where the influence of the white water is completely excluded.

The running waters of the terra firme—standing waters are nonexistent on it—whose valleys lie above the high water line of the main stream are subjected to only relatively slight hydrologic changes during the course of year. Thus the creeks show constant conditions, which are also characteristic of the surrounding forest (Fittkau, 1964, 1967).

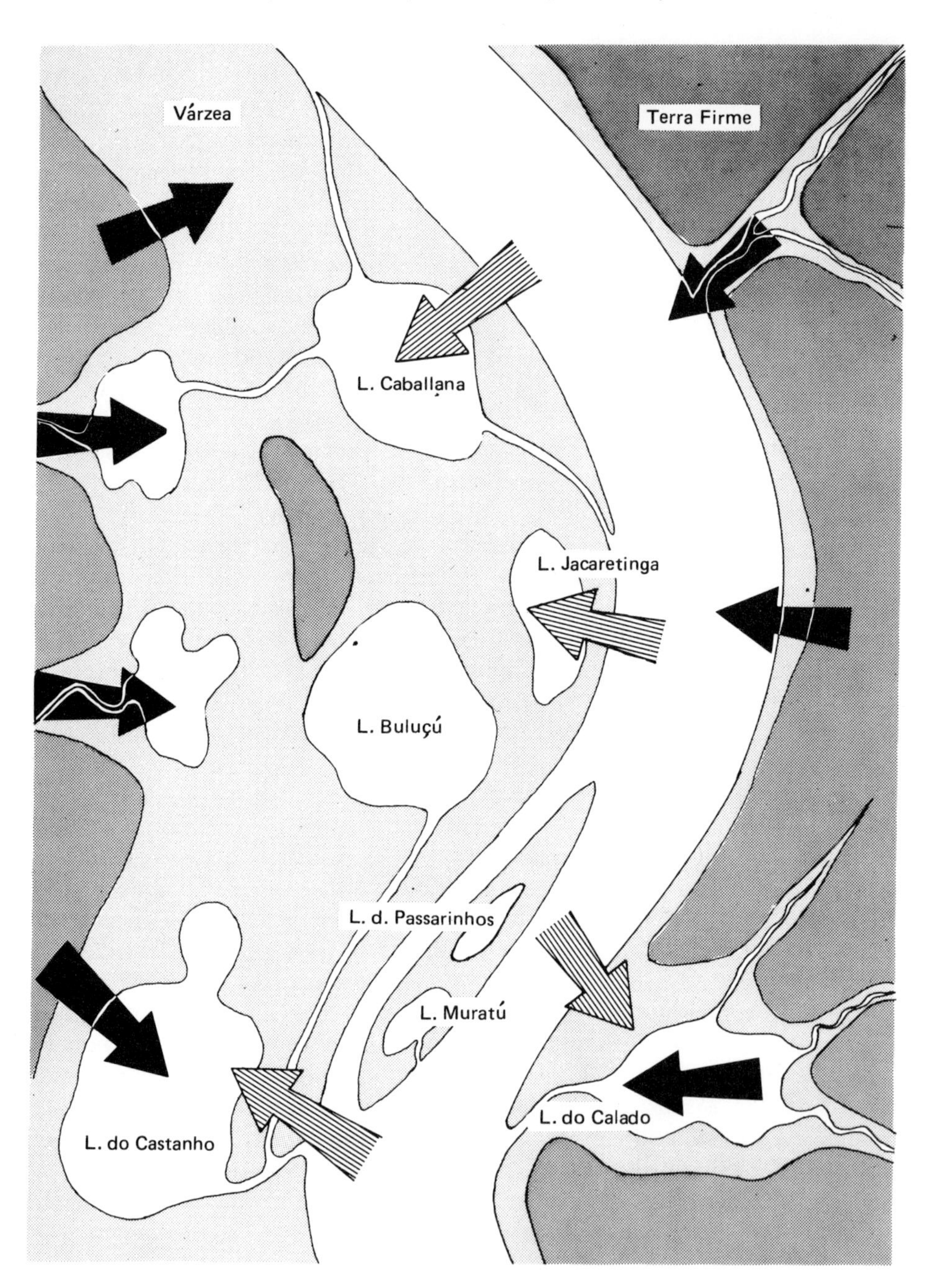


**Fig. 20–2.** Patterns of water movement on the floodplain of the Amazon.

## Primary Production

### Phytoplankton

Phytoplankton production rates were intensively investigated in Lago do Castanho (Schmidt, 1973b). The limnological conditions of this medium-size lake in the floodplain of the Amazon (várzea) in central Amazonia is strongly influenced by the Rio Solimões (Amazonas). Throughout the entire year the lake and river are connected by a channel so that fluctuations in river level are reflected in the lake. This means that nearly every year the lake is almost completely emptied and refilled by river water, resulting in strong seasonal differences in the physical, chemical, and biological conditions. The load of inorganic, suspended material that appears in the lake with the rising water level sinks quickly so that visibility rises, but not to more than about 2 m because of high phytoplankton development and the relatively high concentration of organic material originating from the shore region.

During the low water period, the sediments are resuspended, due to the movement of the water caused by the wind, and the transparency consequently falls to a minimum of about 10 cm. At the same time the concentration of total phosphorus, total iron, and soluble silica (as $SiO_2$) rises. These physico-chemical conditions naturally highly influence the development of fauna and flora in the waters (Schmidt, 1973a).

The primary production of the phytoplankton in Lago do Castanho is characterized by high productivity per unit volume (g C/m$^3$). Because of unfavorable light conditions, the productive zone is usually small. It varies, depending on the water level, between 0.5 and 6 m.

Seasonal differences occur in productivity per unit volume and per unit of surface area. During the low water period from October to November the net production reaches its maximum of more than 2 g C/m$^3$/day. The minimum of less than 0.5 g C/m$^3$/day was obtained from January to May during the inflow of river water containing a high concentration of inorganic suspended material into the lake (Figure 20–3).

The production per unit area reached a maximum of 1.5 g C/m$^2$/day during the high water period from the middle of May until the end of September after the inflow of turbid river water. The average value of the net productivity was 0.8 g C/m$^2$/day. The gross productivity per square meter was about 25 to 40 percent greater than the net productivity.

During the same season the differences in the production from day to day were very small. But the photosynthesizing activity of the phytoplankton varied distinctly throughout one day. During two experiments it was lower between 2:00 and 6:00 PM than between 6:00 and 10:00 A.M. or between 10:00 AM to 2:00 PM.

The proportion of dark fixation of $^{14}C$ varied from 1 to 6 percent of the $^{14}C$ fixation in light bottles.

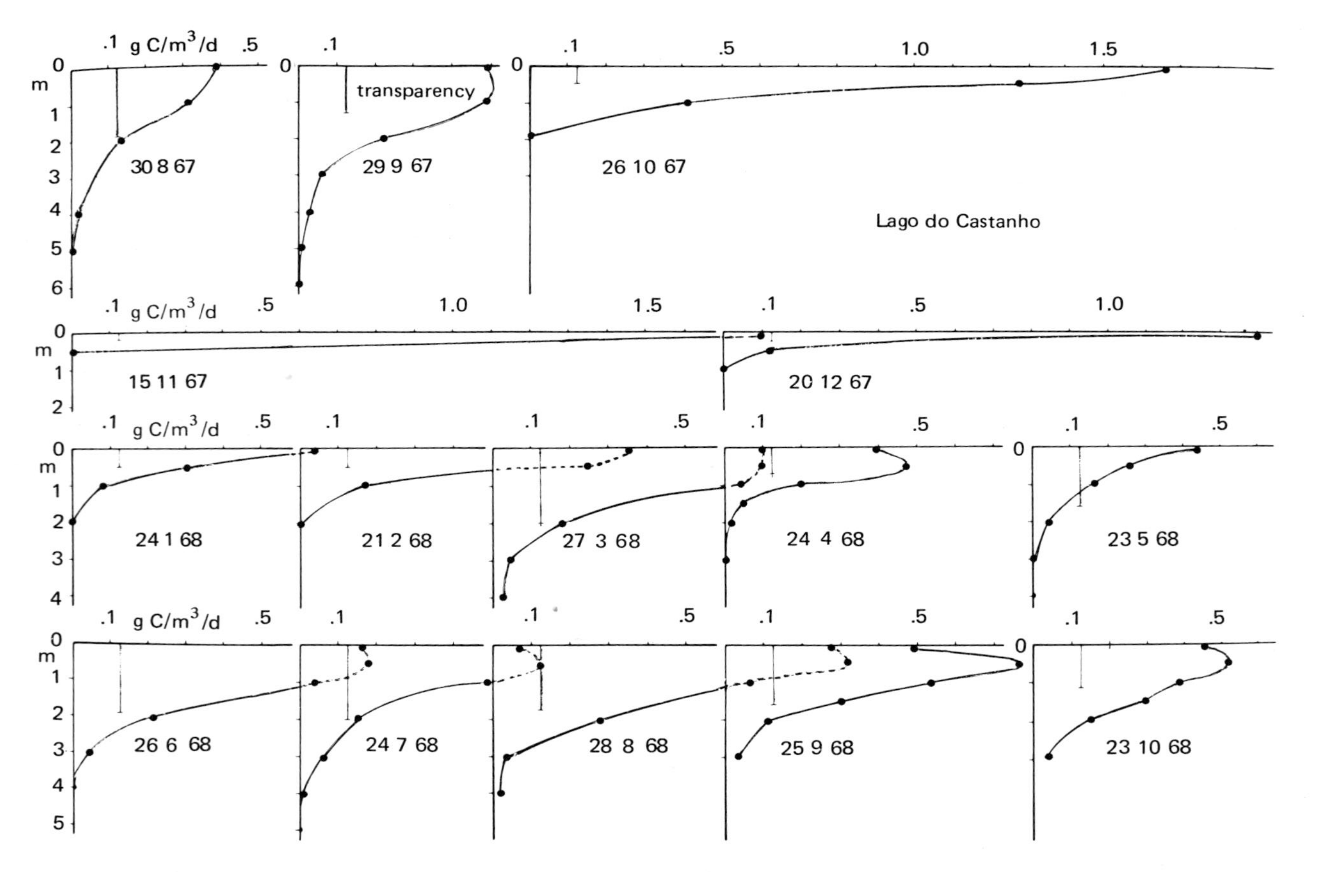


**Fig. 20–3.** Vertical patterns of primary production of phytoplankton and Secchi disk transparency in Lago do Castanho from August 1967 to October 1968.

Between August 1968 and August 1969 the net productivity of phytoplankton was about 3 tons C/ha in Lago do Castanho. The gross productivity was about 3.9 tons C/ha.

From the chlorophyll data of the trophic zone (average 0.052 mg/liter), we calculated the average amounts of C bound to phytoplankton to be about 0.8 mg/liter. The amount of total C bound to other soluble and suspended organic material was about 10 to 15 mg/liter. The average phytoplankton biomass calculated from the chlorophyll data was in the trophic zone 1.9 g $C/m^2$, respectively, 3.8 g $C/m^2$ ash-free dry weight. From this algal biomass and the average gross production of about 1.1 g $C/m^2$, we calculated the average C turnover rate to be about 1.7 days. Altogether the results indicate that in the Lago do Castanho light conditions limit the primary production of phytoplankton more than the nutrient content of the water.

In comparison of the productivity conditions with other tropical inland waters, the Lago do Castanho is a eutrophic lake, but not with as high a productivity as has been found in some other tropical waters eutrophicated by man.

Comparative investigations in the Rio Negro show that the primary productivity of phytoplankton lies about one to two tenth powers below that of Lago do Castanho. This can be attributed to the bad light conditions caused by the high content of humic material and to the considerably poorer nutrient conditions (Schmidt, 1974). In contrast, the Rio Tapajós, a clear water river, the area production can be greater than that of the várzea lakes, despite the poor nutrient supply, because of the favorable light conditions. The plankton density, however, is normally around one-third that of the Lago do Castanho.

## Macrophytes

The second significant source of primary production in the waters of Amazonia are the aquatic and semiaquatic macrophytes. They generally do not occur in black waters presumably because of their low nutrient content and low pH values. In the várzea region, however, they appear in enormous quantities. Floating species have adapted especially well to the ecological requirements of these biotopes since they can easily follow the considerable water level fluctuations and thereby remain in optimal light conditions. Real submerged plants—with the exception of some Utricularia—do not occur for these reasons. Also large areas become dry during the low water period and thus a change between terrestrial and aquatic biotope results. The consequence thereof is that pure aquatic plants are destroyed during the dry phase.

Diverse Gramineae have adapted best to these conditions. Two different types of adaptations can be distinguished. *Paspalum fasciculatum* has its main growth period during the low water phase. It is inundated with rising water, the stems lasting for several months under water; then with low water they begin to grow again. Other species, for example *Oryza*, *Hymenachne*, and especially *Paspalum repens* and *Echinochloa polystachya*, grow higher with the rising water and change over, to a greater or lesser extent, to a floating way of life. Their main growing period is during the high water phase. At low water they dry out and most of them die. A small part, however, adapts to a terrestrial existence and survives the dry season.

The species mentioned have adapted so much to the annual change of high and low water that with the exception of *Paspalum repens* they cannot grow in lakes that do not have these fluctuations in water level. In such lakes *Leersia hexandra*, *Paspalum repens* (Fam. Gramineae), and *Scirpus cubensis* (Fam. Cyperaceae) predominate.

Productivity figures for macrophytes of Amazonia are presently available only for *Paspalum repens*. In the lakes about 6 to 8 mt dry matter/ha are produced per vegetation period. In the Amazon itself the productivity is somewhat lower since the stands do not become as old because of the current. According to the duration of the production period (two to six months), the biomass is around 2 to 6 mt dry matter/ha (Junk, 1970).

This value seems to be relatively low when compared to values given in literature for aquatic macrophytes. Westlake (1963), for example, gives an average value in swamp areas of temperate zones of 45 mt/ha/yr, and even 75 mt/ha/yr for tropical latitudes. But in Amazonia we are dealing mostly with floating species of aquatic macrophytes that do not have to build up supporting elements during the high water phase. With *Paspalum repens*, a water content of the stem of up to a maximum of 95 percent depending upon age was measured; with *Echinochloa polystachya* it amounted up to 84 percent. In contrast, Straskraba (1968) gives a water content of only 37 to 66 percent for *Phragmites communis* stands.

## Secondary Production

### Perizoon

The floating vegetation, with few exceptions, offers excellent living conditions for the aquatic invertebrate fauna. In order to absorb nutrient salts from the water, the Gramineae develop floating root clusters at the nodes, which in connection with the densely intertwining stems and dead leaves serve as a biotope for a large number and great diversity of animals (Dioni, 1967). Planktonic species are favored by the free water regions between the root clusters; sessile species settle and hide in the densely intertwining roots. Due to the varying density of the plant cover, oxygen concentration can range from zero to oversaturation. The large quantities of detritus can serve within limits as direct or indirect food. Emersed plant structures offer protection to imagos of insects whose larvae live in water.

Normally the vegetation is so heavily populated that it must be considered as the richest biotope in the water bodies of the várzea. Those organisms support a division of the vegetation into three distinctly different types. In biotopes with flowing white water (white water type) faunal abundance and composition are influenced predominantly by the current and its load of inorganic suspended matter. As current speed and the load of suspended matter decrease, the number of animals per square meter increases, especially the filter-feeding Phyllopoda. The percentage of total abundance due to Crustacea rises while that due to Insecta declines (Table 20–1).

The total number of individuals fluctuates between 8,000 and 100,000 individuals/m$^2$, and the dry weight between 0.3 and 4.2 g/m$^2$ (Table 20–2), corresponding to a fresh weight of about 1.5 to 20 g/m$^2$. The number of individuals increases from the periphery to the center of the stand because the influence of the current is reduced.

**Table 20–1.** Average percentage of mean total abundance of Crustacea and Insecta in flowing white water.

| Station[a] | Current cm/sec[b] | Trans-parency (cm) | Total Number of Indi-viduals[c] | Crustacea (%) | Insecta (%) |
|---|---|---|---|---|---|
| Paraná do Xiborena | 10–25 | 35 | 15,300 | 39.9 | 52.3 |
| Costa do Baixio marginal region | 5–10 | 40 | 30,100 | 43.6 | 42.9 |
| Costa do Baixio central region | 5 | >40 | 56,600 | 61.4 | 28.1 |

[a] Characterizations of each of the biotopes, as well as discussion of the environmental factors important to the faunal communities are based on Copepoda, Ostracoda, Cladocera, Conchostraca, Ephemeroptera, Trichoptera, and Diptera, since these groups show the clearest relationships. Other groups (Decapoda, Hemiptera, Odonata, Coleoptera, Hydracarina, Mollusca, Oligochaeta) are included in the treatment of total abundance and total biomass.

[b] Current speed was determined by timing floating objects with a stopwatch.

[c] A net designed especially for this purpose was used: mesh width, 223 $\mu$, frame, 50 cm on each side.

In biotopes with sedimented white water (lago type A) the number of aquatic invertebrates and the composition of the fauna are dependent upon a number of local factors, such as density of the substrate, oxygen content, current, and phytoplankton development. Thus every lake (and often every part of a lake) has its special development of fauna in the floating vegetation (Figure 20–4). The number of animals in the vegetation may amount to 780,000 individuals/m$^2$, dry weight between 2.5 and 13.0 g/m$^2$ (12 to 62 g wet weight/m$^2$) (Table 20–2). Usually, faunal densities are about the same in the marginal region and the central region of the stands or decrease by about one-third toward the center.

Under special conditions *Leersia hexandra* and *Scirpus cubensis* form very dense floating mats that can be many years old (lago type B). Anoxia frequently occurs to some extent in the inner part of such stands and hydrogen sulfide can be formed. In the peripheral zone, faunal abundance is about the same as in stands of lago type A. Further inward, the number decreases rapidly (Figure 20–5). Ostracoda seem to be best adapted to these conditions, since

**Table 20–2.** Number of animals and Biomass (dry weight) in the various investigated lakes.

| Locality | Number of Samples | $N$ max./m² | $N$ min./m² | $\overline{N}$/m² | $B$ max. g/m² | $B$ min. g/m² | $\overline{B}$ g/m² | |
|---|---|---|---|---|---|---|---|---|
| Paraná do Xiborena | 9 | 26,640 | 7,360 | 15,330 | 0.984 | 0.344 | 0.610 | White Water Type |
| Costa do Baixio | 22 | 136,220 | 11,220 | 42,155 | 4.255 | 0.545 | 1.402 | |
| Solimões (lotic region) | 7 | 38,790 | 6,960 | 18,785 | 0.923 | 0.317 | 0.642 | |
| Lago do Calado (upper end) | 26 | 207,840 | 34,860 | 107,070 | 6.133 | 2.906 | 4.175 | Lago Type A |
| Lago do Calado (outlet) | 10 | 160,190 | 19,100 | 79,390 | 5.580 | 2.521 | 4.001 | |
| Lago Manacapurú | 26 | 786,500 | 11,890 | 186,880 | 12.966 | 2.448 | 4.907 | |
| Lago do Xiborena (upper end) | 21 | 300,130 | 7,980 | 115,550 | 6.302 | 2.538 | 4.053 | |
| Lago do Xiborena (outlet) | 15 | 197,450 | 9,000 | 86,150 | 5.014 | 2.441 | 3.570 | |
| Costa do Baixio (shore region) | 7 | 384,860 | 67,120 | 156,440 | 6.923 | 3.841 | 5.078 | |
| Lago dos Passarinhos | 4 | 359,100 | 152,600 | 257,620 | 10.239 | 4.855 | 7.244 | |
| Lago Parú | 2 | 307,360 | 195,800 | 251,580 | 8.130 | 5.973 | 7.052 | |
| Lago do Castanho | 1 | —— | —— | 179,160 | —— | —— | 6.179 | |
| Lago dos Passarinhos (Leersia hexandra) (Central region) | 2 | 13,000 | 8,800 | 10,900 | 0.292 | 0.159 | 0.226 | Lago Type B |
| Lago Parú (floating island) | 2 | 0 | 0 | 0 | 0 | 0 | 0 | |
| Lago do Castanho (floating island) | 4 | 0 | 0 | 0 | 0 | 0 | 0 | |

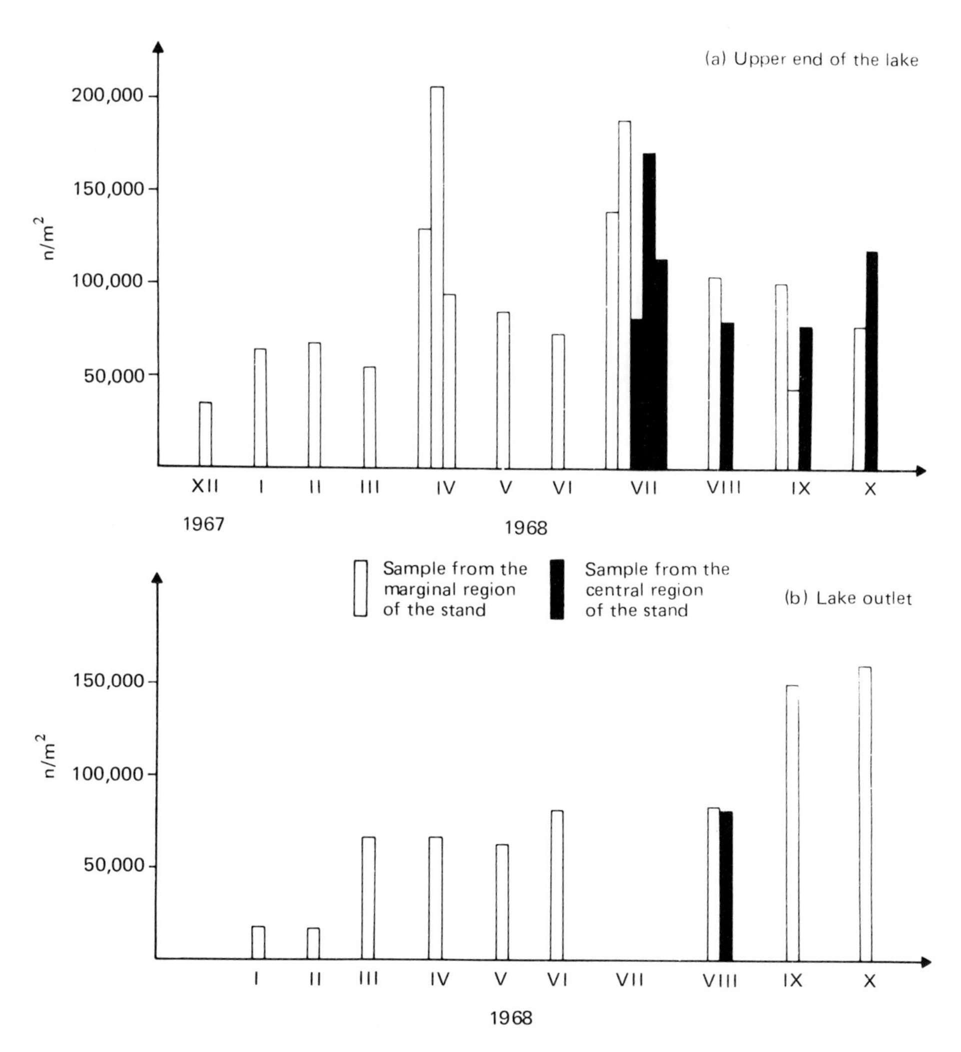


**Fig. 20–4.** Development of aquatic invertebrate fauna (total number) in two different *Paspalum repens* stands in Lago Calado during 1968.

their percentage contribution to the total number increases considerably (Junk, 1973). A strong secondary colonization by nonaquatic plants has the effect of making the floating mats even denser and form floating islands, sometimes with trees. As mentioned, these islands are more or less oxygen free and can be considered totally lacking in animals of the taxa studied here (Table 20–2).

## Lake Macrobenthos

The lakes of central Amazonia are shallow, with a maximum depth of 15 m at high water. To a great extent they are interconnected to the drainage ways

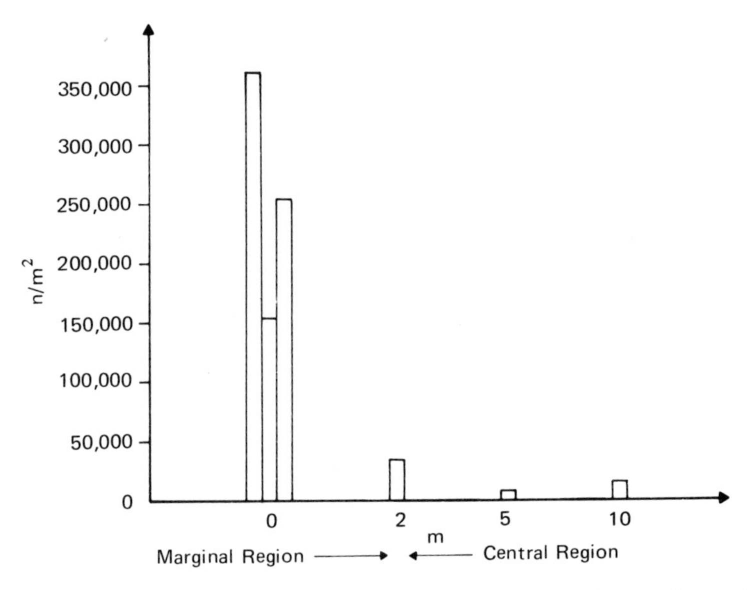


**Fig. 20–5.** Total number of aquatic invertebrates in a pure *Leersia hexandra* stand in Lago dos Passarinhos (lago type B).

and follow the latter's water level fluctuations. At Manaus the 70-year mean range of the annual fluctuations amounts from to 10.12 m (minimum 5.45 m, maximum 14.13 m). All physical, chemical, and biological processes are influenced or determined by these high fluctuations.

Therefore it is important to follow the qualitative and quantitative fluctuations of the macrobenthos of the lakes over a year's period. Investigations were conducted on seven lakes of different water quality in the surroundings of Manaus. Samples were taken at three-week intervals from each lake at fixed positions at the shore and at a central location.

With the exception of Lago Tupé at the lower Rio Negro, which contains black water the whole year around, the lakes were located in the várzea of Rio Solimões (middle course of the Rio Amazonas) and, as already mentioned, are influenced by differing degrees of white and black water. The resulting mixed water body is a codeterminant of the character of the benthic fauna. Figure 20–2 illustrates the position of each lake in the várzea.

The typical black water lake, Lago Tupé, shows the lowest mean annual biomass of the central lake regions (Table 20–3). In addition, Lago Tupé differs in that the distribution of its annual biomass is well balanced. A weak maximum appears in July shortly after highest water and a weak minimum during heavily rising water in March. The impoverished fauna consists of only three groups of organisms (Table 20–3), whereas in other várzea lakes from 6 to 16 groups occur. The three groups—Chaoborinae, Ostracoda, and Hydracarina—also occur at various depths and in the shore region of all studied várzea lakes. They can therefore be considered as the common stock of the

**Table 20–3.** Annual mean biomass and the percentage composition of the constituent groups in the macrobenthic fauna of seven Central Amazonian Lakes.

| Lakes (Central Part) | Mean Annual Biomass, Wet Weight (g/m²) | Percentage Composition | | | | | | | | |
|---|---|---|---|---|---|---|---|---|---|---|
| | | Chaoborinae Larvae (Culicidae) | Ostracoda | Hydracarina | Oligochaeta | Tanypodinae Larvae (Chironomid.) | *Pisidium sterkianum* (Bivaivia) | Hirudinea | *Chironomus gigas* | *Campsurus notatus* (Ephemeropt.) |
| Lago Tupé | 0.14 | 53.7 | 29.5 | 17.0 | — | — | — | — | — | — |
| Lago do Calado | 0.44 | 82.5 | 12.6 | 2.4 | 1.0 | 1.5 | — | — | — | — |
| Lago Cabaliana | 1.01 | 5.8 | 1.3 | <1 | 47.5 | 10.6 | 21.7 | 13.0 | <1 | <1 |
| Lago Buiuçú | 1.15 | 28.3 | 3.1 | <1 | 31.8 | 16.0 | 15.6 | <1 | — | — |
| Lago dos Passarinhos | 2.06 | 1.0 | <1 | 1.4 | 15.4 | 6.6 | — | — | 17.8 | 56.3 |
| Lago Muratú | 2.65 | 4.0 | <1 | <1 | 1.7 | 1.1 | — | — | 65.9 | 27.8 |
| Lago Jacaretinga | 6.20 | 6.3 | <1 | <1 | 4.8 | 1.0 | <1 | <1 | 87.0 | <1 |

macrobenthic fauna of central Amazonian lakes. Since deep water of Lago Tupé has an oxygen concentration below 0.7 mg/liter during the entire year, the representatives of these groups must be very resistant to low oxygen concentration.

Other várzea lakes, in their location near the terra firme border, are influenced strongly by black water and also show relatively low mean biomass values. An example is Lago do Calado (Table 20–3) with only about four times more biomass than Lago Tupé. In the várzea lakes such as Lago Cabaliana and Buiuçú (Table 20–3), where the black water influence is less direct, the mean annual biomass rises roughly eightfold more than in predominantly black water type lakes. An indicator species of such mixed water lakes is the bivalve *Pisidium sterkianum* Pilsbry, which is absent in várzea lakes without a black-water influx. Those lakes, which may be called white water lakes, are located mostly on small islands. Examples are the Lago dos Passarinhos and Lago Muratú. Their mean annual biomass values are 15 to 19 times higher than in black water (Table 20–3). The main portion of biomass is represented by larvae of the large insect species *Chironomus gigas* Reiss (Diptera) and *Campsurus notatus* Needh. and Murphy (Ephemeroptera).

The highest mean biomass value, 44 times higher than in the black water Lago Tupé, was found in Lago Jacaretinga (Table 20–3). This várzea lake is located near the main river but receives water neither directly from the river nor from the adjacent terra firme. At low water there are no inlets or outlets. Because there is no active transport out of the lake during low water period, the sediments of Lago Jacaretinga are unusually rich in organic matter. The quality of the sediments is believed to be the reason for mass development of larvae of the same *Chironomus* species, which, in lower numbers, also occurs in the two white water lakes.

As shown before, the mean annual biomass in the center of lakes ranges from 0.14 g/m$^2$ in a typical black water lake to 6.2 g/m$^2$ in mixed water lakes of the várzea. In addition, striking differences between the shore and central lake areas are also apparent. The mixed water lake Jacaretinga has the highest mean biomass value of 10.4 g/m$^2$ near the shore. Lago Buiuçú, with an average of 0.25 g/m$^2$, shows a lower value than the black water lake Tupé with 0.3 g/m$^2$, which may be due to the strong influence of the fauna of the surrounding forests, which are subject to inundation, on the shore fauna of the lakes. Apparently limnic species of the inundated forests invade the littoral region of the lakes during low water. Also lacustrine species may invade the inundated forests during high water.

The values of mean annual abundance of organisms in the various lakes vary much less than the biomass values and do not show a direct dependence on the water quality. As an example, the abundance for central parts of lakes range from 721 to 2,092 individuals/m$^2$. Corresponding values for the shore region are generally lower. All lakes show a considerable change in the relative dominance of organism groups (percentage of the various groups on total abundance). In the central regions of the lakes the period of strongest dynamics (the highest number of changes of relative dominant groups per time unit, here eight changes per nine weeks) is the period of rapidly falling water levels from mid-August to mid-October. The dynamics of the shore region are even stronger.

In all várzea lakes biomass and abundance are subject to strong seasonal fluctuations. Maxima occur mostly during the low water phase and minima at the high water phase (Figure 20–6). One important factor in these fluctuations is the intensive reduction of oxygen that occurs in water layers close to the bottom during periods of high water. It is likely that additional factors such as the seasonal distribution of zooplankton and phytoplankton and their roles as food for benthic fauna will also have an influence.

The total number of species occurring in the lakes is relatively low, with a maximum of 80. About 25 percent of this number are Chironomidae (Diptera). In lakes of the temperate zone the corresponding number of macrobenthic species is generally higher. According to Mothes (unpublished data) there are more than 650 species in a small lake district in the north German lowlands. Since in the central Amazonian lakes the "Aufwuchs" species are not con-

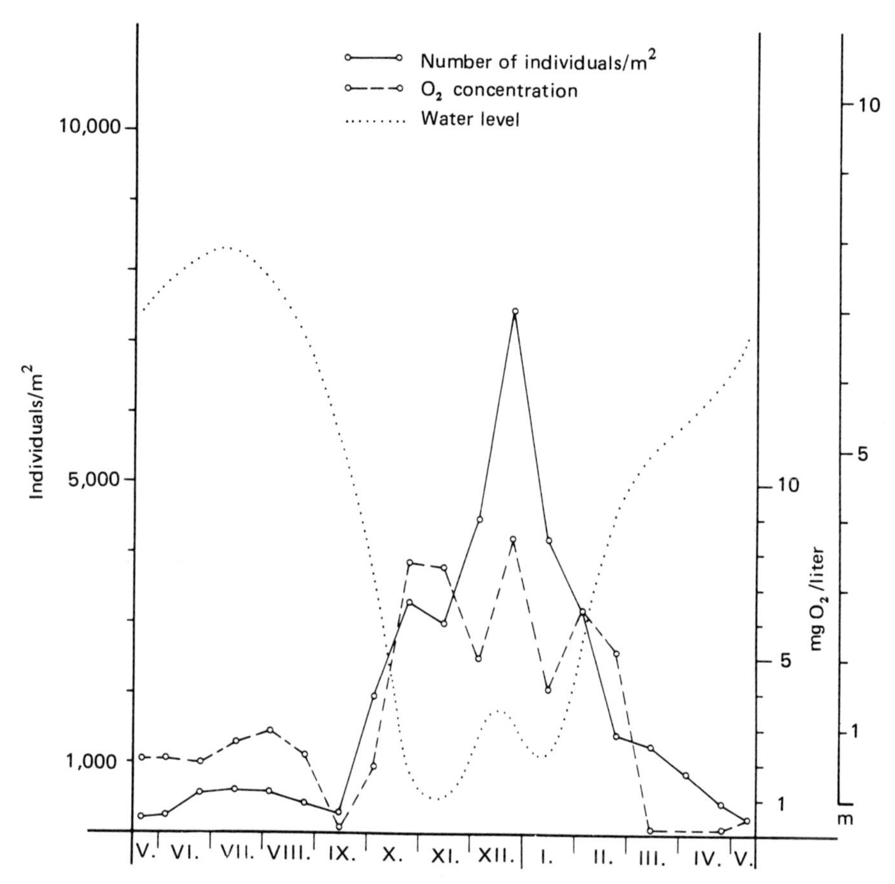


**Fig. 20–6.** Numbers of individuals and oxygen concentration compared to fluctuation in water level in Várzea lakes.

sidered, the above numbers of species are comparable only within limits. Nevertheless these tropical lakes are clearly poor in species. One explanation for this is suggested by the second ecological principle of Thienemann (1918), which states that extreme biotopes have only a few species but large numbers of individuals. The benthic biotopes of central Amazonian lakes can be regarded as extreme as they are exposed to unusual annual variations in water level and to the seasonal influences of two basically different water bodies.

## Macrobenthos of Inundation Forests

Studies were made over one year in inundation forests in a black water area of the lower Rio Negro and in a mixed and a white water area of the middle course of the Amazon River. The marked chemical and physical differences between the two river types near Manaus [white water of the Amazon, black water of the Rio Negro (Sioli, 1968)] naturally imply the existence of different inundation forests and associated aquatic biocoenoses.

In the white water area of the Ilha de Curarí and in the mixed water areas of Lago Janauarí the sediment load of the river and the sedimentation of suspended material along the shores are significant for the bottom fauna (Sioli, 1957b). Over the year at Ilha de Curarí the water transparency fluctuated only between 0.3 and 0.45 m. Here the river carries a high sediment load and deposits predominantly coarse material between 63 and 20 $\mu$ diameter. The bottom is covered with a thick mud layer and only a thin layer of leaf litter at its surface. Therefore, within this area of white water, mainly burrowing animals are found, which live in the mud and filter detritus particles (Sattler, 1967) or which are mud-eaters. As a dominant form in the zone of extremely strong sedimentation, *Campsurus notatus* Needham and Murphy, a mayfly species, which also occurs in the bottom fauna of lakes (Tables 20–3 and 20–4), merits special mention.

In the inundation forest of Lago Janauarí with water that already has lost most of its sediment load, the transparency fluctuates between 0.4 m at the beginning of the inundation period with a strong inflow of white water and 1.5 m at the inflow of black water. Particle size of the bottom sediments is mainly below 2 $\mu$ diameter. In this area only a thin layer of mud covers the leaf litter, which therefore plays an important role for the fauna. Filter feeders that need a firm leaf or branch substrate, and vagile animals, which run on the leaves and eat detritus, predominate. The Bivalvia, principally the genus *Eupera*, is particularly abundant.

The strongly acidic black water of Tarumã Mirím typically is very nutrient poor. Since black waters only carry extremely low amounts of inorganic sediments, the soil surface of these inundation forests consists of a thick leaf layer interspersed with colloidal humic matter. Therefore only very vagile animals that actively search their food can live here, chief among them being Oligochaeta—various genera and species of the family Naididae—and the Decapod *Euryrhynchus burchelli* Calman.

The different soil structures and the different chemical and physical properties of the waters influence the biomass and the production of the bottom fauna. The biotopes influenced by the white waters are distinguished from the black water biotopes by having a higher mean annual biomass. The highest mean annual biomass occurs in the mixed water, since in white water the living conditions are less favorable because of the high sedimentation rate. The values obtained for mixed water and black water inundation forests in the table, however, are too low because of methodological errors. They should really be at least four times greater (Table 20–4).

The mean annual abundance shows a similar relationship to the water quality. In the black water, however, the predominance of the lightweight Naididae results in a disproportionately high abundance. In white water (Ilha de Curarí) the abundance is about 600 individuals/m²/yr, in mixed water (Lago Janauarí) about 1,300 individuals/m²/yr, and in black water (Rio Tarumã Mirím) about 1,500 individuals/m²/yr.

When the changes in biomass over the course of one year are considered, characteristic differences between the black water inundation forests and those influenced by white waters are apparent (Figure 20–7). The latter show two

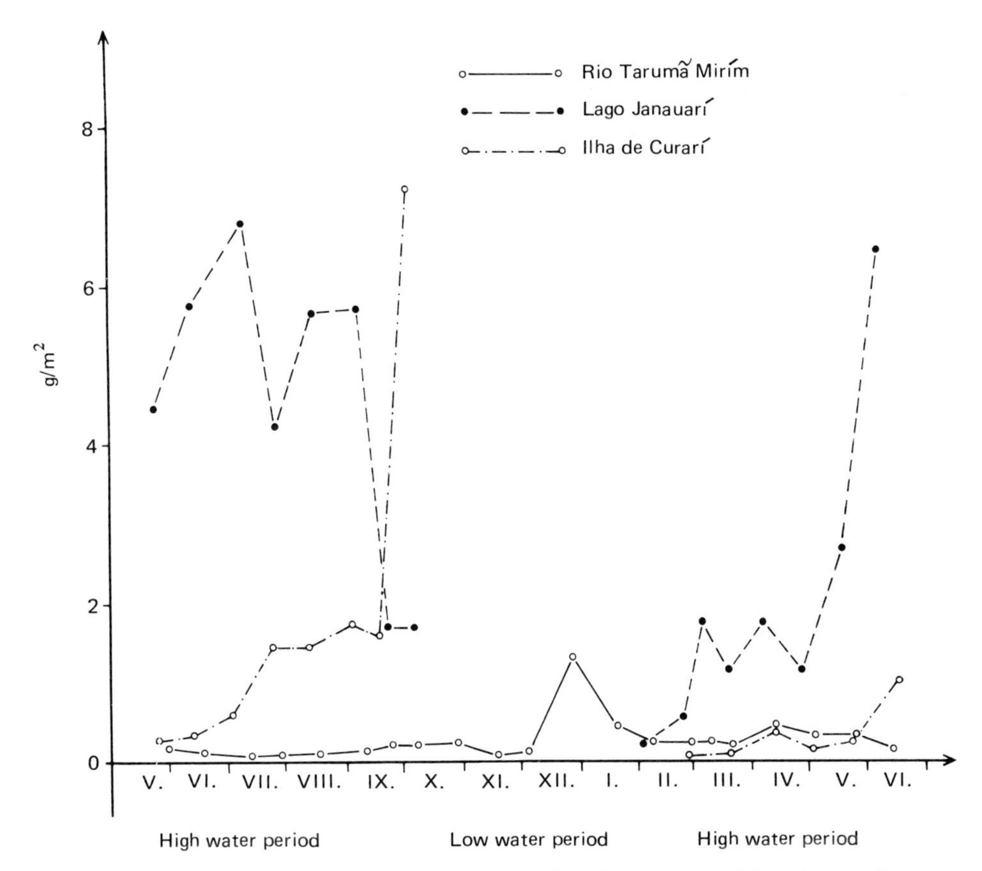


**Fig. 20–7.** Yearly turnover of biomass in the three types of inundation forests.

**Table 20–4.** Annual mean biomass, percentage composition of the constituent groups in the macrobenthic fauna and the conductibility ranges of three Central Amazonian inundation forests.

| Location | Mean Annual Biomass Wet Weight ($g/m^2$) | Percentage of Annual Biomass | | | | | | | Conductibility ($\mu S_{20}/cm$) |
|---|---|---|---|---|---|---|---|---|---|
| | | *Campsurus notatus* (Empemeroptera) | Chironomidae | *Eupera Spec.* (Bivalvia) | Gastropoda | Decapoda | Naididae (Oligochaeta) | Odonata | |
| Ilha de Curarí | 1.2 | 52.5 | 24.6 | 8.1 | 1.1 | — | 1.4 | 2.3 | 27.3–54.2 |
| Lago Janauarí | 3.0 | — | 2.6 | 80.0 | 6.0 | 5.2 | 2.0 | 2.2 | 13.7–65.7 |
| Rio Tarumã-Mirím | 0.2 | — | 10.0 | — | 3.7 | 36.2 | 27.0 | 12.7 | 5.4– 7.5 |

production phases, a short one at the beginning of the inundation period in March/April and a second one after the high water level in June/July. Minimum production during the second phase is greater than that during the first. For the biomass of the bottom fauna in black water inundation forests an almost straight line relationship was obtained. Here, during the low water period the maximum biomass occurs at the river banks, and with increasing water level, the biomass steadily decreases. The two phases in white and mixed water inundation forests occur for different reasons. Whereas in white water inundation forest it is chiefly the sedimentation that induces the development of various successive zoocoenoses, in the mixed water region it is the variation in oxygen concentration in the deep water. The area influenced by white water thus shows a greater degree of fluctuation in its biomass as compared to the black water inundation forest.

This constantly changing environment caused by the variation in water level and in the chemical and physical factors connected therewith demands great adaptability of the fauna to this extreme biotope. Thus migrations and translocations (Schwerdtfeger, 1968) are therefore common occurrences. These movements are also important to the productivity of the inundation forest biotopes because increase in biomass is able to occur at a location other than at which the breakdown takes place. Furthermore, due to reduction of the living space with falling water, animal concentrations can be formed so that the decomposition of the biomass can only take place in a small area in the forest. Migrations occur especially in white water regions where they can be demonstrated for *Campsurus notatus* Needham and Murphy and *Eupera* sp. Here, the expansion, or equally the reduction, in the living space seems to be the decisive factor inducing migration. Translocations were observed mainly in black waters where, due to a strong oxygen deficit in deep water, only a narrow borderstrip is inhabitable for most of the fauna that migrates according to the displacement of the optimal conditions. Figure 20–8 shows the translocation of *Euryrhynchus burchelli* Calman with the changing water level.

Other methods of surviving unfavorable periods are also employed. Molluscs of the genus *Eupera*, for example, survive the dry phase in the inundation forest with a period of dormancy. The animals cease growth even before the dry period and enter a diapause stage. However, according to the length of the dry phase, 15 to 100 percent of the individuals die during this time.

A third possibility, especially for small species, is an extremely rapid development such as is known for Ostracoda and perhaps Culicidae. Here the eggs or larval stages occur in low water areas where, without strong competition from other animals, they develop quickly before the fast rising water changes the living conditions again.

## Fauna of Terra Firme Creeks

The flowing waters of the terra firma, whose valley bottom lies above high water level of the main river system, are characterized by an unusual

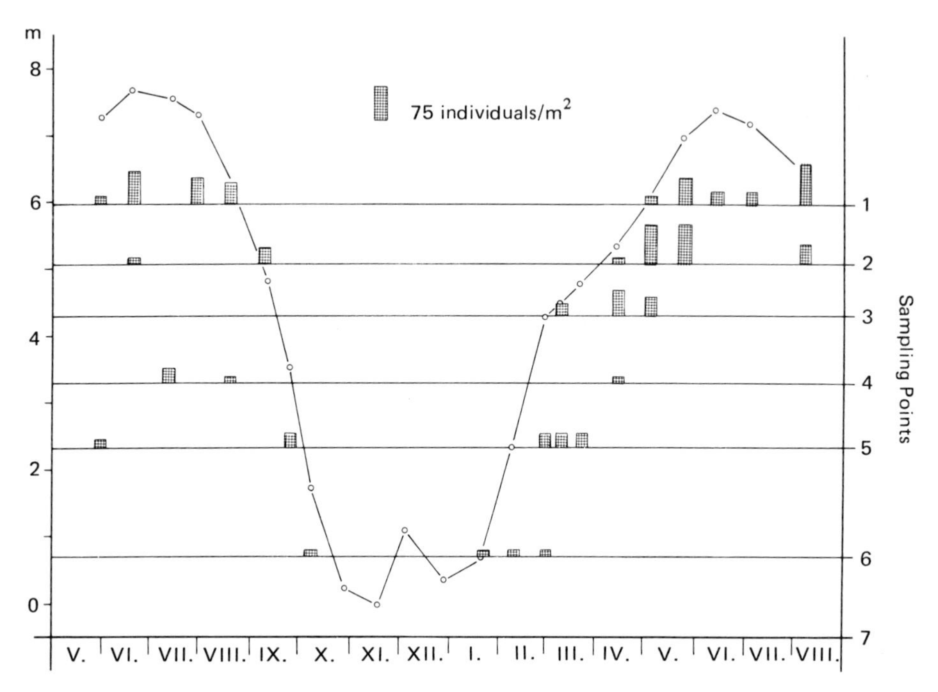


**Fig. 20–8.** Migration of *Euryrhynchus burchelli* Calman (Decapoda, Palaemonidae) in comparison with water level.

uniformity during the course of a year: the temperature and the water chemistry vary only slightly and the fluctuation of the water level is moderate. The ecosystems of these waters are subject to changes when they reach the area liable to be flooded by the main river system. The lotic flowing water biotopes developed during the low water period turn into a lenitic lake-like biotope with rising water, which, according to its geographic position, is determined by dammed white or black water.

Electrolyte and light shortage largely preclude the occurrence of primary producers, especially green algae, in the shady upper river courses. There one can regularly find red algae (*Batrachospermum*), which presumably are not fed on by animals. The only flowering plant in the streams is the semiaquatic species *Thurnia sphaerocephala* (Thurniaceae). The entire food chain is built upon organic matter coming from the forest, the major part of which is made up of leaves. The chief food source, however, is not the leaves as such but the fungi, fungal spores, and bacteria associated with their decomposition (Sattler, 1967).

The animal population is remarkably rich in species. More than 100 fish species can be counted in upper stream courses in the vicinity of Manaus and about 300 macroscopically distinguishable invertebrate animals. A quantitative sample of fish in a creek section 300 m long and 2 to 3 m wide yielded 5 g/m$^2$ made up by 17 species (Fittkau, 1973). Insects and Palaemonidae (Crustacea) comprise the greatest part of the benthic fauna (fishes excluded) both in terms of number of individuals and number of species. Other invertebrate groups are only poorly represented; Isopoda and Amphipoda being completely absent. Important primary consumers include the drift filtering larvae of Trichoptera, Simulidae, and Chironomidae and larvae of other Chironomids and Ephemeroptera, as well as great numbers of larval Odonata, Plecoptera, Megaloptera, beetles (Gyrinidae), and bugs (Heteroptera).

On the basis of many samples the estimated biomass of stream benthos in its upper course amounts of 1 to 2 g/m$^2$ without marked seasonal variations. Depending on the substratum, local maximal values of 20 g/m$^2$ can be reached. These high values appear where a firm substratum, mainly fixed wood or cushions of exposed roots on the bottom or at the sides in water of moderately strong current, becomes available to macro-micro-drift filtering insect larvae.

As the river expands, the areas of free sand increase in extent while the area of firm substratum decreases. At the same time, the number of species and the total biomass of fishes and benthos decreases continuously. Where black water is dammed up in the lower courses of running waters, during the high water phase the benthos consists of a few insect larvae and Oligochaeta, whose biomass amounts to 0.1 to 0.2 g/m$^2$.

## Conclusions

In spite of the relatively high nutritional status of flowing white waters, primary production of phytoplankton in them is unimportant because of the poor light penetration and great turbulence. The remaining types of water may be classified, with regard to phytoplankton production, in the following manner:

| *Water Type* | *Lighting Regime* | *Nutritional Status* | *Phyto-plankton Density* | *Production Per Unit Area* |
|---|---|---|---|---|
| Decanted white and mixed water | Bad to moderately good | Relatively good | High to very high | High |
| Black water | Bad | Very poor | Very low | Very low |
| Clear water | Good | Poor | Moderate | High |

Primary production by aquatic and semiaquatic macrophytes is of great importance in all water bodies with a white water influence but is entirely absent from pure black waters.

The extremely low electrolyte content of the black water is also reflected in a very low secondary production. Production is also inhibited in white water due to a high suspensoid content maintained by the high current speed. Living conditions improve for the fauna as soon as a reduction in current speed causes decantation of the suspended matter. Optimal conditions for the development of secondary producers seem to occur where black water and white water are mixed. This is supported by the fact that in such mixed waters the highest average biomass of both the benthos fauna and the fauna of the floating vegetation of the lakes and inundation forests has been found. Secondary production also is higher in the upper courses of rain forest streams that are influenced strongly by allochthonous matter.

## References


Anonymous. 1969. Os solos da area Manaus-Itacoatiara. Estudos e ensaios. 1. Edição do setor de Relações Públicas. Estado do Amazônas, Secretaria de Estado Produção em colaboração com o Instituto de Pesquisas e Experimentação Agropecuárias do Norte (IPEAN), Manaus, p. 117.

Dioni, W. 1967. Investigación preliminar de la estructura básica de las asociones de la micro y meso fauna de las raices de las plantas flotantes. *Acta Zool. Lilloana* 23:111–137.

Fittkau, E. J. 1964. Remarks on limnology of Central Amazon rain-forest streams. *Verh. Intern. Verein. Limnol.* 15:1092–1096.

———. 1967. On the ecology of Amazonian rain-forest streams. *Atas Simp. sôbre Biota Amazônica (Limnologia)* 3:97–108.

———. 1969. The fauna of South America. In *Biogeography and Ecology in South America.* Vol. II, pp. 624–658. Monographiae Biologicae, 19, The Hague: W. Junk.

———. 1970. Esboço de uma Divisão ecológica da região Amazônica. In Asociacion pro Biologia Tropical. II. *Simposio y Foro de Biologia Tropical Amazonica, Florencia (Caqueta) y Leticia (Amazonas),* Colombia, 1969, 365–372.

———. 1971. Ökologische Gliederung des Amazonasgebietes auf geochemischer Grundlage. *Münster. Forsch. Geol. Paläont.* 20/21:35–50.

———. 1974. Zur ökologischen Gliederung Amazoniens. In preparation.

Gibbs, R. J. 1967. Amazon river: Environmental factors that control its dissolved and suspended load. *Science* 156:1727–1734.

Junk, W. J. 1970. Investigations on the ecology and production-biology of the "floating meadows" (Paspalo-Echinochloetum) on the middle Amazon. I. The floating vegetation and its ecology. *Amazoniana* 2:449–495.

———. 1973. Investigations on the ecology and production-biology of the "floating meadows" on the middle Amazon. II. The aquatic fauna in the root-system of floating vegetation. Amazoniana. 4:9–102.

Santos, U. de M., A. Santos, and W. L. F. Brinkmann. 1971. A composição química do Rio Preto da Eva, Amazônia. Estudo preliminar. *Ciênc. Cult.* 23(5):643–646.

Sattler, W. 1967. Über die Lebensweise, insbesondere das Bauverhalten neotropischer Eintagsfliegenlarven (Ephem. Polymitarcidae). *Beitr. Neotrop. Fauna* 2:89–109.

Schmidt, G. W. 1972a. Chemical properties of some waters in a tropical rain-forest region of Central Amazonia. *Amazoniana* 3:199–207.

———. 1972b. Amounts of suspended solids and dissolved substances in the middle reaches of the Amazon over the course of one year (August 1969–July 1970). *Amazoniana* 3:208–223.

———. 1973a. Primary production of phytoplankton in the three types of Amazonian waters. II. The Limnology of a tropical flood-plain lake in Central Amazonia (Lago do Castanho). *Amazoniana* 4:139–203.

———. 1973b. Primary production of phytoplankton in the three types of Amazonian waters. III. Primary productivity of phytoplankton in a tropical flood-plain lake of Central Amazonia, Lago do Castanho, Amazonas, Brazil. *Amazoniana* 4(4):379–404.

———. 1974. Primary production of phytoplankton in the three types of Amazonian waters. IV. Primary productivity of phytoplankton in a bay of the lower Rio Negro (Amazonas, Brazil). *Amazoniana* (in press).

Schwerdtfeger, F. 1968. *Demökologie, Struktur und Dynamik tierischer Populationen.* Hamburg/Berlin: Verlag Paul Parey.

Sioli, H. 1954a. Gewässerchemie und Vorgänge in den Böden im Amazonasgebiet. *Naturwissenschaften* 41(19):456–457.

———. 1954b. Betrachtungen über den Begriff der "Fruchtbarkeit" eines Gebiets an Hand der Verhältnisse in Böden und Gewässern Amazoniens. *Forsch. Fortschr.* 28(3):65–72.

———. 1957a. Beiträge zur regionalen Limnologie des Amazonasgebietes. IV. Limnologische Untersuchungen in der Region Belém-Bragança ("Zona Bragantina") im Staate Pará, Brasilien. *Arch. Hydrobiol.* 53(2):161–222.

———. 1957b. Sedimentation im Amazonasgebiet. *Geol. Rundschau* 45:608–633.

———. 1963. Beiträge zur regionalen Limnologie des Amazonasgebietes. V. Die Gewässer der Karbonstreifen Unteramazoniens (sowie einige Angaben über Gewässer der anscließenden Devonstreifen). *Arch. Hydrobiol.* 59:311–350.

———. 1964. General features of the limnology of Amazonia. *Verh. Intern. Verein. Limnol.* 15:1053–1058.

———. 1967. The Cururú region in Brazilian Amazonia, a transition zone between hylaea and cerrado. *J. Ind. Bot. Soc.* 46(4):452–462.

———. 1968. Zur Ökologie des Amazonasgebietes. In *Biogeography and Ecology in South America.* Vol. I, pp. 137–170. Monographiae Biologicae, 18, The Hague: W. Junk.

Straskraba, M. 1968. Der Anteil der höheren Pflanzen an der Produktion der stehenden Gewässer. *Verh. Intern. Verein. Limnol.* 14:212–230.

Takeuchi, M. 1962. The structure of the Amazonian vegetation. VI. Igapó. *Fac. Sci. Univ. Tokyo Sect. 3 Bot.* 8:297–304.

Thienemann, A. 1918. Lebensgemeinschaft und Lebensraum. *Naturwiss. Wochenschrift N.F.* 282–290, 297–303.

Westlake, D. F. 1963. Compositions of plant productivity. *Biol. Rev.* 38:385–425.

# CHAPTER 21

# Some Aspects of the Ecology of Tropical Estuaries

GILBERTO RODRÍGUEZ

The main characteristics of estuarine populations are their ability to live in the dynamic conditions of tides and currents and their tolerance of changes in salinity. Daily cycles of tide and currents, salinity, and temperature and rainfall cycles of longer amplitude are superimposed on one another to create an extraordinarily rich and complex environment. There is considerable knowledge of temperate estuaries but very little information on those of the tropical zone, most of which comes from marginal and atypical areas, such as the northern Gulf of Mexico, South Africa, and Australia.

My thesis here is that tropical climatic conditions reflected in estuarine hydrodynamics creates conditions that differentiate tropical from temperate estuaries. This short report is not a complete discussion of extant information nor a complete list of references. Rather; it is an outline of the peculiar condition of tropical estuaries.

## Temperature and the Limits of Tropical Estuaries

The temperature of estuaries is initially determined by the temperature of the river and the sea, and the proportions each forms of the mixture at different stages of the tide. As Francis-Boeuf (1943) and Day (1951) have pointed out, air temperatures are unimportant, but there is a warming by the sun, so that the estuarine temperatures are higher and more variable than the sea. This fact is illustrated by a cross section of temperature in the Maracaibo estuary, Venezuela (Figure 21–1). The mean annual surface in the estuary is about 2°C higher than in the open sea. The annual cycle is well marked, with higher temperatures during the summer, but the fluctuations are small, amounting to approximately 3°C. The relative independency from air temperatures is shown by approximately 1°C mean difference between it and the temperature of the estuary water. The sea has a stabler temperature, so extremes are found at the head of the estuary where there is a greater proportion of river water.

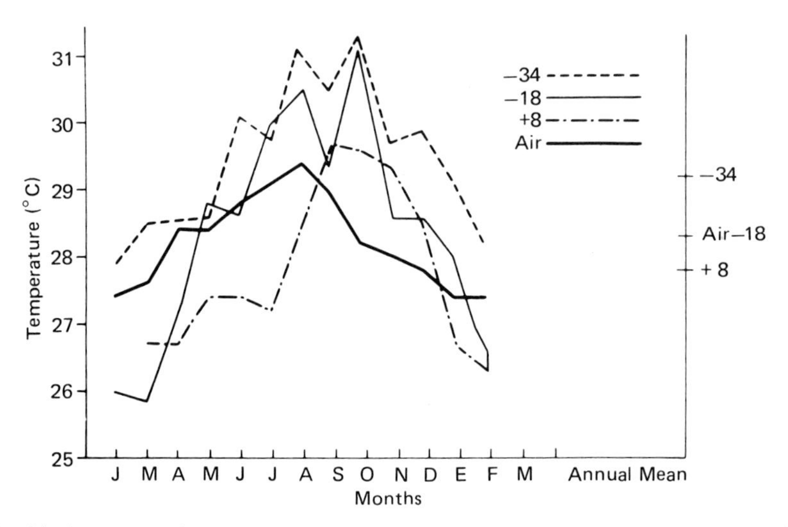


**Fig. 21–1.** Annual fluctuations of the temperature of the water in the Maracaibo estuary, Venezuela, 18 and 34 km seaward from the mouths and 8 km inland, as compared with the air temperature in Maracaibo City.

The fact that land temperatures are relatively unimportant makes them of little use in the definition of "tropical" in relation to estuaries. Here I have employed the criteria of water temperature no less than 20°C in the shallow areas of the adjacent sea and the distribution of mangrove and other tropical organisms to distinguish the limits of tropical estuaries (Figure 21–2).

On the coast of Africa tropical estuaries stretch from the Red Sea down to South Africa, a little south of Durban. On the Asiatic mainland the northern limit lies roughly in the Korean Strait. In the south the limit is on the east and on the west coast of Australia. The tropical region in Australia reaches approximately to Geralton on the west coast and to Cape Sandy near Sydney on the east coast; these two points, with a few exceptions, are the

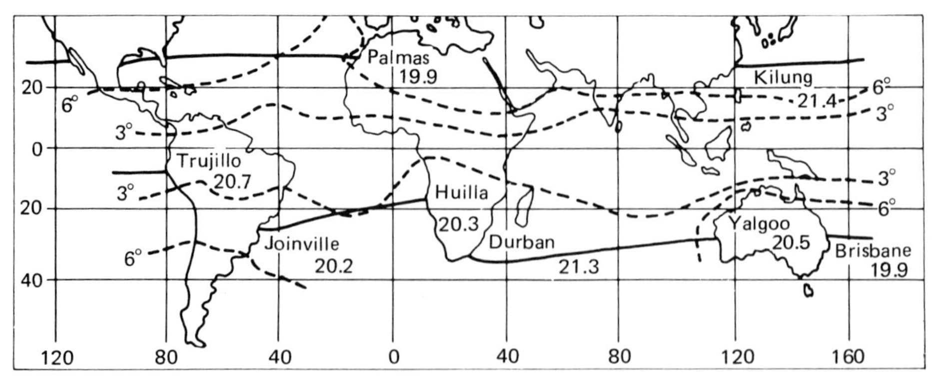


**Fig. 21–2.** Approximate position of the 20°C isotherm on the coast. Broken lines represent mean annual fluctuations of 3° and 6°C.

southernmost limit for many tropical species. New Zealand lies outside the warm water area.

On the America's Pacific the tropical zone includes the California Gulf and the southern tip of lower California. The southern limit lies at a remarkable low degree of latitude: in northern Peru off Point Aguja (6°S), Payta. The shifting of the southern limit of the warm ocean water to the north is due to the cold Humboldt current, whose main part turns west at Point Aguja. This is also the southernmost limit of mangroves.

On the Atlantic side of America there is a marked zoogeographic boundary in northern Florida, possibly near Cape Canaveral, as far as the intertidal marine fauna is concerned (Stephenson and Stephenson 1950). However the northern part of the Gulf of Mexico, from Sanibel Island or Tampa to approximately the US–Mexican border (Hedgpeth, 1953), has a subtropical character. The southern boundary of the zone is more difficult to assess, but presumably it is in the neighborhood of Rio or a little to the south.

The tropical Atlantic region is considerably smaller on the African than on the American side. Here, as in western South America, the reason is that the cool ocean currents are part of the west wind drift, which are forced northward. The limits of mangroves are displaced to the north, somewhere near Benguela. The northern boundary should be located past the mouth of the River Senegal.

The conditions in South Africa are particularly interesting: The east coast of Natal has a tropical biota extending up to the Bashee River south of Durban, while the southwest coast is completely subtropical (Day, 1969).

Obviously these limits are not sharply defined. There is no doubt about areas near the equator, but further north or south there are transitional or mixed zones rather than boundaries in the strict sense. It is relevant here that extreme high temperatures can be attained in tropical estuaries near the equator. Tampi (1969) has recorded 32.5°C in March in coastal lagoons of India, and we have similar values in coastal lagoons of Venezuela (Rodriguez, 1973, unpublished) in August. Tropical species may live at temperatures very near their death temperature. Therefore the effects of heated effluents could be very severe in tropical areas, particularly since few species which could serve as replacement species are adapted to extreme heating. (Naylor, 1965). Thermal pollution in tropical areas would be expected to have severer consequences than in arctic and temperate areas.

## River Flow and Salinity

The salt balance in an estuary is primarily influenced by the fresh water output of the river, which reflects the hydrologic budget of the watershed. In a tropical estuary like the Maracaibo Lake, evaporation is especially important because as much as 25 percent of all the fresh water that enters the watershed evaporates. However, evaporation is almost constant throughout the year. In colder regions evaporation is less pronounced, but seasonal differences are more marked. Consequently, the flow of tropical rivers tends to be governed by the local regime of rainfall, while the flow of rivers from rainy temperate regions, where rainfall occurs fairly regularly throughout the year, tends to be controlled by evaporation with a period of low water in summer when evaporation is most active.

In a tropical area like the northern part of South America (Walter and Medina, 1971) rainfall regimes can be grouped as follows: (1) Very wet regime, without a pronounced dry season; this occurs wherever the Cordilleras tend to create a cul-de-sac that traps the humidity brought by the tradewinds. (2) A bimodal regime with a sharply marked dry season at the beginning of the year and a wet season between April and November. This regime is typical of the Venezuelan llanos, which form a large proportion of the Orinoco watershed. (3) A tetramodal regime, with a dry period at the beginning of the year and another toward the middle of the year, with two wet periods in between. The bimodal pattern is reflected in the regime of the Orinoco River (Figure 21–3). The graph of mean annual flow over 40 years (Alvarez et al., 1964) is very clearly bell-shaped, with a maximum in August. Maracaibo, on the other hand, illustrates the tetramodal regime (Corona, 1964).

The patterns for other tropical rivers are shown in Figure 21–3. The rhythm of the two wet and two less dry periods is reflected in the flow of the Congo. Yet this river is so long and its regime so vast that the pattern is less simple than the one described for the Maracaibo Lake. The basin of the Amazon lies mainly south of the equator. So much water comes from the Brazilian campos, that there is one major flood in May or June.

Patterns of flow similar to the Amazon and Orinoco exist in rivers from monsoonal areas of East Africa and Asia, such as the Zambezi, Hwango Pu, Yangtze, Irrawaddy, Ganges, and Indus.

Output of fresh water is one of the determining parameters of salt penetration in estuaries; consequently, the pattern of salinity fluctuates seasonally with the patterns of river flow. The Atlantic coast of North and South America presents examples of these patterns (Figure 21–4). In temperate areas, as at Beaufort, North Carolina, maximum salinities occur during the summer months. Texas, Florida, and the delta of the Mississippi are approximately at the same latitude (25°–2°N), but while the Mississippi shows a summer maximum typical of more northern latitudes due to the location of its basin, Florida and Texas (Chicken Key and Galveston) reflect the influence of the tropical regime, typical of marginal areas. Salinity fluctuations in Trinidad estuarine areas are very clearly tropical with a maximum during the dry season drought. In southern Brazil the pattern is also tropical, but the maximum occurs during the August-November dry season of the southern hemisphere.

## Types of Tropical Estuarine Areas

The magnitude of the land drainage determines the position of the brackish water in relation to the sea boundary. In the tropical zone three situations may develop.

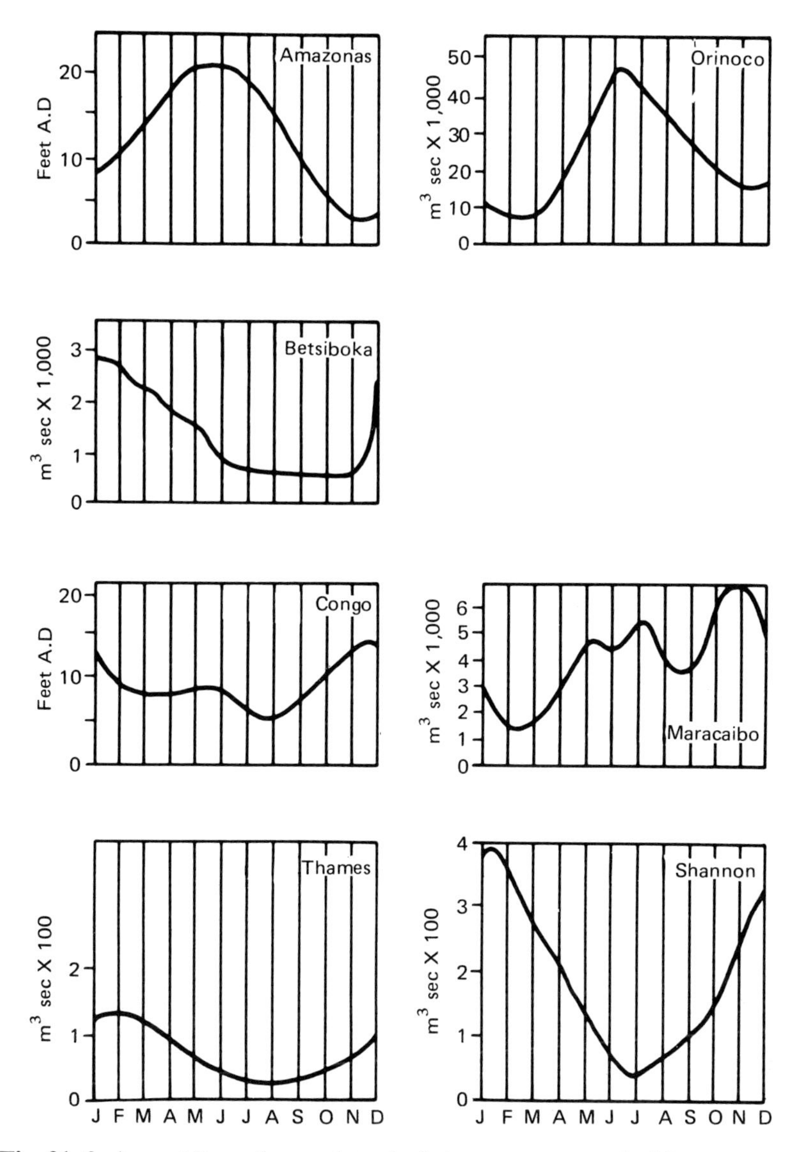


**Fig. 21–3.** Annual flow of several tropical rivers as compared with temperate rivers. The Amazon and Orinoco show a northern hemisphere bimodal regime. The Congo and Maracaibo Lake are also northern hemisphere but with a tetramodel regime. The Betsiboka in Madagascar has a southern hemisphere bimodal regime. The Thames and Shannon, in the British Isles, have a typical bimodal regime with a dry period in summer. (Data from Alvarez et al., 1964; Beckinsale, 1956; Berthois and Crosnie, 1966; Corona, 1964.)

First, in a large watershed, with high volume of discharge even during the dry season, the estuarine zone extends considerably beyond the mouth of the river. All of the ten largest tropical rivers have an extensive estuarine zone outside their mouths. These rivers, with their respective lengths in kilometers and place of origin, are: Amazon, 6,275, Brazil (Atlantic); Congo, 4,600,

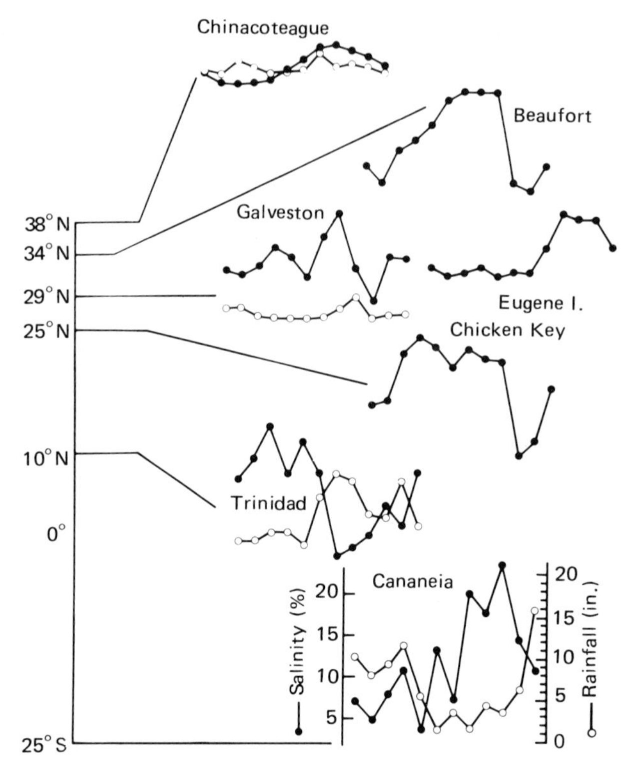


**Fig. 21–4.** Basic patterns of salinity fluctuations in estuaries along the Atlantic coast of America (*dark circles*) and corresponding values of rainfall (*open circles*) over 12-month periods.

Angola (Atlantic); Niger, 4,160, Nigeria (Gulf Guinea); São Francisco, 3,160, Brazil (Atlantic); Orinoco, 2,900, Venezuela (Atlantic); Brahmaputra, 2,700, Bangladesh (Bay of Bengal); Zambezi, 2,570, Mozambique (Indian Ocean); Ganges, 2,480, India (Bay of Bengal); Irrawaddy, 2,000, Burma (Bay of Bengal); Senegal, 1,689, Senegal (Atlantic).

For example, low salinities are found as far offshore as 30 to 40 nautical miles outside the Orinoco delta (Figure 21–5) during the rainy season, including the Gulf of Paria and completely surrounding Trinidad. During the dry season the low salinity water recedes to a narrow strip along the coast and to the western part of the Gulf of Paria (Gade, 1961).

A similar situation is found in the Amazon River, although the area of influence is larger due to the larger flow. Texeira and Tundisi (1967) found surface values of 11.98 o/oo at 180 km from the mouth in April, during the rainy season.

Outside both estuaries the column of water is stratified. The stratification is very steep, from 15 to 36 o/oo in 20 meters in the Orinoco, 180 km from the mouth and from 12 to 34 o/oo in 1.0 meters in the Amazon, 16 km from the mouth. As a result of this stratification the neritic plankton will be subjected,

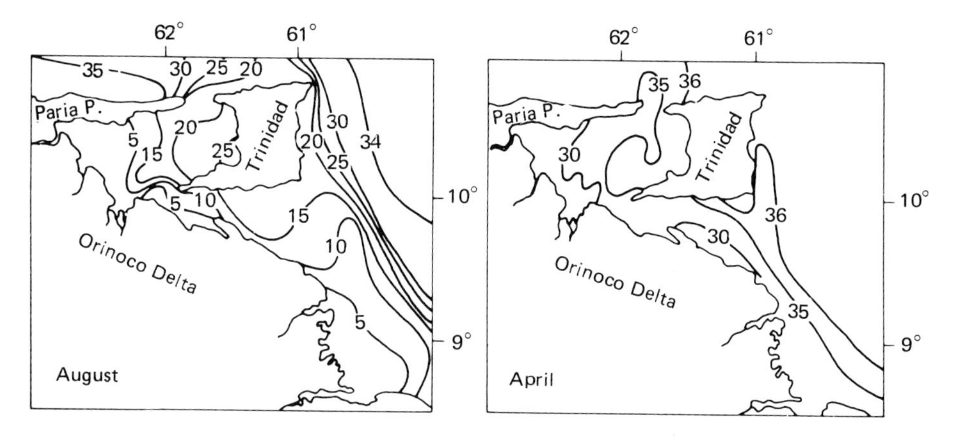


**Fig. 21–5.** Surface salinities in front of the Orinoco delta during the wet (August) and dry (April) seasons. (Data from Gade, 1961).

at least occasionally, to estuarine conditions, but benthic communities and demersal fish will live under salinity conditions very similar to those of more oceanic waters.

Ocean currents move these superficial brackish waters and extend their influence. The Orinoco water moves northward surrounding Trinidad (Gade, 1961); the Amazon moves northwest along the Guiana coast unloading its sediments on the coast of French Guiana (Sioli, 1964). The brackish water in the Gulf of Guinea originates in the Bight of Biafra and on the coast of Liberia and extends during the southern hot season along the coasts of Dahomey, Ghana, and the Ivory Coast (Berrit 1966).

In the eastern tropical Pacific, a tongue of relatively low salinity surface water extends in a very thin layer more than a thousand miles westward from a stretch of coast off Central and South America (Figure 21–6). There are no major rivers in the region and the fresh water originates principally from land drainage from the coast from El Salvador to Ecuador and from precipitation.

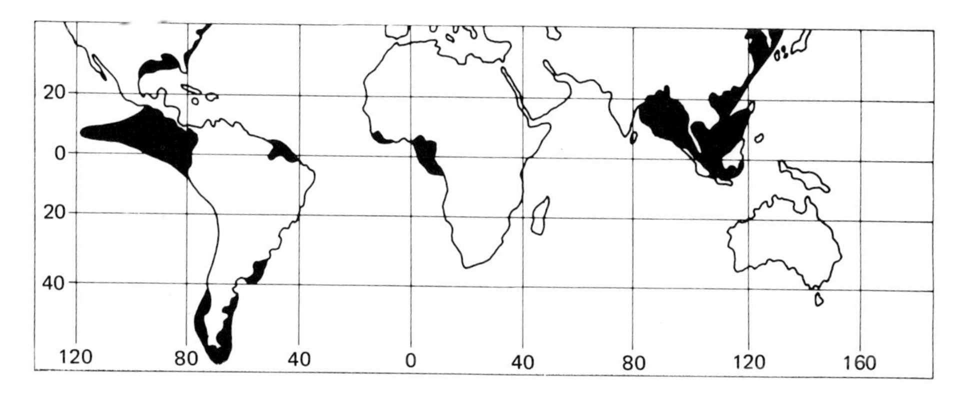


**Fig. 21–6.** Brackish oceanic areas. (Data from McHugh, 1967.)

The discharge of rivers in this area closely follows the seasonal cycle of rainfall. Total runoff is estimated by McHugh (1967) as 300,000 $ft^3$/sec, about half the flow of the Mississippi River.

A second tropical estuarine situation arises in moderate watershed with an appreciable runoff even during dry season. In this case, the estuarine zone extends mainly behind the mouth (circumscribed estuaries). This is the classical estuary, as defined by Pritchard (1967), and is semienclosed in free communication with the open sea where seawater is diluted with the fresh water from land runoff. The estuarine circulation pattern results from vertical and horizonal distribution of densities, and gradients of currents, amplitudes, and other tidal effects.

In this situation the effect of the river is felt outside the mouth, but the magnitude is small. An example is the effect of Sambirano River on the Bay of Ampesidana in Madagascar. According to Magnier and Piton (1972), this bay has a very marked estuarine circulation during the rainy season when the column of water becomes stratified. This influence is weakened during the dry season, and vertical homogenization takes places. Near the bottom the salinity is very stable throughout the year, between 34 and 35 o/oo. Similar results have been obtained in the Gulf of Venezuela into which the Maracaibo estuary discharges (Redfield, 1955; Rodriguez, 1973).

The final case is the small watershed with a pronounced difference between dry and wet seasons. Two situations may develop: First, in a small river on mountain coasts substantial amounts of rainfall may be trapped in small watersheds and flow maintained during the rainy season. However, during the dry season, rainfall and river flow cease. No matter how interesting these ephemeral estuaries are, they have received very little attention.

Another example are blind estuaries (Day, 1951) of smaller rivers that are cut off from the sea for several months during the dry season by sandbars. Rapid evaporation boosts the salinity of these waters to high values. The rivers usually empty into a littoral lagoon and seldom occur outside the tropical and subtropical areas but are very common in the Gulf of Mexico, on the northern South America coast, in Brazil and Africa, and along the Indian Ocean. Childa and Pulicat Lake on the east coast of India, and the chain of lagoons along the coast of Kerala on the west coast, have received considerable attention because of their production of fish (Pillay, 1967b). Lagoons on the coast of Ceylon have a total estimated area of about 350,000 acres. All these Indian lagoons have a typical monsoonal regime with high salinity conditions during the dry season.

In West Africa there is a fairly large number of lagoons from the coast of Liberia to the Niger delta (Pillay, 1967a). Most of the lagoons on the Ivory Coast and western Nigeria are almost fresh and in none of the West African lagoons does there seem to be hypersalinity. In Lagos Lagoon the salinity during the dry season (November-March) is high and during the rainy season (April-October) it falls and almost fresh water conditions prevail, producing mass mortality of many seawater organisms in Lagos Harbor (Hill and Webb, 1958). Similar mass mortalities are produced in Jamaican mangrove lagoons when salinities are less than 20 o/oo after heavy land runoff due to rain in May and October.

## Tidal Action

River flow and tidal currents play a major role in creating the environmental conditions of the estuary. While the net flow is seaward, the direction of flow is reversed once or twice daily according to the tidal regime. Furthermore, when the tidal wave enters a narrow channel like a river mouth it is considerably distorted, although its fundamental characteristics are retained.

In many estuaries the action of tides can be understood according to Redfield (1961). In this theory the tide is considered to be a standing wave resulting from the combined effects of a primary wave entering from the sea and a secondary wave resulting from the reflection of the preceding entering wave at the barrier formed by the head of the embayment. These waves will be in phase to a different degree at different positions along the channel. The theory has been fitted to data from Long Island Sound, the Bay of Fundy, the Juan de Fuca straits, and the Maracaibo estuary.

Tidal reversal is clearly reflected in the regime of salinity, although the absolute values for salinity at flood and ebb will change for each individual estuary depending on the magnitude of the flow and shape of the channel. Under a regime of mixed tides as is characteristic of the Maracaibo estuary, only the highest raising tide is felt, giving a daily pattern of salinity fluctuations in a semidaily tidal regime. This rhythmicity is reflected also in blind estuaries. The tidal regime outside the littoral lagoons in Venezuela is semidaily, but the regime of discharge is daily due to the critical position of the bar that blocks the smaller flood tide of the day.

Progress of the tidal wave can be very complex. Although a complete tidal cycle is the same as that of the sea, the duration of the ebb is greater than the flood. Thus the time of low water at the head of the estuary is later than it is at the mouth. In the enormously long estuary of the Amazon, where the tidal wave has only penetrated to the lower reaches before the next one arrives at the mouth, there are said to be eight tides along the length of the estuary (Day, 1951).

## Planktonic Populations

### Plankton Penetration

Usually there is a surface salinity front within many estuaries near the upper limit of the estuarine zone, related to the magnitude of river runoff and to the amplitude of the tide. In a large estuarine system such as the Chesapeake Bay this front usually occurs at salinities well below 18 o/oo. This critical zone is rich in adults and young of many fish species and in plankton populations.

In the Maracaibo estuary this zone is located near the mouth. Monthly sampling during cycles of 25 hrs have shown that during the hours immediately following the flood tide the number of organisms from the sea increases in the samples. An example is given in Figure 21–7. Note that although the tidal regime is semidaily, the penetration has a predominant daily regime. This phenomenon, associated with the daily regime of salinities and currents, has received little attention.

A summary of the biomass values throughout the Maracaibo system is shown in Figure 21–8 as annual means at each locality. The highest values are attained 5 km from the mouth (40 mg/m$^3$) and toward the west when the flow from the estuary is deflected. These values are higher than those of the open sea (31 mg/m$^3$). There are high values in the mixing zone (17 mg/m$^3$) but they decrease as one progresses into the estuary and become very small (7 and 10) in salinities of 5 o/oo.

## Plankton Mixing and the Stability of Populations

During the ebb tide the flow may not be unidirectional but form a complicated pattern of circulation. A fraction of the neritic plankton that penetrated with the flood tide is left behind and mixes with the outgoing river water. We have estimated the number of organisms that penetrates ($N$) by means of a

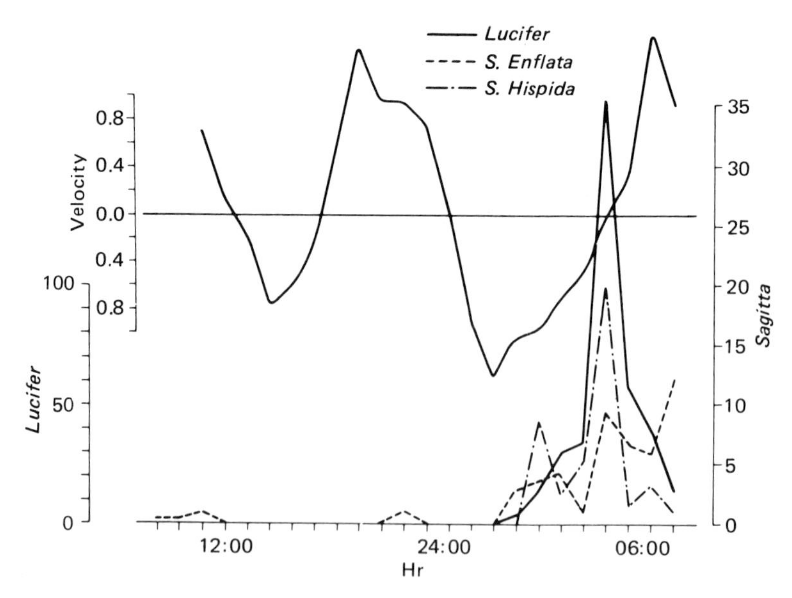


**Fig. 21–7.** Fluctuations in the amount of neritic organisms in the Maracaibo estuary, Venezuela, during a tidal cycle.

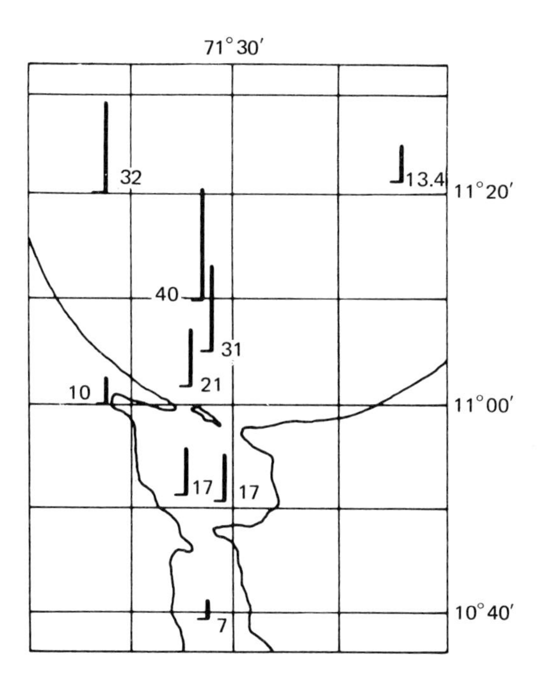


**Fig. 21–8.** Biomass values (mg/m³) through the Maracaibo estuary.

simple mathematical model (Rodriguez, 1969; Lopez, 1968) using the chlorinity (Cl) as a parameter:

$$N = N_o \, e^{-bCl}$$

Where $N_o$ is the concentration of organisms at the origin. But since the chlorinity is a function of the distance ($X$) (Harleman et al., 1967)

$$\mathrm{Cl} = -bx - a$$

we can use the adjusted expression

$$N = N_o e^{-0.16x}$$

The plankton in the mixing zone will be a mixture of the fraction of neritic plankton plus the organisms from the oligohaline waters.

In the oligohaline water of the Amazon, Maracaibo Lake, and several other tropical estuaries, the zooplankton is predominantly composed of rotifers, cladocerans, and a few species of harpacticoid copepods. The copepod *Acartia tonsa* is frequently the most abundant and characteristic component of the oligohaline waters and the zone of mixing in temperate as well as tropical estuaries. Jeffries (1962) found that the number of this species in Yaquina Bay decreased regularly in a seaward direction and were present in small numbers at the mouth only during low water. In Maracaibo estuary it is never present in the mouth but is very abundant through a mean salinity range of 2 to 10 o/oo.

The above-mentioned organisms have to maintain their populations in the seaward drift of the diluted waters. If the circulation were the only process influencing the size of the population throughout the estuary, then distribution would be proportional to that of the fresh water. However, *Acartia tonsa* in Maracaibo (Figure 21–9) populations may be fairly constant throughout the estuaries. Apparently this is achieved by adjustment of the reproduction rate (Ketchum, 1954). Margalef (1967) maintains vertical migration and transport at different stages of the tides are also used. Tundisi and Tundisi (1968) found that copepods migrate toward the bottom during high tide as the light intensity increases and are carried toward the upper reaches of the estuary with the countercurrent.

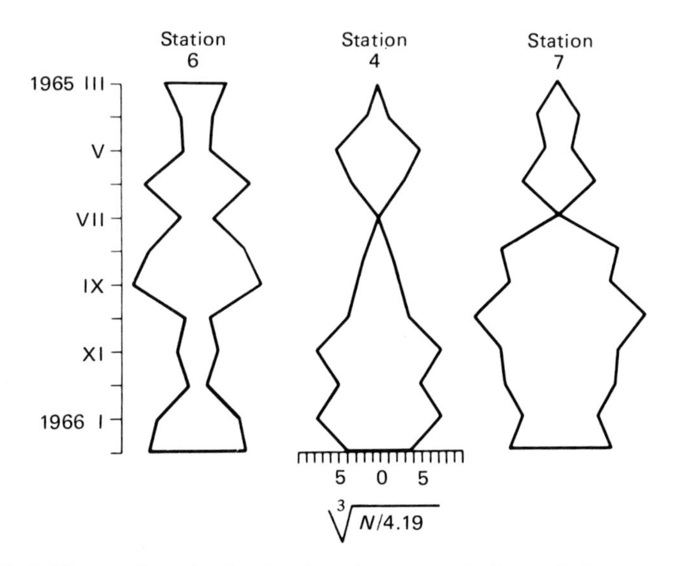


**Fig. 21–9.** Fluctuations in the density of the populations of *Acartia tonsa.*

Recolonization is another mechanism connected with the tidal currents and river flow. A substantial part of the fisheries in tropical countries is dependent on migratory movements of populations between the sea and estuarine areas. This is particularly true of crustacean populations, as demonstrated by Weymouth et al. (1933) for the white shrimp (*Penaeus setiferus*) in Texan estuaries. Similar cycles are being found for other penaerids, for instance *Penaeus schmitti* in Venezuela (Ewald, 1965) and *Penaeus duorarum* in Dahomey, West Africa (Hoestlandtl, 1966). In the latter the postlarval form enters the littoral lagoons from February to March when salinity attains maximum values (30 o/oo), and the largest adults migrate toward the sea when salinities goes down to 1 or 2 o/oo. Blue crabs show a similar cycle but only the females migrate toward the sea for spawning.

In contrast, benthal communities of mollusks and crabs have to be restocked each year with larvae immigrating from the sea or from the lower reaches of the estuary where they have been carried by the flow. Bousfield (1955) has shown that the distribution of cirriped larvae in the Miramiche estuary (USA) is related to the movement of fresh water in the surface and the countercurrent associated with the penetration of the seawater. However, since cirriped larvae are produced mainly in the mixing zone of the estuary, and the distribution is relatively small, these results can hardly be extrapolated to organisms with wide migratory movements such as shrimps or blue crabs. It is possible that some behavioral mechanisms are responsible for these emigrations. In this connection the endogenous rhythmicity is worth exploring.

The mycthemeral rhythm of several adult penaeid prawns has been described. Under experimental conditions *Penaeus duorarum* is active only during the night time, between 19 hr and 23 hr 15′ (Fuss, 1964). This activity is influenced by full moon and temperature (Fuss and Ogren, 1968). Along the Nigerian coast the activity of *P. duorarum* is essentially bimodal, with a maximum 2 hrs after sunset and another 2 hrs before sunrise (Raitt and Niven, 1966).

Rhythms of activity have also been studied in the postlarvae. Tabb et al. (1962) and Baxter (1964), using a plankton net in a channel subjected to tides, have observed that the capture of postlarval *Penaeus* increases during the night and with the rising tide. Hughes (1969) concluded from laboratory experiments that this phenomenon is linked to salinity variations. The postlarvae of *P. duorarum* are scattered through the column of water when salinity is high (high water) and swim to the bottom when the salinity decreases (low water).

## Zonation of the Estuarine Plankton

From what we know of the Maracaibo and other tropical estuaries regarding the penetration and mixing of zooplankton, as well as the fluctuations of biomass along the estuary, four main planktonic zones can be distinguished (Figure 21–10). These zones correspond to the ecological classification of the main species of estuarine holoplankton given by Jeffries (1962).

1. *Oligohaline (limnetic) zone.* Surface salinity rarely is more than 5 o/oo during the dry season. Tidal currents are still felt, but the river flow has the most effect. Low biomass is present (7 to 10 mg/m$^3$ in the Maracaibo estuary). Typical zooplankton comprises Rotifera, Cladocera, a few harpacticoid copepods, and *Acartia tonsa.*

2. *Mixing zone.* Surface salinity is 5 to 20 o/oo during the dry season but rarely is less than 5 o/oo during the rainy season. Tidal currents determine the in-migratory movements of plankters. Modest biomass occurs (17 mg/m$^3$ in the Maracaibo estuary). Typical zooplankters are *Labidocera*, *Acartia tonsa*, and cirriped nauplii.

3. *Critical zone.* Surface salinity is 20 to 10 o/oo during the dry season and rarely is less than 10 o/oo during the rainy season. Hydraulics are determined by the position and topography of the mouth and the periodical penetration of tidal wave. Biomass is high, with maximum values toward the external

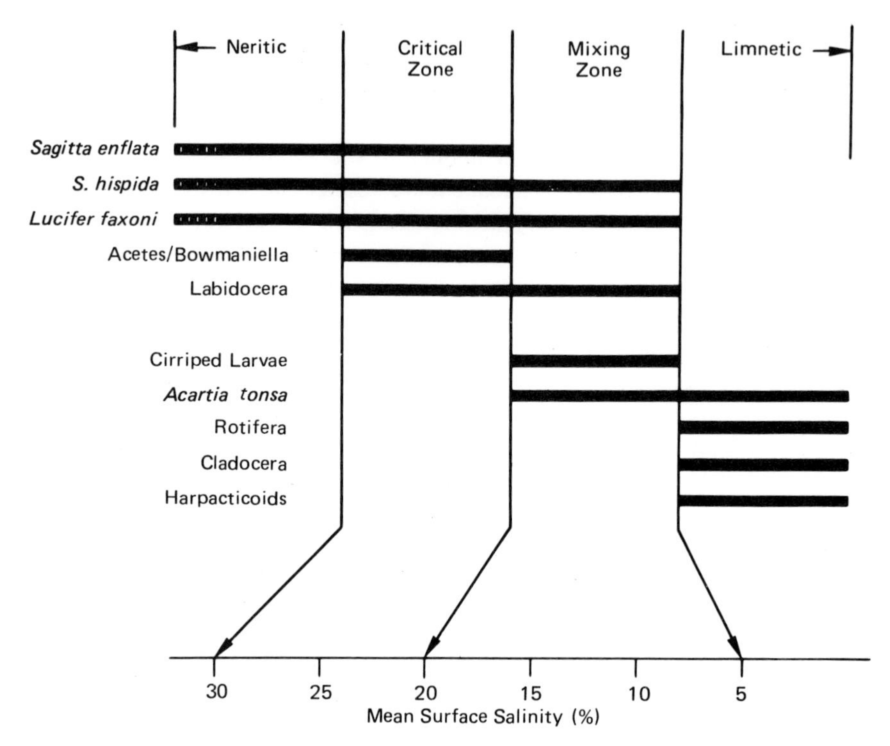


**Fig. 21–10.** Zonation of zooplankters in the gradient of salinities of the Maracaibo estuary, Venezuela.

area of this zone (21 to 31 mg/m$^3$ in Maracaibo estuary). Typical species are neritic plankters periodically introduced by the tide: *Lucifer, Sagitta, Acartia lilljeborghi* or its equivalent. The mysidacean *Bowmaniella* and the decapod crustacean *Acetes* seem to be permanent residents of this zone.

4. *Neritic zone.* Surface salinity is higher than 30 o/oo. The predominant factor is the open sea currents. Between this zone and the critical zone there is a front of high biomass in Maracaibo estuary (40 mg/m$^3$). Species are typically marine neritic with a predominance of *Lucifer.*

## Sea-Estuary Interrelations

The photosynthetic production of organic matter by the phytoplankton in estuaries is made possible by the assimilation of inorganic nutrients from the surrounding water. Most of these substances are present in concentrations greatly in excess of the plants' needs, but some such as nitrogen and phosphorus occur at no more than micromole levels and may be utilized almost to the point of exhaustion by the algae. It is, in fact, the availability of these nutrients that most frequently controls and limits the rate of organic production in natural waters.

These nutrients come from three sources (Ketchum, 1967):

1. Nutrients leached from the land and carried by the river.

2. Nutrients brought in by seawater entering as a countercurrent in estuaries with a two-layered flow. This seawater is drawn from depths below the euphotic zone in the ocean where the concentration of nutrients has not been depleted by the growth of phytoplankton. In some cases organisms grown in the surface layers of the estuary may sink to countercurrent depths where decomposition releases the nutrients that will then be returned again for reuse within the estuary. Nutrients can thus become trapped within the estuary and build up unusually high concentrations.

3. Local pollution within the estuary, which may be very diverse according to each particular situation.

Phosphate is a convenient index of organic pollution. Its analysis is far easier than analysis of other chemical nutrients, and it persists when other products of organic decomposition such as nitrogenous compounds have disappeared from solution. It is easy to conclude that phosphate is the causative agent or algal growth, euthophication, and other adverse effects associated with organic pollution; however, some authors, particularly Ryther and Dustan (1971) do not support this conclusion. Phosphate is present in normal mud as interstitial phosphate, which could be released by stirring the mud (it is physically absorbed on the mud particles, as in soils and bound in some insoluble compound) (Rochford, 1950). There is considerable evidence of the exchange of phosphate between the sediments and the overlying water of shallow estuaries. This process, according to the experiences of Pomeroy et al. (1965) in a Georgia estuary, consist of a two-step ion exchange between clay, mineral, and water, plus an exchange between intertidal microorganisms and water. In this situation the exchange tends to maintain a concentration of phosphate in the water of 1 $\mu$mole of phosphate/liter. The process has been confirmed in the Vellar estuary in India (Balasubrahmanyan, 1961) and in the Heeia mangrove swamp in Hawaii (Walsh, 1967).

Ryther and Dustan (1971) show that nitrogen is the critical limiting factor to algal growth and eutrophication in estuaries of Long Island. This is a matter worth investigating in tropical estuaries in view of the increasing nutrient enrichment both from agriculture and detergent-containing sewage. These authors point out that removal of phosphate from detergents is not likely to slow the eutrophication of coastal marine waters, and its replacement with nitrogen-containing nitrilo–triacetic acid may worsen the situation.

Tropical rivers usually drain heavily forested basins. For this reason organic detritus is considerably more important than in temperate rivers. In some parts of the Maracaibo estuary 75 percent of the total dry weight of the plankton tows is made up of vegetable matter. Also in many tropical estuaries the water is colored orange to brown. As Prakash and Hodgson (1966) have demonstrated, humic substances added in small quantities to cultures of dinoflagellates have a marked stimulatory effect on its growth, and thus may influence primary production of phytoplankton, especially during the rainy season.

There are very few measurements of primary production taken with the $^{14}C$

method or oxygen method in tropical estuarine waters. Most of them come from lagoons (see Table 21–1). Although we speculate that production can be high due to a accelerated rate of turnover, the data are not adequate for a firm conclusion.

### The Effect of Estuaries on the Oceanic Ecosystem

As we have already suggested, tropical rivers might have a considerable influence on the adjacent ocean, and it is of significance to determine if water runoff may affect the organic productivity of the ocean. The discharge of the Amazon accounts for approximately 18 percent of the total river discharge in the world, and the discharge of the second largest tropical river, the Congo, is more than twice that of the Mississippi. However, according to Ryther et al. (1967), the concentration of nitrate, phosphate, and phytoplankton organisms is lower, while the levels of silicate are appreciably higher in the surface water influenced by the Amazon River compared with the surrounding seawater. The direct overall effect of the river, therefore, is to decrease the fertility of the ocean into which it flows.

This is not always true, however. For example, a region of high biological activity exists off the coast of the Guianas, resulting from geostrophic uptilting of water, bringing nutrient-rich subsurface water into the euphotic layer. Ryther et al. (1967) suggest that this could be caused or enhanced by the weight of the accumulated river water offshore. To this nutrient-rich water, and not to the Amazon water, is attributed the abundance of diatoms found by Hulburt and Corwin (1969) in the coastal water of the Guianas. Wood (1966) found considerably greater quantities of phytoplankton in the shallow coastal waters than in the deep oceanic water seaward of the continental shelf, and Texeira and Tundisi (1967) described a change from abundant diatomaceous plankton near shore to less abundant coccolithophorid plankton offshore from the Amazon River. This mass of water seems to be responsible also for the three-fold increase in zooplankton biomass observed in this area by Calef and Grice (1967) during the wet season.

The paradoxical effect of such a large river impoverishing the oceanic water is explained by Ryther et al, (1967) in the following manner: In regions where

**Table 21–1.** Estimates of primary production of tropical estuaries by oxygen on $^{14}C$ methods.

| Locality | Author | Daily Production |
|---|---|---|
| Laguna de Unare, Venezuela | Gessner and Hammer (1962) | 2.6–2.9 ($^{14}C$) g $C/m^3$ |
| Cananeia, Brazil | Tundisi (1970) | 1.13–9.71 ($O_2$) g $C/m^3$ |
| Mandapan, India | Tampi (1969) | 0.08–0.12 ($O_2$) g $C/m^3$ |
| Tropical ocean water | Sorokin and Kliashtorin (1963) | 0.1–0.2 g $C/m^3$ |

nutrients are plentiful, solar energy is limiting and vertical mixing is adequate to carry plants below their critical depth, the increased stability provided by fresh water drainage might be expected to enhance the growth of phytoplankton. In the tropics, on the other hand, the supply of nutrients rather than light is the common limiting factor to primary production; thermal stratification is normally well developed, and further stability of the surface layers by fresh water outflow suppresses the resupply of nutrients from below and thereby inhibits phytoplankton growth.

## The Benthic Biocenose

Benthic communities resemble each other in estuaries of different parts of the world. If we compare the composition of the infauna of the Vellar estuary in India (Balasubrahmanyan, 1961) with that Maracaibo estuary (Rodriguez, 1973) the main elements are:

1. *Phoronis architecta*, which is common in both estuaries.
2. Several especies of errant polychaetes, with three genera in common in both estuaries.
3. Three species of *Pelecypods* in each estuary.

The abundance of organisms is also very similar.

In addition there are differences that stress the relative isolation of estuaries. For example, several species of polychaetes and two species of pelecypods are endemic in the Maracaibo estuary. Among these endemic pelecypods is *Polymesoda*, a genus that is distributed along the Atlantic coast of the Americas but segregated in several species with small areas of distribution. Similar endemism occurs among the fishes: several species in the families Sciaenidae, Engraulidae, and Potamotrygonidae are restricted to the Maracaibo estuary and do not occur in the nearby Magdalena estuary, which is only 400 km away.

Biomass values seems to be modest. In the Maracaibo estuary they rarely are higher than 0.1 g/m$^2$.

## The Mangrove Community

The physiology of mangrove has been the object of many studies since the time of Schimper (1891). At present, this is possibly the best known estuarine community and is outside the scope of the present report. It is, however, of some interest of examine briefly the role of mangrove in the economy of the tropical estuaries.

Mangrove forest is established on most tropical coasts. The general structure and the process of succession have been described by several authors (see, for instance, Macnae, 1967). In the marginal zone of the tropics, however, the structure of the mangrove undergoes some changes. In temperate Australia (Fisher, 1940) it is reduced to two species at the latitude of New South Wales, *Avicennia officinalis* L. and *Egiseras majus* Gaert., other tropical forms, such as *Rhizophora*, *Bruguiera*, *Cerops*, and *Acanthus*, being excluded. In South Africa (Day, 1969) *Avicennia marina* reaches to the estuary of Kabomgaba, but *Rhizophora mucronata* and *Bruguiera gymnoryza* reach only to the Ungozaba and Kei rivers, respectively.

Mangroves influence the general economy of the estuary in the following aspects:

1. The most obvious influence of the mangrove is its contribution of organic matter. Golley et al. (1962) estimated in a Puerto Rican mangrove that the tide carried 1.14 g C/m$^2$ day from the mangrove. There seems to be very little work done on the degradatory processes that this particular matter undergoes.

2. There is evidence that the presence of mangrove influences the seasonal cycle of the phytoplankton (Tundisi and Tundisi, 1972). This could be due to the introduction of growth or inhibitor factors (Smayda, 1970) other than the normal nutrients from river discharge and runoff.

The combined effect of these factors and the nutrient input by land drainage, mainly during the summer, could affect the timing of the phytoplankton growth and the selection of the nanno- and microplankton (Tundisi, 1970).

3. A very important role of mangrove is the multiplication of niches created within its structure. Mangroves are not peculiar to estuarine situations, but may be developed in shallow oceanic lagoons. However, there is a fundamental difference in the associated fauna of estuarine mangrove and those from sea coasts. Mangroves of mesohaline and oligohaline waters commonly support fewer species on the roots than those of euryhaline waters and the fauna of the mud is also relatively impoverished (Rodriguez, 1963; Walsh, 1967).

An example of the multiplication of niches created by the mangrove could be found in the discrimination of species of fiddler crab (*Uca*). In each mangrove situation three to four species of this genus occur, depending on the type of sediment and trees. The fiddler crabs ingest the sediments. The species on clean sand have special spoon-tipped hairs that are used to scrape the periphyton from the sand grains. Those from muddy substrata lack the spoon tips, and those from sediment with intermediate grain size have ill-developed tips. This correlation shows that there is a real discrimination of niches.

## Conclusions

We conclude that tropical estuaries are not only a geographic entity but could be characterized as well by their peculiar hydrographic regime, dominated by the seasonality of river flow with a concomitant salinity regime and uniform high temperatures. A substantial part of the estuary in many large tropical rivers is located outside the mouth.

Several biotic characters, such as the effect of the input of organic matter derived from the bordering vegetation, migrations, and rhythmicities, seem to be peculiar to or at least more relevant than in temperate areas.

The idea of a sizable contribution of large tropical estuaries to the fertility of the sea is worth reexamining. Even so, since tropical oceanic waters in general are low producing it is worth increasing the effort to utilize the relatively richer estuarine areas for fishing development.

## References


Alvarez, R. J., A. Klanke, and J. M. Volcan. 1964. Aforos del Rio Orinoco. *Rev. Fac. Ing.* 8:1–177.

Balasubrahmanyan, K. 1961. Studies in the ecology of the Vellar estuary. 1. A preliminary survey of the estuarine bottom. *J. Zool. Soc. Ind.* 12:209–215.

Baxter, K. N. 1964. Abundance of postlarval and juvenile shrimp. *Circular U. S. Bur. Comm. Fish.* 183:23–29.

Beckinsale, R. P. 1956. *Land, Air and Ocean.* London: Gerald Duckworth and Co.

Berrit, G. R. 1966. Les eaux desalées du Golfe de Guinée. *Document Centre Rech. Ocean. d'Abidjan* 9:1–15.

Berthois, L., and A. Crosnier. 1966. Etude dynamique de la sedimentation au large de l'estuaire de la Betsiboka. *Cahiers Orstom Ser. Oceanogr.* 4:49–130.

Bousfield, E. L. 1955. Ecological control of the occurrence of barnacles in the Miramichi estuary. *Bull. Natl. Mus. Can.* 137:1–69.

Calef, G. W., and G. D. Grice. 1967. Influence of the Amazon River outflow on the ecology of the western tropical Atlantic. II. Zooplankton abundance, copepod distribution with remarks on the fauna of low salinity areas. *J. Mar. Res.* 25: 84–74.

Corona, L. 1964. *Balance Hidrológico del Lago de Maracaibo.* Mimeo. Rept. 2 vol. Caracas: Instituto Nacional de Canalizaciones.

Day, J. H. 1951. The ecology of South African estuaries. I. A review of estuarine conditions in general. *Trans. Roy. Soc. S. Afr.* 33:53–91.

———. 1969. *A Guide to Marine Life on South African Shores.* Cape Town: A. A. Balkema.

Ewald, J. J. 1965. *La biología y pesquería del camarón en el occidente de Venezuela.* Caracas: IVIC.

Fisher, P. H. 1940. Notes our le peuplements littoraux d'Australie. III. Sur la faune de la mangrove australienne. *Mem. Soc. Biogeogr.* 8:315–329.

Francis-Boeuf, C. 1943. Physico-chimie du milieu fluvio-marin. *Compt. Rend. Sommaire Seances Soc. Biogeogr.* 169–170:19–26.

Fuss, C. M. 1964. Observations on burrowing behaviour of the pink shrimp, *Penaeus duorarum* Burkenroad. *Bull. Mar. Sci.* 14:62–73.

———, and L. M. Ogren. 1968. Factors affecting activity and burrowing habits of the pink shrimp, *Penaeus duorarum* Burkenroad. *Biol. Bull.* 130:170–191.

Gade, H. 1961. On some oceanographic observations in the south eastern Caribbean sea and the adjacent Atlantic ocean with special reference to the influence of the Orinoco River. *Bol. Inst. Oceanogr.* 1:287–342.

Gessner, F., and L. Hammer. 1962. La producción primaria en la Laguna Unare, XII Convención Anual, Associación Venezolana Av. Cienc.

Golley, F., H. T. Odum, and R. F. Wilson. 1962. On structure and metabolism of a Puerto Rican red mangrove forest in May. *Ecology* 43:9–19.

Harleman, D. R., L. Corona, and E. Partheniades. 1967. Analisis de la distribución de la salinidad en el canal de Maracaibo. Caracas: Instituto Nacional de Canalizaciones. *Publi. Técn. DI-1.*

Hedgpeth, J. 1953. An introduction to zoogeography of the Northwestern Gulf of Mexico with reference to the invertebrate fauna. *Pub. Inst. Mar. Sci.* 3: 107–224.

Hill, M. B., and J. E. Webb. 1958. The ecology of Lagos Lagoon II. Topography and physical features of Lagos Harbour and Lagos Lagoon. *Phil. Trans. Roy. Soc. B.* 241:319–333.

Hoestlandtl, H. 1966. Premieres recherches sur le cycle biologique de *Penaeus duorarum* en Afrique Occidental (Dahomey) *Mem. Inst. Fond. Afr. Noire* 77:478–497.

Hughes, D. A. 1969. Responses to salinity change as a tidal transport mechanism of pink shrimp. *Penaeus duorarum. Biol. Bull.* 136:43–53.

Hulburt, E. M., and N. Corwin. 1969. Influence of the Amazon river out flow on the ecology of the western tropical Atlantic. III. The planktonic flora between the Amazon river and the windward islands. *J. Mar. Res.* 27:55–72.

Jeffries, H. P. 1962. Succession of two *Acartia* species in estuaries. *Limnol. Oceanogr.* 7:354–364.

Ketchum, B. H. 1954. Relation between circulation and planktonic populations in estuaries. *Ecology* 35:191–200.

———. 1967. Phytoplankton nutrients in estuaries. In *Estuaries*, G. F. Lauff, ed. AAAS Publ. 83, pp. 329–335.

Lopez, H. 1968. La penetración de zooplankton marino al estuario de Maracaibo durante la estación seca. Caracas: Instituto Nacional de Canalizaciones. *Publ. Téc. DI-4.*

Macnae, W. 1967. Zonation within mangroves associated with estuaries in North Queensland. In *Estuaries: A Perspective*, G. F. Lauff, ed. AAAS Publ. 83, pp. 432–441.

Magnier, Y., and B. Piton. 1972. La circulación en Baie d'Ampasindava (Madagascar) et ses implications biochimiques. *Cahiers ORSTOM Ser. Oceanogr.* 10:75–97.

Margaleff, R. 1967. The food web in the pelagic environment. *Helgol. Wiss. Meersumters* 15:548–559.

McHugh, J. L. 1967. Estuarine nekton. In *Estuaries*, G. F. Lauff, ed. AAAS Publ. 83, pp. 581–620.

Naylor, E. 1965. Effect of heated elements upon marine and estuarine organisms. *Advan. Mar. Biol.* 3:63–103.

Pillay, T. N. R. 1967a. Estuarine fisheries of West Africa. In *Estuaries*, G. F. Lauff, ed. AAAS Publ. 83, pp. 639–646.

———. 1967b. Estuarine fisheries of the Indian Ocean coastal zone. In *Estuaries*, G. F. Lauff, ed. AAAS Publ. 83, pp. 647–657.

Pomeroy, L. R., E. E. Smith, and C. M. Grant. 1965. The exchange of phosphate between estuarine water and sediments. *Limnol. Oceanogr.* 10:167–172.

Prakash, A., and M. Hodgson. 1966. Physiological ecology of marine dinoflagelates. *Bedford Inst. Oceanogr. Fifth Ann. Rept.* 93–94.

Pritchard, D. W. 1967. What is an estuary: Physical viewpoint. In *Estuaries*, G. F. Lauff, ed. AAAS Publ. 83, pp. 3–5.

Raitt, D. F. S., and D. R. Niven. 1966. Exploratory prawn trawling in the waters off the Niger Delta. *FAO Fish. Rept.* 51:59–60.

Redfield, A. C. 1955. The hydrography of the Gulf Venezuela. *Pap. Mar. Biol. Oceanogr. Deep. Sea Res. Supple.* 3:113–115.

———. 1961. The tidal system of Lake Maracaibo, Venezuela. *Limnol. Oceanogr.* 6:1–12.

Rodriguez, G. 1963. The intertidal estuarine communities of Lake Maracaibo, Venezuela. *Bull. Mar. Sci. Gulf Carib.* 13:191–218.

———. 1969. Seasonal fluctuation and penetration of the zooplankton in the Maracaibo estuary, Venezuela. *Mem. Simp. Intern. Lagunas Costeras.* UNAM-UNESCO. pp. 591–600.

———. 1973. *El Sistema de Maracaibo.* Caracas: IVIC.

Rochford, D. J., 1950. Studies in Australian estuarine hydrology. I. Introductory and comparative features. *Australian J. Mar. Freshwater. Res.* 2:1–116.

Ryther, J. H., P. W. Menzel, and N. Corwin. 1967. Influence of the Amazon River outflow on the ecology of the western tropical Atlantic. I. Hydrography and nutrient chemistry. *J. Mar. Res.* 25:69–83.

Ryther, J. H., and W. M. Dustan. 1971. Nitrogen, phosphorus and eutrophication in the coastal marine environment. *Science* 171:1008–1013.

Schimper, A. F. 1891. Die indo-malayische Strand flora. *Bott. Mitt. Tropen* 3:1–280.

Sioli, H. 1964. General features of the delta of the Amazon. Scientific problems of the humid tropical zone deltas and their implications. *Proceedings of the Dacca Symposium.* pp. 381–390. UNESCO.

Smayda, T. J. 1970. Phytoplankton studies in the lower Narrangansett Bay. *Limnol. Oceanogr.* 2:342–359.

Sorokin, Y. I., and L. B. Kliashtorin. 1963. Primary production in the Atlantic ocean. *Fish. Res. Board Can. Serv. 421* (transl.). *Trudy Usesoinzuogo Gidrob. Obsch.* 11:265–284.

Stephenson, T. A., and A. Stephenson. 1950. Life between tidemarks in North America. I. The Florida Keys. *J. Ecol.* 38:354–402.

Tabb, D. C., D. L. Dubrow, and A. E. Jones. 1962. Studies on the pink shrimp, *Penaeus duorarum* Burkenroad, in Everglades National Park, Florida. *Florida State Board Conservation Tech. Ser.* 37:1–30.

Tampi, P. R. S. 1969. Productivity of a saline lagoon near Mandapam (India). Lagunas costeras. *Mem. Simp. Intern. Lagunas Costeras* 479–484.

Texeira, C., and J. Tundisi. 1967. Primary production and phytoplankton in equatorial waters. *Bull. Mar. Sci.* 17:884–891.

Tundisi, J. G. 1970. O Plankton estuarino. *Contr. Inst. Oceanogr. Univ. Sao Paulo Ser. Ocean. Biol.* 19:1–22.

———, and T. M. Tundisi. 1968. Plankton studies in a mangrove environment. V. Salinity tolerances of some planktonic crustaceans. *Bolm. Inst. Oceanogr. Sao Paulo* 17:57–65.

———. 1972. Some aspects of the phytoplankton in tropical inshore waters. *Cienc. Cult.* 24:189–193.

Walsh, G. E. 1967. An ecological study of a Hawaiian mangrove swamp. In *Estuaries*, G. F. Lauff, ed. AAAS Publ. 83, pp. 420–431.

Walter, H., and E. Medina. 1971. Caraterización climática de Venezuela sobre la base de estaciones particulares. *Bol. Soc. Venez. Cienc. Nat.* 29:211–240.

Weymouth, F. W., M. J. Linder, and W. W. Anderson. 1933. Preliminary report on the life history of the common shrimp *Penaeus setiferus* (Linn.) *Bull. U. S. Bur. Fish.* 48:1–26.

Wood, E. J. F. 1966. A phytoplankton study of the Amazon River region. *Bull. Mar. Sci.* 16:102–123.

# CHAPTER 22

## Diurnal Rates of Photosynthesis, Respiration, and Transpiration in Mangrove Forests of South Florida

Ariel E. Lugo, Gary Evink, Mark M. Brinson, Alberto Broce and Samuel C. Snedaker

The mangrove forests of south Florida extend over an area of 1,750 $km^2$, located primarily within the boundaries of the Everglades National Park and within the region known as the Ten Thousand Islands (Figure 22–1). Davis (1940) described the zonation and succession in some of these areas, and more recent descriptions appear in Craighead (1971). The majority of the literature on mangroves is descriptive; only recently have investigators studied ecosystem function and the role(s) of mangroves in a regional ecosystem. The work of Golley et al. (1962) established baseline information on the primary productivity and respiration for a red mangrove stand in Puerto Rico, and demonstrated its role in exporting organic matter to the adjacent estuarine ecosystem. The regional role of mangrove forest ecosystems in Florida has been documented by Heald (1969) in terms of detritus export, and by Odum (1969) in terms of the role of detritus in the food chains of estuarine and commercial fisheries.

Few measurements of photosynthesis, respiration, and transpiration of mangrove trees are reported in the literature. Of those available, the most comprehensive are the studies of Golley et al. (1962) in Puerto Rico and of Miller (1972), who has developed a model of productivity of south Florida mangroves in Key Largo near Turkey Point (Figure 22–1). Relatively few data exist for seedlings (Brown et al., 1969; Chapman, 1962) and transpiration in leaves (summarized by Walter, 1971 and Macnae, 1968).

Our objective in this study was to measure carbon dioxide exchange and transpiration for all four of the south Florida mangrove-forest tree species (*Rhizophora mangle, Avicennia nitida,* Laguncularia racemosa,* and *Conocarpus erecta*) by compartment (trunks, prop roots, pneumatophores, seedlings, and shade and sun leaves) on a diurnal basis in order to establish magnitudes and possible zonation of photosynthesis, respiration, and transpiration rates.

* Synonymous with *A. germinans.*

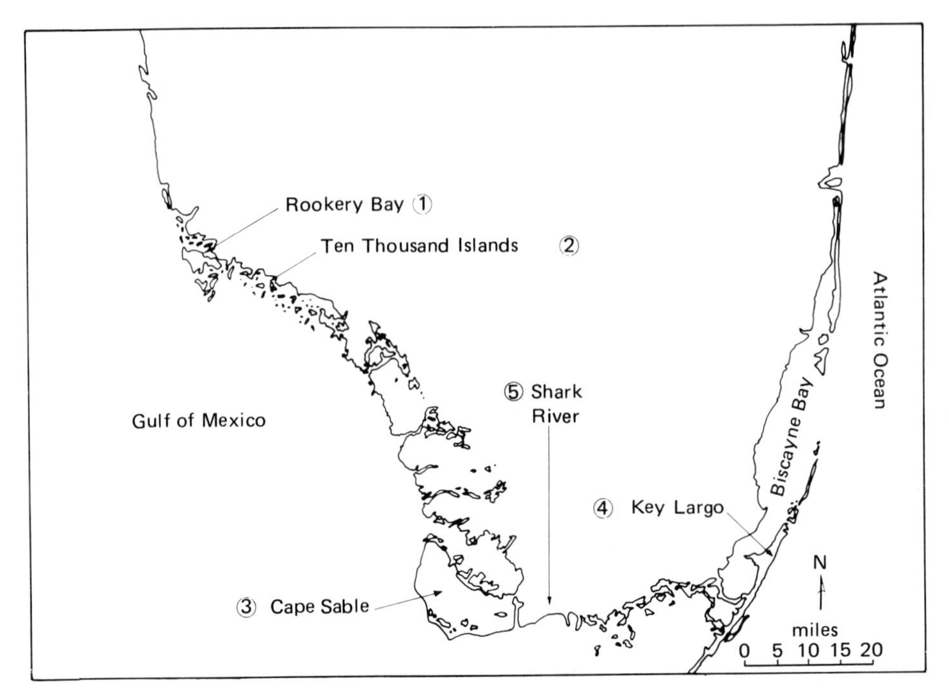


**Fig. 22–1.** Map of south Florida showing the sites where this study was conducted (Rookery Bay and Ten Thousand Islands), and those where others have made metabolism determinations (Key Largo), detritus export measurements (Shark River), and observations of mangrove vigor (Cape Sable).

## Methods

Carbon dioxide gas exchange was measured in open systems with a Beckman Infrared Gas Analyzer Model 215. Simultaneously, water vapor exchange with the atmosphere was measured with a Hygrodynamics Lithium Cloride Hygrometer with wide range temperature and relative humidity sensors. The complete apparatus—consisting of sensors, pumps, valves, recorders, and timers—has been described in Lugo (1969) and Odum et al. (1970). Switching between ambient and chamber air flows allowed three or four compartments to be monitored during the same day at 20-min intervals. All measurements were made *in situ*. Results were expressed in grams of carbon fixed or respired and/or grams of water transpired per square meter of leaf or stem surface area per hour, and converted to total surface of ground of forest by calculating leaf, trunk, and prop-root area indices. Data on ecosystem standing crop (Snedaker, unpublished) were available for relating estimates based on compartment biomass to surface area. The experiments were conducted in a mixed-mangrove forest on the Rookery Bay during the months of August 1971 and January-February 1972. Over 30 diurnals were analyzed during this period.

Data for periphyton productivity were provided by Delbert Hicks of the Environmental Protection Agency. He determined net carbon fixation utilizing the $^{14}C$ method by duplicating determinations on glass slides and red mangrove root surfaces. Incubation time was 3 hr, and the experiments were conducted near the Rookery Bay mangrove forest.

The mangrove forest at Rookery Bay consists of a zone of red mangroves (*Rhizophora mangle*) facing the bay landward from which is a stand of black mangroves (*Avicennia nitida*) interspersed with individual red mangroves and white mangroves (*Laguncularia racemosa*). The buttonwood mangroves (*Conocarpus erecta*) occur on higher ridges above the zone of tidal influence. Areas of regeneration occur where hurricane damage has created openings in the otherwise closed canopy. The regeneration in such areas is exclusively red mangrove, whereas red mangrove seedlings are not present beneath the closed canopies. Figure 22–2 is a map of this site and indicates where the metabolism studies were conducted.

## Results

Unless otherwise stated, the values reported in this section are expressed as a function of the surface area of the compartment (left, trunk, prop root, and pneumatophore) of interest. To express these values on a per-unit-land-area basis, each value must be multiplied by the appropriate area index (that is, total compartment surface area-per-unit-land area). The land surface conversions are introduced in the discussion section.

Net daytime photosynthesis rates were higher in the canopy leaves as compared to those of shade leaves (Table 22–1). Shade leaves, however, had higher rates of night respiration. The red mangrove leaves exhibited higher daytime net photosynthesis rates and lower nighttime respiration rates than black mangrove leaves. *Laguncularia* had the lowest rates of the three species (Table 22–1). A diurnal curve for red mangrove canopy leaves is shown in Figure 22–3 to illustrate the trend of metabolism. The top part of the figure shows the diurnal measure of carbon dioxide in the leaf and in the air. Net photosynthesis and respiration are calculated from the diurnal curve, and are shown in the center of the figure.

The daily magnitudes of trunk metabolism are ten times higher as compared to leaves (Table 22–1). The hourly rates (Figure 22–4) are much lower, however. Trunk respiration also was similar during the daytime compared to nighttime (Figure 22–4). Red mangroves had greater trunk respiration rates than either black or white mangroves (Table 22–1).

The respiration of the pneumatophores was five times higher than that of the prop roots (Table 22–1), suggesting a greater energy expenditure for adaptation to anaerobic conditions. The magnitudes of hourly respiration in this compartment (aerial roots) are ten times lower than those observed for trunks and 100 times lower than those of leaves.

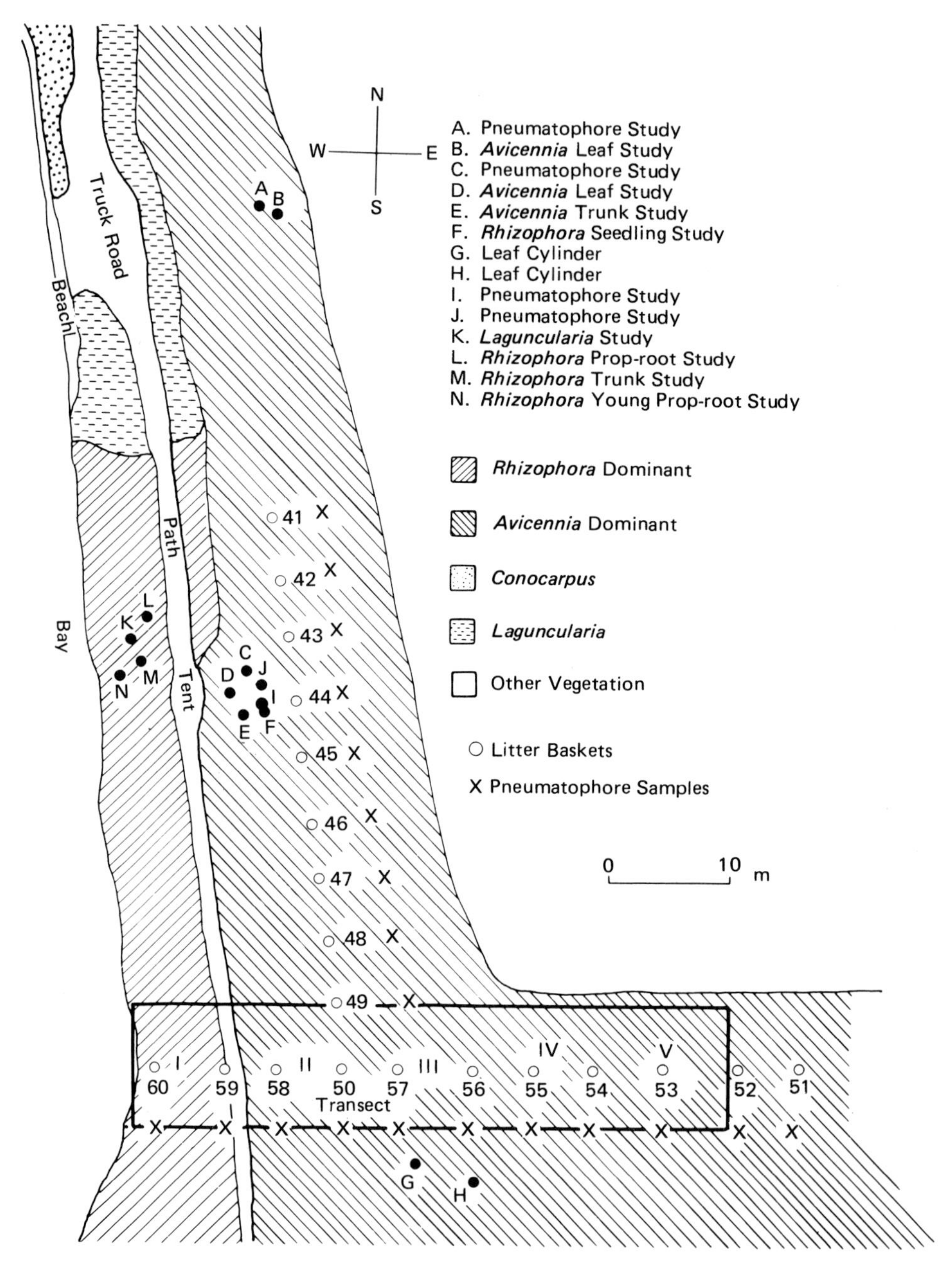


**Fig. 22–2.** Diagram of the Rookery Bay forest where metabolism observations were made. Determinations are shown for the location for August, 1971. Also shown are the tree growth plots, leaf fall and pneumatophore biomass stations, and the location of the instrument tent.

**Table 22–1.** Calculation of net daytime photosynthesis, nighttime respiration, and their ratio for several mangrove ecosystem components. Values are based on one surface area.

| Ecosystem Component | Species | Number of Diurnals | Mean of Diurnals P | Mean of Diurnals R | Mean of Diurnals P/R |
|---|---|---|---|---|---|
| Canopy leaves | *Rhizophora mangle* | 5 | 1.38 | 0.23 | 6.0 |
| | *Avicennia nitida* | 7 | 1.24 | 0.53 | 2.3 |
| | *Laguncularia racemosa* | 3 | 0.58 | 0.17 | 3.4 |
| Lower story leaves | *Rhizophora mangle* | 1 | 0.54 | 0.61 | 0.9 |
| | *Avicennia nitida* | 1 | 0.72 | 2.31 | 0.3 |
| Trunks | *Rhizophora mangle* | 3 | — | 3.62 | — |
| | *Avicennia nitida* | 3 | — | 2.19 | — |
| | *Laguncularia racemosa* | 1 | — | 1.23 | — |
| Prop roots | *Rhizophora mangle* | 2 | — | 0.10 | — |
| Pneumatophores | *Avicennia nitida* | 2 | — | 0.54 | — |
| Seedlings | *Rhizophora mangle* | 4 | 0.31 | 1.89 | 0.1 |
| Periphyton | | | 1.90 | — | — |

In Figure 22–5 the metabolism of a red and a white mangrove growing adjacent to one another is compared. Measurements were made simultaneously. Whereas the magnitudes of hourly rates of net daytime photosynthesis and nighttime respiration are the same, the red mangrove had a higher total daytime net photosynthesis for the same light intensity (2.86 versus, 1.56 g C/m$^2$/day). Subsequent comparative studies in January 1973 confirmed these differences in metabolism among the tree species. The results suggest that as one proceeds from the riverine condition and into the swamp, the rates of net photosynthesis in individual leaves decrease. Within each zone, the leaves of the species characteristic of that environment exhibit higher rates of net photosynthesis than those from other zones growing with them. Thus, in the black mangrove zone, leaves of *Avicennia* exhibit a higher net photosynthetic rate than those of *Rhizophora.* However, leaves of red mangroves growing adjacent to the bay had a higher rate of net daytime photosynthesis than those of *Avicennia* inside the swamp.

Considering compartments, sun leaves have higher net daytime photosynthesis and lower nighttime respiration than shade leaves (Table 22–2). Due to their high nighttime respiration, shade leaves seem to have a higher gross photosynthesis than sun leaves. This high gross production may be possible because these leaves are not in true dark rain forest shade but are frequently exposed to sun flecks penetrating the relatively open canopy (LAI = 3.5). Miller (1972), in his analysis of mangrove productivity, found a high net photosynthesis in the shade leaves of the upper canopy. On a surface area basis, trunks have as much respiration as shade leaves, and prop roots and pneumatophores have the least respiration (Table 22–2). To understand the relative contributions of these compartments to the forest as a whole, values must be standardized on the basis of ground area. These estimates will be presented in the discussion section.

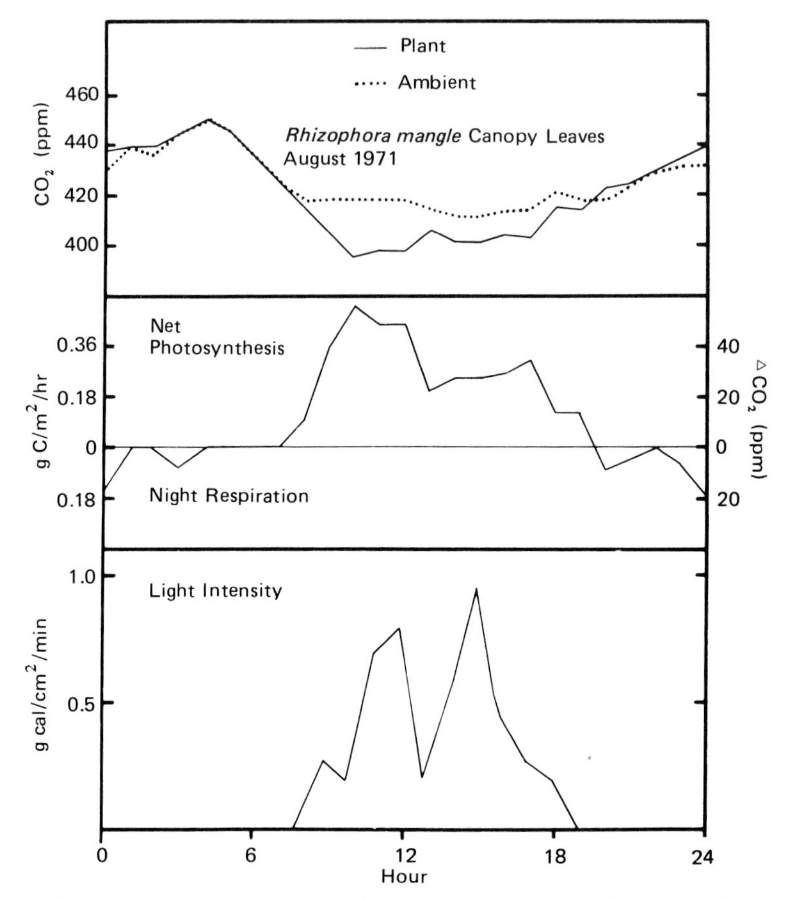


**Fig. 22–3.** Diurnal rates of net daytime photosynthesis, nighttime respiration, and incident total radiation for canopy leaves of the red mangrove *Rhizophora mangle*. Total daytime net production and nighttime respiration (g C/m$^2$), their ratio, and total incident radiation (kcal/m$^2$) were, 2.87, 0.43, 6.6, and 2,946.

**Table 22–2.** Vertical distribution of metabolism of the Rookery Bay mangrove forest. Values represent the means, by compartment, of all diurnals and are expressed on one surface area basis.

| Ecosystem Compartment (All Species) | Number of Diurnals | Mean Daytime Photo-synthesis | Mean Nighttime Respiration | Mean Gross Photo-synthesis | 24-Hr Respiration |
|---|---|---|---|---|---|
| Canopy leaves | 15 | 1.1 | 0.34 | 1.44 | 0.68 |
| Shade leaves | 2 | 0.6 | 1.4 | 2.0 | 2.8 |
| Tree stems | 7 | — | — | — | 2.6 |
| Prop roots | 2 | — | — | — | 0.1 |
| Pneumatophores | 2 | — | — | — | 0.5 |
| Seedlings | 4 | 0.3 | 1.9 | 2.2 | 3.8 |
| Periphyton | 2–3 hr | 1.9 | — | — | — |

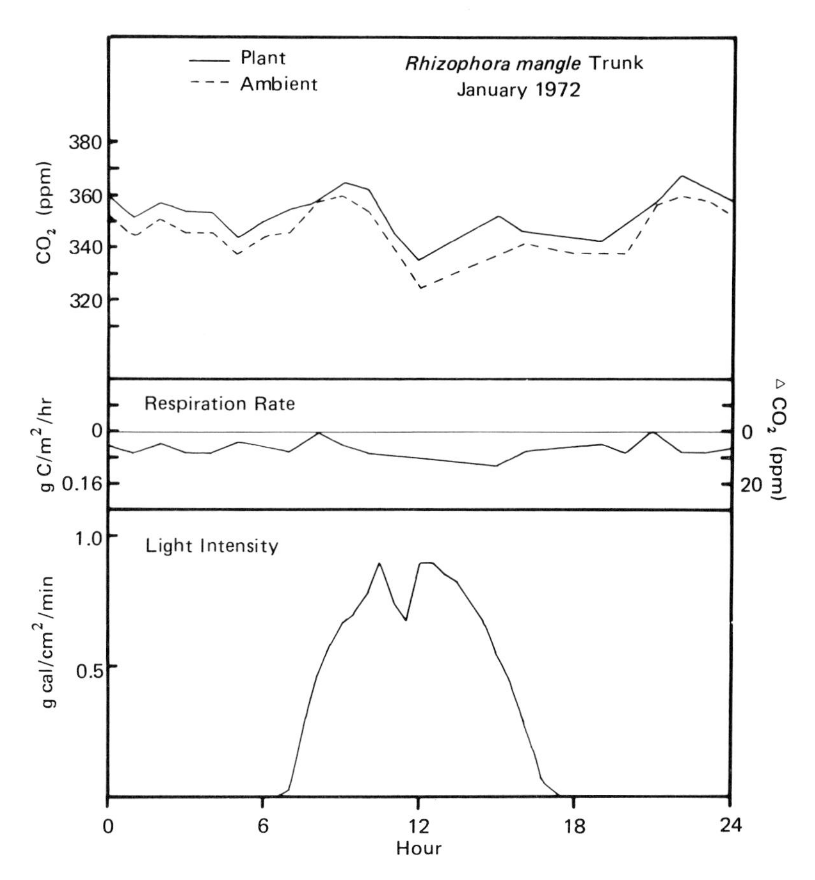


**Fig. 22–4.** Diurnal rates of respiration for trunks of the red mangrove *Rhizophora mangle*. Total respiration and daytime respiration (g C/m²) and total incident light (kcal/m²) were, respectively, 1.8, 1.2, and 2,659.

The net photosynthetic rate of periphyton growing on red mangrove roots is reported in Table 22 3. The productivity was greater in periphyton growing on roots as compared to that of periphyton growing on inert substrates (glass slides). This suggests there may be a symbiotic relationship between the periphyton and the mangrove root, involving an exchange of carbon dioxide from

**Table 22–3.** Net photosynthesis of periphyton growing on red mangrove prop roots.[a]

| Periphyton | Net Photosynthesis (g C/m²/hr) Experiment 1 | Experiment 2 |
|---|---|---|
| Growing on roots | 0.189 | 0.128 |
| Growing on slides | 0.035 | 0.076 |

[a] Experiments done by Delbert Hicks of the Environmental Protection Agency at Naples, Florida.

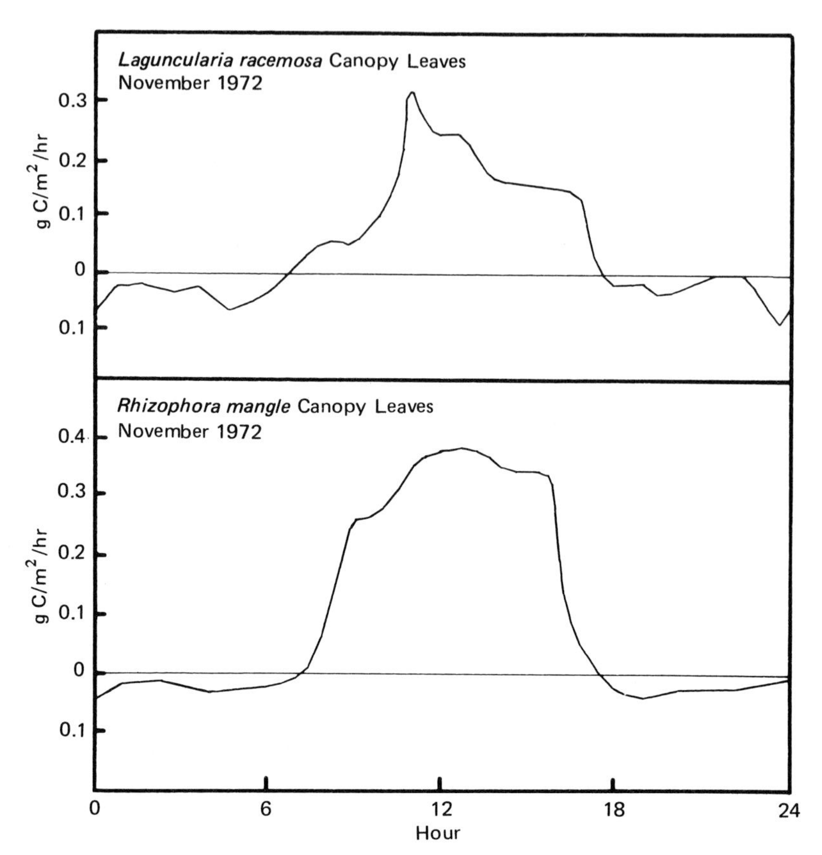


**Fig. 22–5.** Diurnal rates of net daytime photosynthesis and nighttime respiration for canopy leaves of the white mangrove *Laguncularia racemosa* (top) and the red mangrove *Rhizophora mangle* (bottom). These determinations were made simultaneously in a riverine forest located in the lower reaches of the Fakahatchee Strand by the Environmental Protection Agency, as part of their south Florida environmental study. Total net daytime photosynthesis, nighttime respiration (g $C/m^2$), and their ratio were, respectively, white: 1.56, 0.52, and 3.0; red: 2.86, 0.52, and 5.4. Total incident radiation (kcal/$m^2$) was 2,365.

the root to the producer and possibly oxygen from producer to the roots. In any event, measurements made on glass slides apparently underestimate periphyton productivity.

Diurnal curves of transpiration were also made for forest species and components. A diurnal for canopy leaves of the red mangrove shown in Figure 22–6 for illustration of the results. The transpiration data were more variable than the data for photosynthesis and respiration, but species and zonation differences are evident. The data are summarized in Table 22–4. Red mangrove canopy leaves have higher water-loss rates than were observed in any other species or compartment. Other canopy leaves (*Avicennia* and *Laguncularia*) have lower transpiration rates than shade leaves. Trunks had lower rates than leaves.

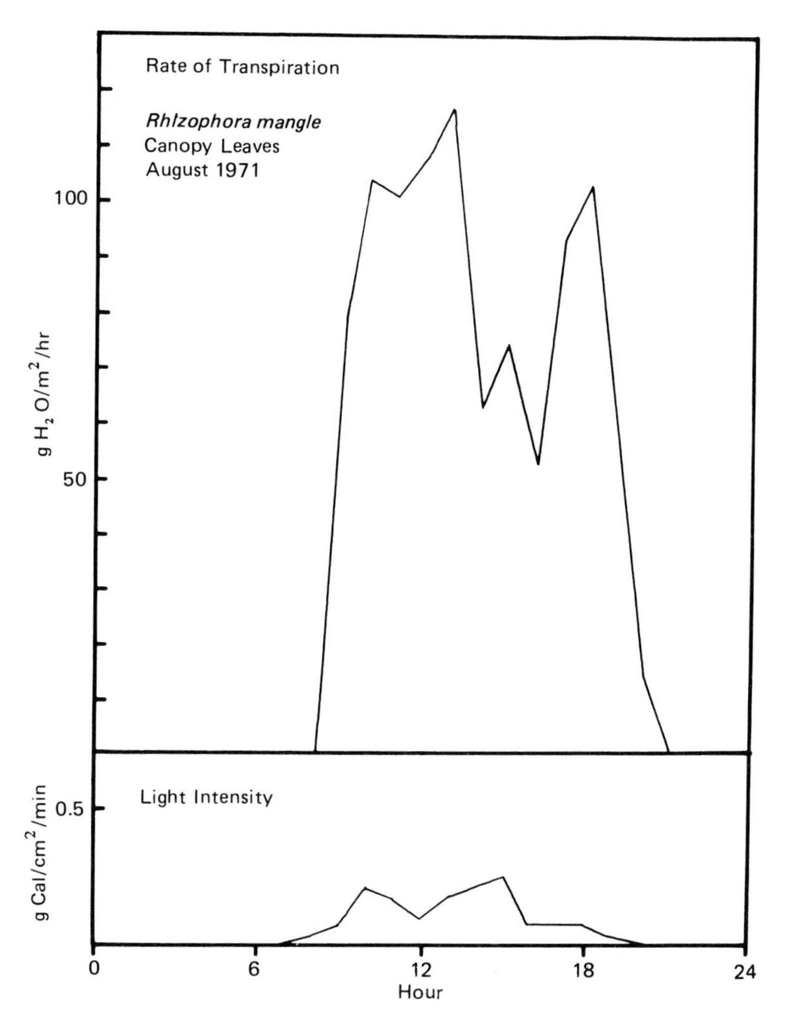


**Fig. 22–6.** Diurnal rates of transpiration for canopy leaves of the red mangrove *Rhizophora mangle*. Total water loss (g $H_2O/m^2$) and incident radiation ($kcal/m^2$) were, respectively, 936 and 882.

## Discussion

The rates of photosynthesis, respiration, and transpiration reported in this paper for the species and compartments of the mangrove forest in south Florida agree with literature reports for other tropical forest ecosystems. These include previous studies on red mangroves by Golley et al. (1962) and the studies in the rain forest by Odum et al. (1970) and Odum and Jordan (1970). While considerable day-to-day variations in the total net daytime photosynthesis and nighttime respiration were observed, due to the expense and logistics of the operation measurements had to be taken at the times when

**Table 22–4.** Diurnal transpiration rates for mangrove species. Values are expressed on one surface area basis.

| | | |
|---|---|---|
| *Rhizophora mangle:* | | |
| Canopy | 3 diurnals | 168–1,037 |
| Shade | 1 diurnal | 704 |
| Seedings | 4 diurnals | 0–520 |
| *Avicennia nitida:*[a] | | |
| Canopy | 3 diurnals | 79–238 |
| Shade | 1 diurnal | 715 |
| Pneumotophores | 1 diurnal | 78 |
| *Laguncularia racemosa:* | | |
| Canopy | 1 diurnal | 184 |
| *Conocarpus erecta* | 1 diurnal | 332 |
| Trunks | 2 diurnals | 134–154 |

[a] Salt-excreting plant.

all measuring systems were operational. Therefore some diurnals were determined on rainy, sunny, or cloudy days, and the results varied accordingly. Nevertheless, diurnal determinations of metabolism and transpiration have an advantage over short-term observations made in the field or the laboratory with potted or excised plant parts because they cover the span of environmental conditions characteristic of the locality. Diurnal measurements made under different atmospheric conditions yield a wide statistical variation among the day-to-day comparison, but the natural range of metabolism and transpiration is better understood through the mean values, as shown in Table 22–1. The difference in metabolism among species may be compared in the field using simultaneous measurements under the same natural conditions (Figure 22–5).

## Vertical Distribution of Metabolism

The vertical distribution of metabolism is similar to the trends and magnitudes observed in Puerto Rico, with the exception of the respiration of red mangrove prop roots, which was higher in Puerto Rico (Table 22–5). This is because the Puerto Rican forest is a well-developed monospecific stand of red mangroves and has a higher prop-root area than the Rookery Bay forest. The leaf area index is also higher for the red mangrove zone in Puerto Rico, which also helps to explain the higher leaf respiration (Table 22–5). The Rookery Bay forest is characterized by $> 1$ ratio of gross photosynthesis to total respiration (P/R) in contrast to the Puerto Rican forest, which had a ratio of 0.8. This difference, however, may be due to the availability of more data for the Rookery Bay forest. The measurements in Puerto Rico were few and of short duration. Mangrove forests are characterized by their exports of organic matter (Golley et al., 1962; Heald, 1969); thus their ratios of P/R should be greater than one, unless a portion of the respiration is subsidized by organic inputs from upstream ecosystems.

**Table 22–5.** Mean estimates of gross photosynthesis ($P_g$), 24-hr respiration ($R_t$) (g C/m²/day) and transpiration ($T$) (g $H_2O$/m²/day) of mangrove ecosystems. Values have been corrected for leaf, trunk, and root area indices.

| | LAI | Leaves | | | Stems[a] | | Surface[b] Roots | | Peri-phyton | Soil | Losses | Total for the System | | | |
|---|---|---|---|---|---|---|---|---|---|---|---|---|---|---|---|
| Ecosystem | | $P_g$ | $R_t$ | $T$ | $R_t$ | $T$ | $R_t$ | $T$ | $P_{(net)}$ | $R_t$ | Export | $P_g$ | $R_t$ | $T$ | $P_g/R_t$ |
| Rookery Bay: | | | | | | | | | | | | | | | |
| Red mangrove stand (fringe) | 3.5 | 5.63 | 1.61 | 2,500 | 0.23 | 10 | 0.06 | 60 | 1.1 | — | — | 6.73 | 1.90 | 2,570 | 3.5 |
| Black mangrove stand (inside swamp) | 5.1 | 9.02 | 5.40 | 1,482 | 0.14 | 8 | 0.54 | 78 | — | 0.20 | — | 9.02 | 6.28 | 1,568 | 1.4 |
| Puerto Rico: | | | | | | | | | | | | | | | |
| Red mangrove stand | 4.4 | 8.23[d] | 5.39[d] | — | — | — | 2.03 | — | — | 0.37 | 1.37 | 8.23 | 9.16 | — | 0.8 |
| Key Largo, Fla.: | | | | | | | | | | | | | | | |
| Red mangrove stand | 2.4[c] | 5.34[d] | 6.05[e] | 736[e] | — | — | — | — | — | — | — | 5.34 | 6.05 | 736 | 0.8 |

[a] Mean trunk area index was 0.66 m²/m².
[b] Mean root area index was 0.67 m²/m.²
[c] Mean of three determinations.
[d] Includes seedlings.
[e] Mean of four determinations for sunny and cloudy days.

The biomass production by periphyton may represent a mechanism for the removal of nutrients from the waters passing through the mangroves. In these mangroves, the productivity of periphyton producers, on a forest-area basis, was 1.1 g $C/m^2/day$ (Table 22–5). This appears to be a significant contribution to the overall productivity of the forest and reinforces the idea of their role in nutrient uptake and concentration of materials from passing waters.

## Metabolic Basis of Zonation

While each species of mangrove is capable of similar hourly rates of photosynthesis, respiration, and transpiration in corresponding compartments, it is clear from Table 22–1 that the integrated total rates of these processes show consistent differences between species. Simultaneous measurements, such as the one in Figure 22–5, confirm that on a given day leaves of the red mangrove exhibit the highest total net daytime photosynthesis, followed by those of black, white, and buttonwood mangrove in descending order. This gradient of metabolic efficiency suggests a metabolic basis for the zonation of mangrove species. Gross photosynthesis and transpiration are highest in the outer zones and decrease toward the landward zone. Within each zone the characteristic species has maximized, apparently, photosynthesis and thus exhibits a higher metabolic rate than any invading species, which would be at a competitive disadvantage due to its lower metabolic efficiency in that habitat. This is consistent with Lotka's (1956) and Odum and Pinkerton's (1955) principle of maximum power, which states that each species maximizes power output in the environment to which it is adapted. For this reason, mangrove zonation should be explored in relation to those factors that most affect the photosynthetic output of each species.

In the area of study, buttonwood mangroves on the landward side grow on high ground over limestone or shell low in nutrients and with low moisture-holding capacity. The metabolic and transpiration rates are correspondingly low. White mangrove is a minor component of this ecosystem, and its metabolism is also low. However, high rates of metabolism were observed in a riverine condition where nutrient-rich terrestrial runoff was high (Figure 22–5). Similarly, the high values observed for the red mangrove were observed in areas where terrestrial runoff and tidal flushing were high. The black mangroves grow in protected areas where terrestrial runoff and tidal exchange were low. Possibly these factors (nutrients and terrestrial runoff), more than salinity, are responsible for the observed zonation of metabolism rates.

One would expect a greater net energy flow in habitats with low environmental stress and a lesser net flow in areas of more stress. Table 22–1 demonstrates this to be true when comparisons are based on metabolism per unit surface area of leaves. Red mangrove leaves exhibited greater net photosynthesis and lesser respiration than black mangrove leaves. In this habitat, the tidal flushing may alleviate severe anaerobic conditions in the roots, while nutrient runoff accelerates photosynthesis. In the black mangrove zone, standing water and a high peat content resulting from low tidal flushing and runoff produce a more anaerobic condition which may stress root function (no significant salinity differences were observed). This might also explain why pneumatophores exhibited a greater flux in carbon dioxide exchange than prop roots, but it does not explain why more respiration was observed in red mangrove trunks. The fact that red mangrove leaves exhibit less photosynthesis when growing in the black mangrove zone suggests a lack of adaptation for these extreme conditions. Pneumatophores, with greater area per unit ground, seem to be the more effective mechanism for adapting to anaerobic soils.

## Comparison of Stands

In Table 22–5 we have computed the relative contribution of each compartment to the total metabolism of the Rookery Bay forest. These data are compared with similar data reported for a red mangrove stand in Puerto Rico and in Key Largo, Florida. The black mangrove zone of the Rookery Bay forest is more productive than either of the other two forests. The red mangrove zone is similar to the Puerto Rican forest and is more productive than a similar stand in Key Largo. The Rookery Bay forest also has a higher rate of water loss than the Key Largo forest.

Highest respiration of leaves was observed in Key Largo, followed by Puerto Rico, and then by the study reported here. The trend in net photosynthesis, however, was the reverse. It appears that this study site exhibits a greater net energy flow (5.8 g C/m$^2$/day) than either Puerto Rico (3.65 g C/m$^2$/day) or Key Largo (2.31 g C/m$^2$/day). Key Largo is by far the least favorable mangrove environment of the three. It has a very low terrestrial runoff and is a relatively dry site. La Parguera, in Puerto Rico, is in the dry section of the island but does receive heavy terrestrial runoff during a short wet season.

By comparing stands on the basis of their gross productivity the same trends discussed among sites are not as clear. This is because the red mangrove stand in the Rookery Bay, although growing faster than the one in Puerto Rico, does not exhibit as high a gross productivity (Table 22–5). In addition, it appears that black mangrove stands have a higher rate of gross productivity than red mangrove stands. These differences can first be explained on the basis of the structural complexity of each stand. The Puerto Rican mangrove forest had a leaf area index of 4.4, higher than any measured by us in eight mangrove stands where the red mangrove was dominant. The greater leaf area index may be indicative of an older stand or one that has not been disturbed by a hurricane as recently as the ones we studied. Recall also that the prop roots are also more developed in the Puerto Rican forest as compared to those in Florida.

The structural differences of larger trees and the higher leaf area index in black mangrove stands represent functional differences as well, because each species is growing in a unique environment. The black mangrove zone is on a more isolated habitat, with less runoff and less flushing to the bays compared with the red mangrove zone. This results in organic material accumulation in this part of the swamp, greater oxygen deficits in the root zones, and perhaps more dependency on efficient nutrient recycling to maintain high rates of photosynthesis. Structural adaptations to these conditions may maximize energy flow by accelerating energy capture as well as respiration. Our data have demonstrated the greater rates of leaf and pneumatophore respiration in this zone as well as the higher gross photosynthesis for the forest as a whole. In the red mangrove zone there is greater nutrient input and also more flushing from tides, which removes organic materials from the floor and aerates roots. Tidal flushing may be an auxiliary energy source that subsidizes the energy budget of the red mangrove forest, thus reducing selective pressure in this zone for complex structural development as in the black mangrove zone. Thus the greater availability of nutrients in this zone is not reflected by a higher gross photosynthesis of the ecosystem but does represent a subsidy that maintains high net production and export to adjacent bays.

### Comparison of Transpiration Rates

The rates of transpiration in Table 22–4 are higher than those reported in the literature by Miller at Key Largo (Table 22–5). However, most literature values are based on short-term measurements and, as we have discussed, the diurnal method is a better indicator of normal fluctuations in rates. Mangroves are characterized by low transpiration rates (Walter, 1971), and the total transpiration from the Rookery Bay forest is similar to the value reported by Odum and Jordan (1970) for a tropical rain forest (2,136 g $H_2O/m^2$/day). The rain forest in Puerto Rico also had a low transpiration rate as compared to temperate forests during the dry and hot summer days. Kramer and Kozlowski (1960) report transpiration rates for this time about 4,700 g $H_2O/m^2$/ day. Odum and Jordan explained the low transpiration in the rain forest in terms of the low saturation deficit characteristic of the atmosphere of that montane forest. The mangrove forests occur in lowlands where saturation deficits are large and wind velocities high. Here, the lower transpiration rates are probably not related to saturated deficits but to the energetic cost of maintaining at all times sap pressures of —35 to —60 atm (Scholander et al., 1965). In so doing, mangroves reverse osmosis, take up fresh water from the sea, and exclude salt water from their transpiration stream. Some species allow some salt to enter the transpiration stream, thus having an added energetic cost of excreting salt through specialized glands on their leaves (Walter, 1971). One expects that the energetic expenditure of ultrafiltration and salt excretion will limit the rate of transpiration in mangrove stands.

## Conclusions

It appears that the zonation of mangrove species also involves zonation in their photosynthesis, respiration, and transpiration rates. The species of each zone and their symbionts have adaptations that take advantage of auxiliary energy sources, such as tidal flushing and nutrient runoff, and in so doing maximize the capture and retention of solar energy for useful work. It could be that salinity adaptations represent the energetic cost of tapping the nutrient-rich habitat in which mangroves occur.

## Acknowledgment

This report is part of a more comprehensive study of the structure and function of mangrove ecosystems, their relationship to regional water quality, and their role in the maintenance of regional fisheries. The initial studies were supported by the University of Florida's Division of Sponsored Research and are being carried to completion by the U.S. Department of Interior, Bureau of Sport Fisheries and Wildlife (Contract No. 14–16–008–606). The Environmental Protection Agency and the University of Miami's Rookery Bay project provided some data and essential on-site logistic support. The authors also acknowledge the assistance of members of the University of Florida's Department of Botany, Department of Environmental Engineering, and the Center for Aquatic Sciences.

## References

Brown, J. M. A., H. A. Outred, and C. F. Hill. 1969. Respiratory metabolism in mangrove seedlings. *Plant. Physiol.* 44:287–294.

Chapman, V. J. 1962. Respiration studies of mangrove seedlings. I, II. *Bull. Mar. Sci. Gulf Carib.* 12:137–167, 245–263.

Craighead, F. C. 1971. *The Trees of South Florida.* Miami: Univ. Miami Press.

Davis, J. H. 1940. The ecology and geologic role of mangroves in Florida. *Papers Tortugas Lab.* 32:305–412.

Golley, F. B., H. T. Odum, and R. F. Wilson. 1962. The structure and metabolism of a Puerto Rican red mangrove forest in May. *Ecology* 43: 9–19.

Heald, E. J. 1969. The production of organic detritus in a south Florida estuary. Ph.D. dissertation, Univ. Miami, 111 pp.

Kramer, P. J., and T. T. Kozlowski. 1960. *Physiology of Trees.* New York: McGraw-Hill.

Lotka, A. J. 1956. *Elements of Mathematical Biology.* New York: Dover.

Lugo, A. E. 1969. Water, carbon, and energy budgets for a granite outcrop ecosystem. Ph.D. dissertation, Univ. North Carolina, Chapel Hill.

Macnae, W. 1968. A general account of a fauna and flora of mangrove swamps and forests in the indo-west-pacific region. *Advan. Mar. Biol.* 6:73–270.

Miller, P. C. 1972. Bioclimate, leaf temperature, and primary production in red mangrove canopies in south Florida. *Ecology* 53:22–45.

Odum, H. T., A. Lugo, G. Cintron, and C. F. Jordan. 1970. Metabolism and evaporation of some rain forest plants and soils. In *A Tropical Rain Forest*, H. T. Odum and R. F. Pigeon, eds. Chap. 1–8. Oak Ridge, Tenn.: AEC, U.S. Div. Tech. Inf. Education.

———, and C. F. Jordan. 1970. Metabolism and evapotranspiration of the lower forest in a giant plastic cylinder. In *A Tropical Rain Forest*, H. T. Odum and R. F. Pigeon, eds. Chap. 1–9. Oak Ridge, Tenn.: AEC, U.S. Div. Tech. Inf. Education.

———, and R. C. Pinkerton. 1955. Time's speed regulator: The optimum efficiency for maximum power·output in physical and biological systems. *Am. Sci.* 43:331–343.

Odum, W. E. 1969. The structure of detritus based food chains in a south Florida mangrove system. Ph.D. dissertation, Univ. Miami.

Scholander, P. F., H. T. Hammel, E. D. Bradstreet, and E. A. Hemmingsen. 1965. Sap pressure in vascular plants. *Science* 148:339–346.

Walter, H. 1971. *Ecology of Tropical and Subtropical Vegetation*. Edinburgh: Oliver & Boyd.

# Island Ecosystems

The International Biological Program has stimulated research on ecosystems throughout the world. However, tropical ecosystems have been studied mainly in Africa and Asia. In the Americas, ecosystem studies of tropical forests usually have not been sponsored by the IBP, but rather by the USAEC in Puerto Rico and Panama (Odum and Pigeon, 1970; Golley et al., 1974) or by the Smithsonian Institution at Barro Colorado Island (see several papers in this volume). The major exception has been Hawaii where the US/IBP supports an island ecosystem study. Mueller-Dombois, program director for this area, discusses some of the basic concepts and conclusions.

In IBP ecosystem studies the focus has often been on the structural character and the dynamics of the type rather than on the variation in these characteristics from one place to another within the type. Mueller-Dombois argues that ecological concepts developed on continental areas cannot be transferred wholesale to islands, and that islands must be studied individually since they have unique characteristics that influence their ecosystem structure and function or relationships between species populations. Besides this point, Mueller-Dombois also stresses that there is a middle ground between the study of ecosystems and populations. He terms this "study of the general niche," a concept that is fundamentally similar to that of the component in ecosystem analysis or to plant-animal groups in the study of interactions. Thus study of the ecology of tropical islands permits us to see some of the processes discussed in earlier papers in a new and different light.

## References

Golley, F. B., J. T. McGinnis, R. G. Clements, G. I. Child, M. J. Duever. 1974. *Mineral Cycling in a Tropical Moist Forest Ecosystem.* Athens: Univ. Georgia Press. In press.

Odum, H. T., and R. F. Pigeon. 1970. *A Tropical Rain Forest.* USAEC,, Div. Tech. Inf.

# CHAPTER 23

## Some Aspects of Island Ecosystem Analysis[1]

DIETER MUELLER-DOMBOIS

I have assumed that one of the reasons for emphasizing the subject matter of tropical ecology is a general realization that ecological principles developed through the study of temperate ecology are not always sufficiently complete to explain important ecological relations in the tropics. My thesis in this presentation is that island ecology differs enough from continental ecology that ecological principles developed from the study on continental ecosystems are not always sufficiently complete to explain important ecological relations on islands.

The peculiarities of island life have received considerable attention by evolutionary biologists since Darwin (1859) developed his theory of speciation from his observations of the Galapagos finches. Carlquist (1965, 1970) presented comprehensive reviews together with theories that may explain the often rather bizarre life forms developed on isolated islands. Fosberg (1963, 1966, 1967) and others (for example, Dorst, 1972) have written conceptual papers dealing with the precarious balance and fragility of island ecosystems.

The object of this paper is to point out how biological evolution enters into the study of ecosystems analysis on islands as currently carried out by the Island Ecosystems Integrated Research Program of the U.S. International Biological Program (IBP).

## Three Important Features of Island Systems

The first and most important feature is the geographic isolation of island ecosystems. In this respect the Hawaiian Islands are perhaps one of the best examples. The islands are 2,000 mi away from the nearest continent to the east, North America. The distance to the Philippines is 5,300 mi, to Japan 2,500 mi, to Australia 4,000 mi. The nearest island is Johnson Island, 450 mi SW.

[1] Contribution No. 22, Island Ecosystems IRP/US IBP Hawaii, supported by NSF grant GB23230.

The second factor is size. Different opinions exist with regard to defining Australia as an island or continent but, as a rule, islands are small landmasses. A third factor is age. So-called continental islands like Ceylon exist that are as old geologically as the continental landmass from which they originated. However, many oceanic islands are of volcanic origin. If they are still mountainous, they are, as a rule, geologically of relatively recent origin. The age of the oldest parts of the high Hawaiian Islands is estimated as 11.3 million years (Macdonald and Abbott, 1970).

These three features have important bioecological consequences. Geographic isolation has a "screening effect" on the biota that can reach the island. Only those capable of long-distance transfer could become established before the arrival of man. While the present native flora of Hawaii has been estimated by Fosberg (1948) to consist of a little over 1,700 species and varieties, it developed from only 272 original arrivals that became successfully established. A few of these have not changed morphologically—the indigenous species. Others have changed morphogenetically, or became fractured into many distinct species and varieties, forming the endemic taxa. Endemism in the vascular flora is estimated to be in excess of 95 percent, meaning that over 1,600 of the native Hawaiian species and varieties do not occur anywhere else in the world.

While geographic isolation is responsible for the unique biota of Hawaii, it is not simply distance over water and disseminule size that have determined the composition of the native flora. If this were so, the Hawaiian flora should show the greatest similarity to the North American temperate and subtropical flora. Instead, 71 percent of its flora is of tropical, 19 percent of temperate, and 10 percent of unknown origin. The high tropical component is most likely a reflection of habitat selection for successful establishment of plants preadapted to tropical climates. Of the 71 percent native vascular Hawaiian plant taxa of tropical origin, 40.1 percent show Indo-Malayan, 18.3 percent tropical American, and 12.5 percent pan-tropical affinities. Of the 19 percent plant taxa of temperate origin, 16 percent show Australian and only 3 percent north temperate origin. Thus the most successful invasions came from the southwest in spite of the greater distance to major landmasses in that direction.

The biological implication of "small land size" is that population sizes of perennial organisms are also small. Even in mountainous islands, which usually have much greater landmasses than flat coral islands, the recurrence of similar habitats is limited in comparison to most continental mountain ecosystems. Small areas for populations imply limitations in gene flow as well as number of individuals. Both factors contribute to a greater fragility of island populations.

The bioecological consequence of "recent geological age" is that tropical island communities are much younger than most tropical continental communities. For example, forests on continents may have undergone more or less uninterrupted succession from giant equisetum-lycopod and seed fern forests to primitive gymnosperm and angiosperm forests, and then to the modern angiosperm forests. In contrast, the origin of most existing volcanic island forests is within the modern angiosperm era. As mentioned before, the oldest parts of the high Hawaiian Islands are estimated as being only 11 million years old. Fosberg (1948) estimated that only one arrival form was required to become successfully established every 20,000 to 30,000 years to account for today's native flora of more than 1,700 taxa. The shorter geological time available for community development may in part account for a lower alpha diversity in tropical island communities as compared to tropical mainland communities.

It may be argued that the three factors mentioned are not entirely unique to islands. However, the biological consequences of these conditions are most clearly developed on islands. In particular, these conditions have caused a different pattern of evolution in island ecosystems. Furthermore, on islands many ecological relationships involve the interactions among native and exotic species. Many of the latter were introduced by man within the last 200 years.

The peculiarities of island ecology are studied in three areas of ecosystem analysis through the island ecosystems Integrated Research Program of the US/IBP. These three areas are (1) spatial distribution of island biota, (2) community structure and niche differentiation, and (3) successional phenomena.

The ecological relationships in these three areas have to do mostly with species interactions in an ecosystem context rather than with metabolic process research of ecosystems as pursued in the US/IBP Biome Studies. Because of the unique evolution of the island biota, we are concerned primarily with the interaction of native and exotic species.

## Spatial Distribution of Island Biota

### Ecological Amplitudes

In contrast to most tropical continental vegetation, the native forest vegetation of the Hawaiian Islands is remarkably simple. There are, in fact, only two dominant, tall-growing (25 to 30 m maximum) native tree species complexes: *Metrosideros collina* and *Acacia koa*. These are wide-ranging dominants whose spatial distribution can be ecologically characterized as follows (Fig. 23–1):

On the most southern island Hawaii (the so-called Big Island with an area of 4,038 sq mi), *Metrosideros* ranges from sea level to 8,200 ft (2,500 m) elevation on Mauna Loa (13,677 ft = 4,170 m). In terms of climatic parameters this means that the tree grows in a year-round warm-tropical climate of 23°C mean air temperature all the way up into a cool-tropical, high-altitude climate of 8°C mean air temperature. The upper boundary of *Metrosideros* is marked approximately by the year-round nocturnal ground-frost boundary (Mueller-Dombois, 1967). *Acacia koa* has a more limited range on Mauna Loa, from 4,000 to 6,700 ft (1,220 to 2,043 m), within the range of *Metrosideros*. On Mauna Kea (13,796 ft = 4,206 m) and on the island of Maui, *Acacia koa* occurs in a similar altitude range as on Mauna Loa, but it also occurs in the lower montane belt together with *Metrosideros*. On the islands of Oahu and Kauai, *Acacia koa* only occurs below the range of *Metrosideros*, and the two

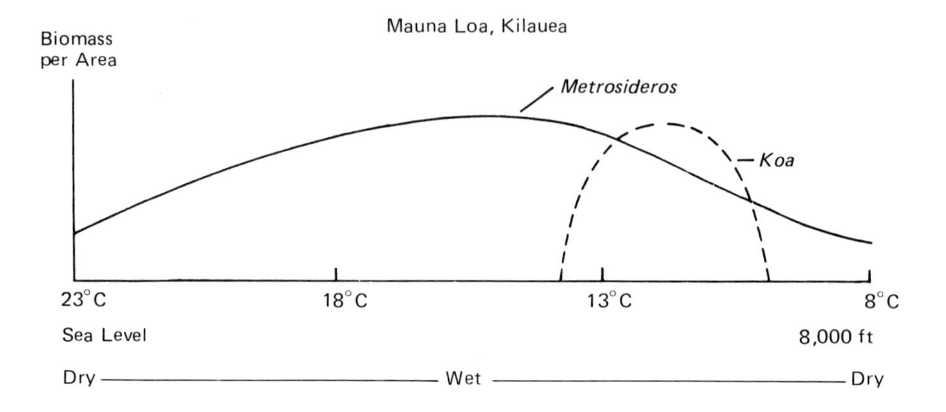


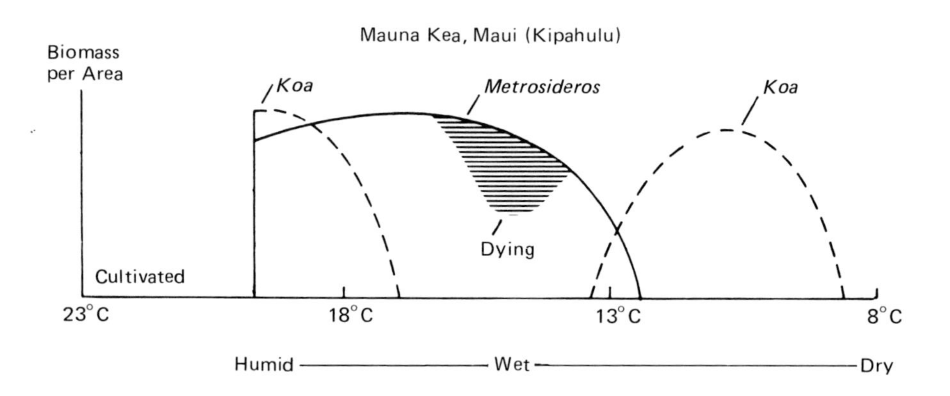


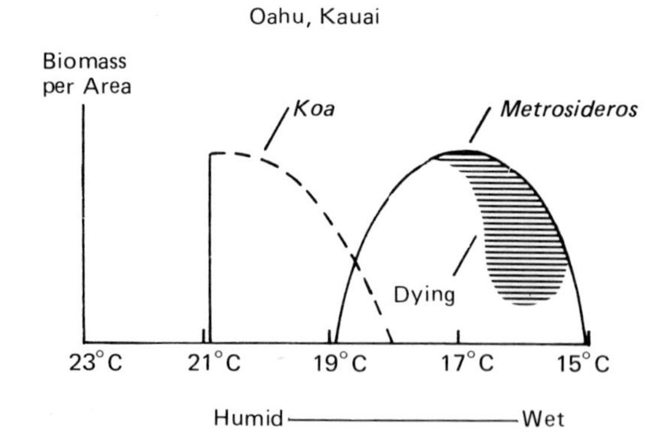


**Fig. 23–1.** General ecological amplitudes of the two most important native tree species (*Metrosideros collina* and *Acacia koa*) in the mountainous Hawaiian Islands.

species rarely form mixed stands. However, a few widely scattered or rare *Metrosideros* individuals of shrub stature do occur down to 500 ft (150 m) on Oahu and to sea level on Kauai.

The change in range of these species is correlated with substrate age. Mauna Loa is still an active volcano whose currently exposed lava rock substrate ages range from probably post-Pleistocene to present. In contrast, most of the Mauna Kea surfaces are from late Pleistocene (from 600,000 to about 15,000 years ago). The substrates on Oahu and Kauai are from 1 to 6 million years old (Macdonald and Abbott, 1970). The substrate age effect is primarily one of changing soil-water regimes, from xeric to hydromorphic, in the midelevation rain forest range. The dying *Metrosideros* in the wet areas on the older substrates is often correlated with poor drainage. However, correlation with substrate age and soil moisture regime does not mean that those factors are also the causes of these distributions. The causes are as yet unknown.

Several other native woody life forms of small tree or shrub stature have similarly wide-ranging (but often interrupted) distributions, for example, the trees *Myoporum sandwicense, Sophora chrysophylla* and the shrub *Styphelia tameiameiae* [which may be split into low (*S. tameiameiae*) and high elevation (*S. douglasii*) species]. Other species of similar life form have narrow ranges. Such species may belong, however, to wide-ranging genera, such as the shrub *Cyrtandra*, which occurs with 118 species on Oahu (St. John, 1966). These fractioned taxa are responsible for another ecological peculiarity, found mostly on the older volcanic islands, namely, that physically similar habitats may be occupied by quite dissimilar communities in terms of species composition. Thus the pattern of recurrence of similar species combinations in similar habitats varies between islands and generally decreases with island age.

## Hypotheses of Species Distribution

Based on a working hypothesis that endemic birds, insects, and other life forms evolved primarily in adaptation to the dominant native plant species, we are testing the degree of spatial association of native biota along environmental gradients. Our test gradient is a 22-mi-long transect on the east flank of Mauna Loa, which cuts through 12 structurally well-defined vegetation types, from alphine scrub to *Metrosideros-Cibotium* (tree fern) rain forest. Both tree species, *Metrosideros collina* and *Acacia koa*, form the structural dominants in these ecosystems, and we are determining if native ecosystem stability is related to these two native tree species.

So far, this hypothesis appears to be supported by the distribution of phytophagous insects (Gagne, 1972) but not by the soil arthropods (Radovsky, 1972). In the former group many endemic species are found where these two tree species are most vigorous, while among the soil arthropods most species are exotics. The latter show wide distributions, quite unrelated to the vigor or distributional variations of the native tree species and ecosystems.

There are currently four hypotheses about species and community distribution along environmental gradients. These have been stated by Whittaker (1970, p. 35) as follows:

1. "Competing (dominant) species exclude one another along sharp boundaries. Other species evolve toward close association with the dominants and toward adaptation for living with one another." This species distribution would result in distinct zones.

2. "Competing (dominant) species exclude one another along sharp boundaries, but (other species) do not become organized into groups with parallel distributions." This would result in zones with overlapping species ranges.

3. "Competition does not (usually) result in sharp boundaries (that is, zonation) between species populations. Evolution of species toward adaptation to one another will, however, result in the appearance of groups of species with similar distributions." This would result in a pattern of typical communities separated by ecotones.

4. "Competition does not usually result in sharp boundaries between species populations, and evolution to one another does not result in the formation of groups with similar distributions. Instead, centers and boundaries of species populations are scattered along environmental gradients." This would result in no recognizable zonation and absence of typical communities.

All four spatial distribution patterns seem possible. Whittaker's studies in continental temperate mountain ecosystems supported the last hypothesis. Our studies in island tropical mountain ecosystems seem to support the third hypothesis as far as the native biota are concerned.

Moreover, Whittaker (1970) holds that the highest degree of integration is accomplished by a high beta diversity. High beta diversity implies accommodation of a large number of species with restricted distributions along a given environmental gradient as opposed to a few wide-ranging species on the same gradient. Low beta diversity appears to be a characteristic on altitudinal gradients on Hawaii. In the tropics this is a peculiarity. Whether this also means poor integration needs further examination. An increase of beta diversity through exotic species invasion appears to show a decrease in integration.

## Community Structure and Niche Differentiation

For the purpose of distinguishing ecological peculiarities in island communities and ecosystems, we may recognize three levels of community structure.

1. Gross structure at the level of the formation or biome
2. Fine structure at the level of the species assemblage
3. Intermediate structure at the level of species groups with similar gross morphologies (life forms) and/or similar general ecological adaptations

### Biome Structure

Island communities and ecosystems are not unique in terms of biome structure. For example, on the island of Hawaii alpine tundra, open and closed scrub, evergreen rain forest, evergreen seasonal forest, savanna, grassland, desert, and other formations are represented. Moreover, several of these formations show similar relations as in continental tropical lowland and mountain ecosystems. In addition, one can recognize even certain gross taxonomical relationships among equivalent island and continental formations. To give a general altitudinal outline: The leeward lowlands have a winter rainfall and a summer drought climate. The typical community is the perennial grass savanna. The woody plants may be scattered trees or shrubs, or shrubs variously clumped. Among trees, members of the Leguminosae are important. Open seasonal evergreen forest occurs above this lowland vegetation on the leeward side. Where well preserved, this is a taxonomically quite rich tree vegetation, with at least 20 native tree species (Wirawan, 1972). In the montane environment evergreen rain forest predominates with Myrtaceae (*Metrosideros*, *Eugenia*, also with introduced taxa, *Psidium* spp.) and native tree ferns (*Cibotium* spp.). Above the rain forest occurs mountain parkland or savanna. There, legume trees are again important (*Acacia, Sophora*). Above this, subalpine open forest and scrub occurs. The scrub includes heath, composed of shrubby members of Ericaceae (*Vaccinium* spp.) and those of a closely related family, Epacridaceae (*Styphelia*). One also may find here individuals of the peculiar tree-like hapaxanthus plants, the silverswords (*Argyroxiphium* spp.). Their homologs in the tropical Andes are the giant espeletias, on Kilimanjaro, the tree-like senecios. Above these is an alpine tundra or stone desert with a sparse moss (*Rhacomitrium lanuginosum*) and/or crustose lichen (for example, *Lecanora* spp.) growth. This general sequence of altitudinal formations is quite similar to several continental tropical mountains (Troll, in Walter, 1971). Thus, on the level of gross structure, island ecosystems are not at all unique.

## Niche Differentiation

At the level of fine structure, however, island communities are quite unique because not only do we find different species assemblages in habitats comparable to continental ones but we may also encounter quite different spatial relations along environmental gradients. But the degree of departure of island ecosystems in comparison with continental ecosystems is probably best recognized at the level of intermediate structure. In plant ecology this level is conceived as "synusia" (Gams, 1918; Lippmaa, 1939). In animal ecology there exists no direct equivalent concept; however, on a functional level there is the term "general niche" (Miller, 1967).

The concept of niche can be defined as "the space occupied by an organism and the functional role that it assumes in that space." It is useful to distinguish between "specific" and "general" niche. A *specific niche* refers to a species population within an ecosystem. *General niche* refers to a group of species of closely similar ecological characteristics. A functional plant ecological concept for the species occupying a general niche is "ecological group" (Ellenberg, 1956), which also can be applied to animals and microorganisms that share a general niche.

In the Island Ecosystems Research Program of the US/IBP, we have integrated our studies on the basis of ecological groups and the general niches that these groups occupy. This, we believe, will form the best basis for biogeographic and ecological comparisons of the structure and function of ecosystems.

We are currently studying a variety of ecological groups along an altitudinal ecosystem gradient and in a 200-acre plot of a large homogeneous montane rain forest. On the gradient the groups include phytophagous insects (sap-sucking insects, leaf miners, and so on), bark and wood beetles, soil arthropods, fungi, algae, and so on. On a similar basis we are studying the plant synusiae in the rain forest community, including the canopy and emergent trees, the herbaceous ground-rooting plants, and the mosses and liverworts on the forest floor.

In the context of the general niche we observe that the canopy tree synusia has a very low diversity, while there is an absence of native forms in the herbaceous ground-rooted synusia. Here, only ferns have become established, and only in low quantities. Exotic plants invade this synusia, an otherwise completely native community. However, the reason is not merely the poor occupation of this general niche by native herbaceous species but also the active pigs that disturb the forest floor. (Pigs were first introduced by the Hawaiians 1,000 to 2,000 years ago.) Furthermore the number of native species of plants and their quantitative occupation of the general niches may provide a useful index to predict the resistance offered against invasion of exotic biota of similar ecological adaptation. One general plant niche practically unfilled on the island is the vine, or liana, synusia. In this forest there is only one thin-stemmed woody vine, *Alyxia olivaeformis*, and the native *Rubus hawaiiensis* shrub, which sometimes grows as a vine but with upward growth limited at a 5- to 8-m height. An introduced vine from tropical South America, *Passiflora mixta*, recently has spread vigorously in another forest (about 40 mi away on Mauna Kea), which has been strongly disturbed by cattle and pigs. The vine has assumed epidemic proportions by forming dense curtains from the ground to the canopy of the tallest (25 to 30 m) *Acacia koa* trees, killing many members of the subordinate tree and shrub synusia. The exotic vine is so successful because it fills a practically empty niche. In addition, the original community structure was partly destroyed by introduced herbivores that interfered with the reproduction of woody plants.

## Successional Phenomena

Succession refers to a directional rather than rhythmic change in the biotic community on the same site that continues for a period of at least several years. In island community succession the interaction of exotic and native species can bring about important ecological consequences. Replacement by exotic species and recovery trends of native species have been studied in this project.

## Replacement by Exotic Species

A closed perennial grassland occurs in the mountain parkland ecosystem on the east flank of Mauna Loa (4,500 to 6,700 ft, or 1,370 to 2,040 m, elevation). It has a simple composition of about ten herbaceous species. The dominant grasses are the native *Deschampsia australis* and the European *Holcus lanatus.* The latter was first collected in 1903 (Whitney et al., 1939). During a detailed reconnaissance in 1965, I noticed that *Holcus lanatus* appeared to be commoner on areas that were disturbed by feral pigs (Mueller-Dombois, 1967). From 1971–1972 a detailed experimental study was made to determine if pig-digging aided in the spread of the introduced grass (Spatz and Mueller-Dombois, 1972b). While both grasses participated in the invasion of freshly scarified soil, the rate of invasion of *Holcus* exceeded that of *Deschampsia.* For example, a place that contained about 60 percent *Deschampsia* and only 25 *Holcus* before ground disturbance had after one year a plant cover of 30 percent, of which about 15 percent was *Holcus* and less than 5 percent was *Deschampsia.* Areas with no noticeable ground disturbance are still dominated by the native grass. It appears that the invasion of the exotic *Holcus lanatus* is decidedly favored by pig-digging.

Another example of grassland establishment seems to have more serious ecological consequences. An eastern North American grass, *Andropogon virginicus* (broomsedge), was noticed on Oahu, Hawaii, in 1932 (Rotar, 1968). This grass spread, and today it forms the dominant herbaceous cover in all denuded lowland (up to about 800 ft, or 250 m, elevation) rain forest habitats on windward Oahu. The grass is clearly a fire-adapted bunchgrass that accumulates dead standing yellow foilage in a few years. In addition, it goes into dormancy during the winter months just when the rainfall increases. At that time of the year it forms a straw-like mulch that reduces evapotranspiration from the surface, in an area where excess water is a problem. Transpiration studies have shown that the *Andropogon* grass cover removes only a fraction (about 20 to 25%) of the incoming rain water during October through April (Mueller-Dombois, 1972a). On the same habitat, evergreen trees remove considerably more water because they maintain a much greater quantity of active leaf material on a square meter basis (2.24 kg/m$^2$ green foliage for the trees versus 0.58 kg/m$^2$ green blades for the grass). The habitat shows considerable erosion and runoff, much of which can be indirectly attributed to the introduced grass. The dominance of *Andropogon* is maintained by periodic fires.

## Recovery Trends of Native Species

### *Succession on New Volcanic Surfaces*

Plant invasion and recovery after a volcanic eruption were studied for nine years in a rain forest location (Smathers and Mueller-Dombois, 1972) on a number of new volcanic habitats at the Kilauea Crater in Hawaii. On a pahoehoe lava rock habitat, algae, mosses, and ferns arrived in the first year. Lichens arrived in the third year and seed plants in the fourth. The seed plants consisted of five species, four of which were native woody plants (*Metrosideros collina* seedlings, *Vaccinium reticulatum*, *Dubautia scabra* and *Hedyotis centhranthoides*); the other was an exotic species (*Lythrum maritimum*). In time the frequency of the four native seed plants increased more uniformly over the new surface, but they were always found in the rock crevices. The exotic species disappeared, but other exotics became established, in particular the grasses *Paspalum conjugatum*, *P. dilatatum*, *Andropogon virginicus*, and *Setaria geniculata*. These remained only in a very restricted area in moist microhabitats.

Next to an undisturbed rain forest, on ash, several exotic woody plants (*Buddleja asiatica*, *Rubus rosaefolius* and *Rubus penetrans*) became established, mostly at the base of standing snags where the moisture relations were favorable. However, native woody seed plants also became established. After a few years considerable mortality was observed among the exotic woody plants (*Buddleja*, in particular), but not among the native species. Moreover, some individuals of the exotic *Rubus* shrubs were replaced by native shrubs (*Dubautia scabra* and *Vaccinium reticulatum*), indicating competitive replacement. Here exotic grass species also arrived but they remained confined to moist microhabitats.

The results show that, in general, native pioneer species are better adapted to the new, edaphically extreme habitats. However, exotic species participate in primary succession. Exotics that receive no or little competition from native plants appear to be most successful. For example, no pioneer grasses have evolved in the Hawaiian rain forest. Exotic grasses, therefore, fill a vacant niche in primary succession on lava and ash surfaces in rain forest climates.

### *Recovery Following Experimental Herbivore Displacement*

The mammalian herbivore niche was not filled in the course of island evolution of Hawaii. However, several large hervibores were introduced to the islands by the Europeans. Among these, goats became particularly abundant on the island of Hawaii, where they were introduced about 200 years ago. Concentrations occur in the seasonally dry climates in the mountain parkland ecosystem and the coastal lowland of Hawaii Volcanoes National Park.

The impact of goats on the vegetation has been studied using several goat exclosures in these two areas. A 10-m by 100-m exclosure was built in the mountain parkland in 1968. The exclosure was established in the grassland adjacent to an *Acacia koa* (koa) stand. Koa reproduces in the mountain parkland from root suckers, while in the rain forest it reproduces primarily from seed.

In 1971 a quantitative analysis (Spatz and Mueller-Dombois, 1972a) showed that the exclosure was stocked with a dense sapling stand of koa suckers from 10 cm to 2 m tall, while hardly any suckers of this size occurred outside the exclosure. However, a very large number ($> 3/m^2$) of small

(< 5 cm) herbaceous root suckers were found outside. A survey of koa throughout the mountain parkland showed that the observations were representative. Very few taller woody saplings were observed, and among these about 50 percent were girdled and defoliated, had broken stems, or were dead and still standing.

These results showed that koa has a tremendous capacity to resprout when its small herbaceous suckers are browsed, but has no capacity to resprout when the woody saplings are girdled or broken from browsing. The capacity to resprout from roots appears to be an adaptation to grow competitively within closed grassland. Sprouting may also be encouraged by burning or ash fallout from volcanic explosions. Fire appears to be a natural stress factor in this island environment (Mueller-Dombois and Lamoureux, 1967; Vogl, 1969). However, the very dense resprouting after clipping of herbaceous suckers is undoubtedly a new response. When goats are removed after heavy browsing, koa saplings grow in dense thickets in which each individual becomes exposed to high intraspecific competition.

In contrast, the inability of woody saplings to resprout or form new leaders shows that koa is not really well adapted to browsing pressure. In a woody vegetation where browsing is a long-established factor, as for example in the monsoon forest-scrub vegetation of Ceylon, nearly all tree species were observed to respond to browsing by formation of new branches (Mueller-Dombois, 1972b).

In 1968 a similar goat exclosure was constructed in the coastal lowland, in a summer drought climate with 800 mm rainfall. Here the vegetation was dominated by the exotic, pan-tropical annual grass *Eragrostis tenella.* Within two years the plant cover changed completely. Perennial bunchgrasses (*Sporobolus africanus, Rhynchelytrum repens*) became established. Among these was *Heteropogon contortus*, a grass introduced long ago by the Hawaiians. Also, woody chamaephytes (*Waltheria indica*, *Indigofera suffruticosa*, and *Cassia leschenaultiana*) began to grow. But most surprising was the appearance of a native legume vine, *Canavalia kauensis*, that was identified as a new endemic species (St. John, 1972). In the third year *Canavalia* covered more than 50 percent of the surface area of the exclosure (Mueller-Dombois and Spatz, 1972).

Both examples show that native plant recovery is, at least in part, still possible when the introduced stress factor is removed. The Park Service is now fencing large areas in the lowland in an effort to recreate at least a partially native Hawaiian ecosystem.

## Conclusions

Island ecology differs from continental ecology in several ways at the level of species interactions. Spatial distributions are often different because only a few well-adapted species may exist in certain life form groups. Their range or amplitude can be unusually broad because of an absence of competitors. Certain general niches, for example that of the phytophagous insects, are well filled and diverse. In those niches there is greater stability in the sense that penetration by exotics is much more limited. In poorly occupied niches, such as that of the soil arthropods or the grasses in disturbed rain forest habitats, sweeping invasion by exotics may occur. Even ecologically ill-adapted exotics may in such cases retain dominance, for example, the temperate *Andropogon virginicus* in disturbed rain forest habitats. However, in most cases so far observed among plants, exotic species do not penetrate easily into established native communities, unless new niches are created by direct disturbances, for example, by cattle grazing in forests, digging in grassland, or by goat grazing and gradual elimination of certain species from the area. In these cases, exotics that are better adapted to such stresses may take over, for example, the annual *Eragrostis tenella* in the overgrazed lowland. On the other hand, removal of these man-introduced stresses still offers chances of partial recovery of native vegetation and associated biota.

Contrary to common opinion, many endemic island species are strong competitors. They would not be displaced or eliminated from their niche in the island ecosystems if it were not for the new stress factors introduced by man. The island species evolved with stress factors associated with volcanism, fire in seasonally dry habitats, and occasional hurricanes. The effects of volcanism resulted in superior adaptation of many native species to extreme edaphic conditions existing on volcanic rockland, where soil-water regimes fluctuate almost instantly in direct relation to rainfall.

To preserve the unique remaining island biota it is necessary to eliminate the recently man-introduced stress factors that were not part of the spectrum of evolutionary stress factors of the isolated, oceanic island ecosystems.

The effect of evolution that is so clearly apparent in many ecological phenomena in island ecosystems is, of course, important also in continental ecosystems. However, ecological studies are concerned usually with processes and relationships that do not need a historical or evolutionary explanation. Thus ecology and evolution are traditionally quite separate areas of research. In any fundamental study of island ecology, however, the evolutionary background cannot be ignored.

Moreover, future contributions to ecosystem theory will have to come from ecological studies other than ecosystem process studies at the gross structural level. At this level the answer will merely substantiate Tansley's (1920, 1935) concept of an ecosystem as a quasiorganism. Also, contributions to an understanding of ecosystems will hardly be expected from studies of species populations out of context of their general niche, since such studies describe ecosystems merely as a collection of parts. However, studies of species interactions within the context of their general niches and within the context of the ecosystem as a whole will provide an understanding of the degree and preciseness of the web-like ramifications that make a community and its habitat an interacting system.

## References

Carlquist, C. 1965. *Island Life. A Natural History of the Islands of the World.* Am. Mus. Natural Hist. Garden City, N.Y.: The Natural History Press.

———. 1970. *Hawaii. A Natural History.* Am. Mus. Natural Hist. Garden City, N.Y.: The Natural History Press.

Darwin, C. R. 1859. *On the Origin of Species by Natural Selection, or the Preservation of Favoured Races in the Struggle for Life.* London: John Murray.

Dorst, J. 1972. Parks and resources on islands. Paper presented at Second World Conf. on National Parks. USDI, National Park Service.

Ellenberg, H. 1956. *Aufgaben und Methoden der Vegetationskunde.* Stuttgart: Verlag Eugen Ulmer.

Fosberg, F. R. 1948. Derivation of the flora of the Hawaiian Islands. In *Insects of Hawaii,* E. C. Zimmerman, ed. Vol. 1, pp. 107–110. Honolulu: Univ. Hawaii Press.

———. 1963. The island ecosystem. In *Man's Place in the Island Ecosystem,* F. R. Fosberg, ed. pp. 1–6. Honolulu: Bishop Museum Press.

———. 1966. The oceanic volcanic island ecosystem. In *Galápagos. Proc. Galápagos Intern. Sci. Proj.* R. I. Bowman, ed. pp. 55–61. Univ. California Press.

———. 1967. Opening remarks. Island ecosystem symposium. Presented at XI Pac. Sci. Congr., Tokyo. *Micronesia* 3:3–4.

Gagne, W. C. 1972. Species and community distribution along an island ecosystem gradient. Foliar insect communities. Abstract. Nov. 10, 1972. Workshop Symp. Island Ecosystems IRP. Honolulu, Hawaii.

Gams, H. 1918. Prinzipienfragen der Vegetationsforschung. Ein Beitrag zur Begriffserklärung und Methodik der Biocoenologie. *Vierteljahrsschrift Naturforsch. Ges. Zürich* 63:293–493.

Lippmaa, T. 1939. The unistratal concept of plant communities (the unions). *Am. Midland Naturalist* 21:111–145.

Macdonald, G. A., and A. T. Abbott. 1970. *Volcanoes in the Sea. The Geology of Hawaii.* Honolulu: Univ. Hawaii Press.

Miller, R. S. 1967. Pattern and process in competition. In *Advances in Ecological Research.* Vol. 4, pp. 1–7. New York: Academic Press.

Mueller-Dombois, D. 1967. Ecological relations in the alpine and subalpine vegetation on Mauna Loa, Hawaii. *J. Ind. Bot. Soc.* 96(4):403–411.

———. 1972a. A non-adapted vegetation interferes with soil water removal in a tropical rain forest area in Hawaii. Dept. Bot. Univ. Hawaii. *US/IBP Island Ecosystems IRP Tech. Rept. No. 4.*

———. 1972b. Crown distortion and elephant distribution in the woody vegetations of Ruhuna National Park, Ceylon. *Ecology* 53(2):208–226.

———, and C. H. Lamoureux. 1967. Soil-vegetation relationships in Hawaiian kipukas. *Pac. Sci.* 21(2):286–299.

———, and G. Spatz. 1972. The influence of feral goats on the lowland vegetation in Hawaii Volcanoes National Park. Dept. Bot. Univ. Hawaii. *US/IBP Island Ecosystems IRP Tech. Rept. No. 13.*

Radovsky, F. J. 1972. Species and community distribution along an island ecosystem gradient. Distribution and abundance of soil arthropods. Abstract, 2 tables, 1 figure. Nov. 10, 1972. Worshop Symp. Island Ecosystems IRP. Honolulu, Hawaii.

Rotar, P. P. 1968. *Grasses of Hawaii.* Honolulu: Univ. Hawaii Press.

Smathers, G. A., and D. Mueller-Dombois. 1972. Invasion and recovery of vegetation after a volcanic eruption in Hawaii. Dept. Bot. Univ. Hawaii. *US/IBP Island Ecosystems IRP Tech. Rept. No. 10.*

Spatz, G., and D. Mueller-Dombois. 1972a. The influence of feral goats on koa (*Acacia koa Gray*) reproduction in Hawaii Volcanoes National Park. Dept. Bot. Univ. Hawaii. *US/IBP Island Ecosystems IRP Tech. Rept. No. 3.*

———. 1972b. Succession patterns after pig digging in grassland communities on Mauna Loa, Hawaii. Dept. Bot. Univ. Hawaii. *US/IBP Island Ecosytems IRP Tech. Rept. No. 15.*

St. John, H. 1966. Monograph of *Cyrtandra* (Gesneriaceae) on Oahu, Hawaiian Islands. *Bernice P. Bishop Mus. Bull. 229.*

———. 1972. *Canavalia kauensis* (Leguminosae), a new species from the Island of Hawaii. Hawaiian plant studies 39. *Pac. Sci.* 26(4):409–414.

Tansley, G. A. 1920. The classification of vegetation and the concept of development. *J. Ecol.* 8:118–149.

———. 1935. The use and abuse of vegetational concepts and terms. *Ecology* 16: 284–307.

Tomich, P. Q. 1969. *Mammals in Hawaii.* Bernice P. Bishop Mus. Special Publ. 57. Honolulu: Bishop Mus. Press.

Vogl. R. J. 1969. The role of fire in the evolution of the Hawaiian flora and vegetation. *Proc. Ann. Tall Timbers Fire Ecology Conf.* 5–60.

Walter, H. 1971. *Ecology of Tropical and Subtropical Vegetation.* Edinburgh: Oliver and Boyd.

Whitney, L. D., E. Y. Hosaka, and J. C. Ripperton. 1939. Grasses of the Hawaiian ranges. *Hawaii Agr. Exptl. Sta. Univ. Hawaii Bull. No. 82.*

Whittaker, H. 1970. Communities and ecosystems. In *Current Concepts in Biology Series.* London: Macmillan.

Wirawan, N. 1972. Floristic and structural development of native dry forest stands at Mokuleia, N. W. Oahu. M. Sc. thesis, Univ. Hawaii, Honolulu.

# Applications

It is very apparent that the science of ecology has application to many of the problems man faces in maintaining a highly productive and satisfying environment. In a sense, ecology is to the environmental management sciences as physics is to the engineering sciences. This point is well made in the first paper by Terborgh, who reasons from the quite esoteric studies of island biogeography to the optimum size of preserves for wildlife. This extremely practical question is fundamental to the preservation of the vast genetic diversity inherent in the tropics, which man has used with great success in the past and will undoubtedly continue to utilize in the future. We wonder if the administrators of tropical lands realize the size of reserves needed to preserve their fauna.

Ecology contributes to many subjects other than wildlife management. Industrial development, transportation, agriculture, and forestry all impact the natural environment. Kreb's study of the changes in volcanic soils as they are converted from forest to permanent agriculture is an example of the type needed for realistic evaluation of the environmental impact of development on the land.

# CHAPTER 24

# Faunal Equilibria and the Design of Wildlife Preserves

JOHN TERBORGH

A recurrent theme in science is that the utilitarian harvest from new theoretical developments comes after some delay and often in unanticipated quarters. So it can be said of modern island biogeography whose founders, MacArthur and Wilson (1963), perceived the simplicity and analytical tractability of isolated ecosystems. Now it is becoming increasingly apparent that the methods and the way of thinking they developed are extensible to a much larger range of situations, including the design of faunal preserves.

Because my own experience has been largely with birds, my arguments will pertain especially to them, but there is no evident reason why the principles should not apply equally well, with appropriate modification, to other groups of animals. This presentation is organized into three sections. First, we will consider evidence that tropical forest bird species, with few exceptions, have very limited dispersal and colonization abilities in relation to their temperate counterparts. Second, we will examine the kinetics of species loss from forested islands that have been cut off from the normal interchange of species with adjacent forested regions. Last, the conclusions to be drawn from these results will be applied to the problem of how to optimize the design of faunal preserves.

## Response to Habitat Disruption

By the time of World War II the eastern half of the United States had been so thoroughly logged that less than 1 percent of the virgin forest remained. Yet this systematic deforestation resulted in the extinction of at most two species of birds, the ivory-billed woodpecker and the Carolina parakeet, and the proximal cause of extinction of the latter is somewhat in doubt (Greenway, 1967). Others, such as the wild turkey, pileated and red-cockaded woodpeckers, and swallow-tailed kite, suffered great loss of numbers and have retreated from parts of their original ranges but are not seriously threatened. This fauna's high level of resistance to extinction is due to a number of factors, both biological and historical.

First, we must remember that the cutting of the eastern forests, although thorough, was gradual, taking nearly 300 years. During this entire period the total area in forest was never much below half of the original amount, and today the extent of forest is greater than it was 70 years ago and still increasing. Harvesting of the primeval timber was thus done in a way that buffered its impact on native wildlife.

The biological properties of a great majority of temperate bird species provide them with a high degree of resiliency to local habitat disruption. A glance at the breeding bird list of any forested locality in eastern North America shows that two-thirds or more of the species are partially or wholly migratory. Moreover, a great majority of forest-nesting species will accept middle-aged second growth as suitable habitat. Finally, species nesting at temperate latitudes typically possess high reproductive rates and maintain dense populations. These properties that confer a notable degree of resistance to extirpation following habitat destruction are just those that, in general, distinguish temperate from tropical nesting species.

The outstanding work of Skutch (1954, 1960) and others on the nesting of tropical bird species shows that most lay two or three eggs at a time instead of four, five, or six, and additionally, because of high predation, the ratio of young fledged to eggs laid is lower. This implies very slow population recoveries following disturbances.

At least three lines of evidence indicate that tropical forest birds are poor at invading second growth. One is that many regions possess a large complement of nonforest species that seem intrinsically better adapted for the early and intermediate stages of second growth. In the Amazon basin these are the birds characteristically inhabiting the margins of rivers, particularly unstable alluvial meander belts where periodic flooding maintains large areas in a successional condition. This floodplain vegetation is very similar in both structure and species composition to the second growth that develops in clearings on higher ground. Thus it is not surprising that riparian bird species contribute disproportionately to the faunas of second growth (Terborgh and Weske, 1969).

A second reason for the poor showing of forest birds in successional vegetation derives from the importance of vertical structure in habitat subdivision by competing species. Virgin forests in the Amazon basin commonly contain upward of 150 resident bird species, many of which are locked into delicately balanced competitive complexes. One frequent manifestation of these competitive interactions is compression of foraging zones (Figure 24–1). Species confined to narrow foraging zones become extremely dependent on the precise way that the vegetation is structured. The reduced stature and altered vertical organization of second growth necessarily mean that some species will be eliminated, since the full complement of interacting species cannot be spatially accommodated. Just which species will succeed in invading second growth can be predicted with a fair degree of accuracy, since we have found that within any vertically interactive group of species, those having the greatest variance in

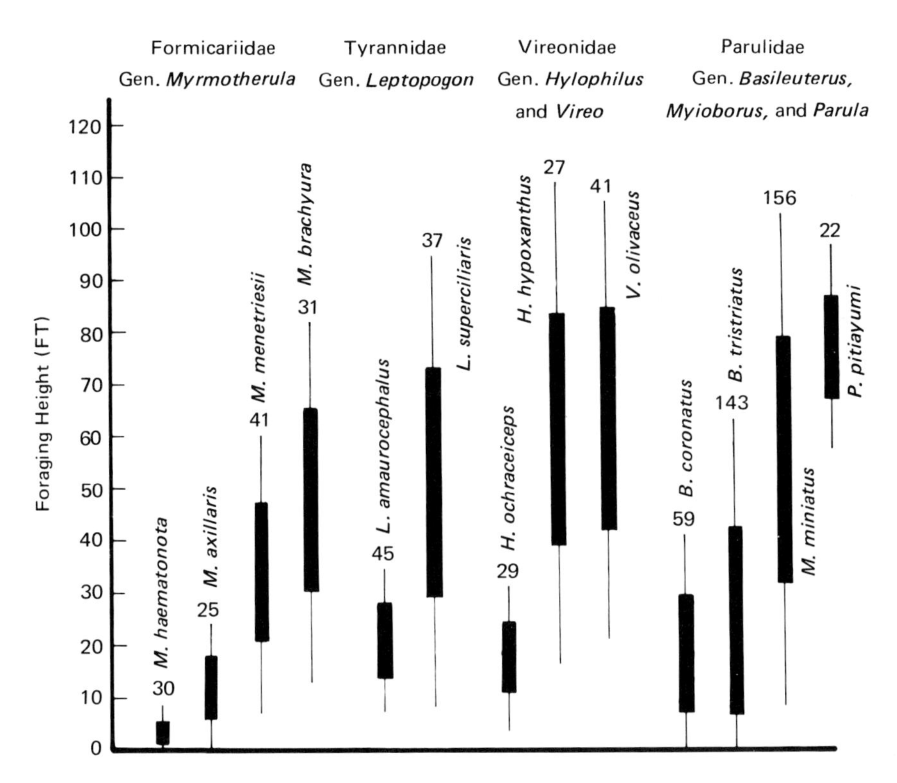


**Fig. 24–1.** Foraging heights of four vertically interactive species complexes in Peruvian forests. Heavy bars indicate a range of ±1 standard deviation about the mean foraging height. Thin bars indicate the range encompassed by two standard deviations. The number of observations is given at the top of each bar. Only one species in each group readily invades second growth. These are *Myrmotherula brachyura, Leptopogon amaurocephalus* (low elevations), *L. superciliaris* (higher elevations), *Vireo olivaceus*, and *Myioborus miniatus*. The remaining species enter second growth at greatly reduced density or not at all.

foraging height usually display the greatest versatility in accommodating to vegetation of altered structure (Figure 24–2).

Extremely limited dispersal ability gives tropical birds a third handicap in surviving the effects of habitat destruction. We learned a number years ago, to our considerable surprise, that the species composition of scattered patches of disturbed vegetation depended as much on their proximity to sources of colonists as on their structure (Terborgh and Weske, 1969). Further investigation has amply confirmed the initial finding (Figure 24–3). Patches of young second growth near or adjacent to a river and a kilometer or more from forest contain nothing but riparian bird species. Very similar patches, however, which are surrounded by forest and some distance from the nearest river, contain large numbers of forest species and a greatly reduced component of riparian species. Merely a kilometer or two of unsuitable habitat seems sufficient to screen out about half the expected species. Moreover, it should be noted

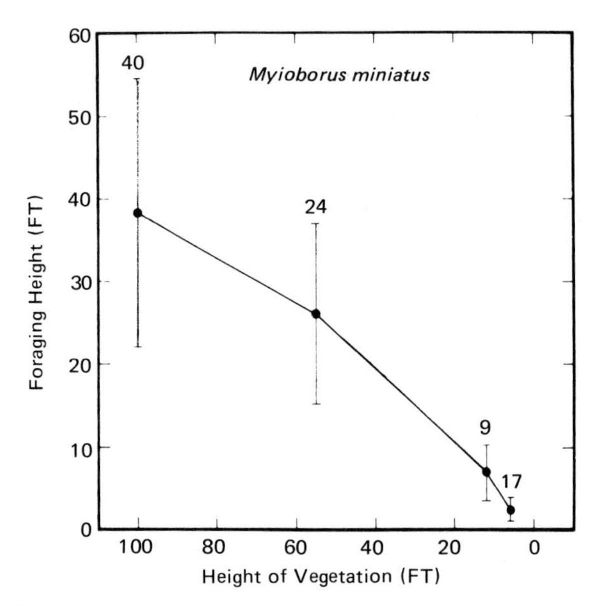


**Fig. 24–2.** Effect of reduced stature of vegetation on the foraging behavior of *Myioborus miniatus*, a versatile colonizer of second growth. The bars indicate ±1 standard deviation and the number of observations is given above each bar. The control data for mature forest (100 ft) were taken in a different locality from those in Figure 24–1 and are somewhat lower. Note that colonizing success requires a flexible behavorial response and an ability to maintain a high rate of prey capture despite contraction of the variance in foraging height.

that these experiments test the dispersal abilities of riparian bird species, which, by virtue of living naturally in shifting and successional vegetation, should be more adept at dispersal than forest species.

## Nature's Island Lesson

From the results just presented, one would readily conclude that numerous extinctions of forest bird species would accompany wholesale destruction of the virgin rain forest in South America or in other humid tropical regions. Let us now, then, examine evidence of a kind that intimates at the degree of success to be anticipated in preserving species in forest reserves or parks.

Such evidence can be derived from experiments that nature performed for us at the conclusion of the Pleistocene. Geologists are now in satisfactory agreement that the present sea level is approximately 100 m higher than it was during the peak of the most recent glaciation (Flint, 1957; Broecker, 1966; Mercer, 1972). We may thus confidently suppose that for an extended period of thousands of years prior to about 10,000 years ago, large areas of continental shelf were emergent and presumably supported well-developed vegetation. Reversal of this situation commenced some 10,000 to 14,000 years ago

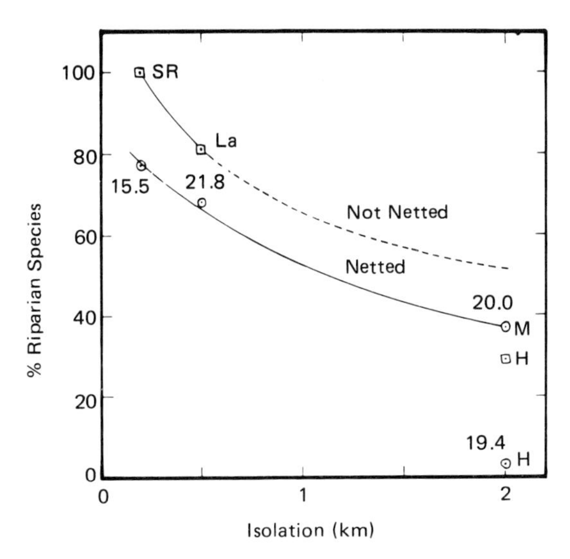


**Fig. 24–3.** Effect of distance from source habitat on the species composition of faunas colonizing patches of second growth. The curves show a parallel decline of riparian species in the netted (understory) and not-netted (canopy dwelling) components of the colonizing faunas. In each instance the remainder of the fauna was composed of forest species. The plots consisted of young second growth 20 to 40 ft high, and were separated from the nearest riparian vegetation by unsuitable habitat, either open fields or forest. All of the plots were adjacent to or surrounded by undisturbed forest and were thus accessible to the full forest fauna. Figures beside the points give the number of equally common species in the netted samples and the abbreviations refer to localities in the Apurimac Valley of Peru: *SR*, Rio Santa Rosa; *La*, Hacienda Luisiana; *M*, Rio Mapitunari; *H*. Huanhuachayoc. The latter is the site of an abandoned coffee plantation and is isolated from riparian vegetation both by distance and by being 600 m higher in elevation.

when glacial melting brought about a rapid rise that restored the sea level to nearly its present position at a time estimated at 6,000 to 8,000 years ago. An interesting consequence of the fluctuating sea levels was that a number of present-day islands were connected to the mainland at a known time in the past. Such islands can be identified on hydrographic charts as those lying inside the 100-m contour. Within the Neotropical region these include Trinidad, Tobago, Margarita, Coiba, and the Perlas archipelago.

Comparison of the species-area curve for these islands with that for the West Indies, which have not had a recent land connection, shows that the land bridge islands carry many more bird species, especially the larger ones (Figure 24–4). It can be argued that the richer faunas are due to the greater proximity of these islands to the mainland, but this interpretation is countered by the fact that nearby islands which are separated by deep channels (Grenada, Cozumel, Bonaire, La Orchilla, Blanquilla) have species numbers that agree well with the West Indian curve. Thus the most probable explanation for the result is that the land bridge islands have lost varying numbers of species since their separation from the mainland in inverse proportion to their areas. This

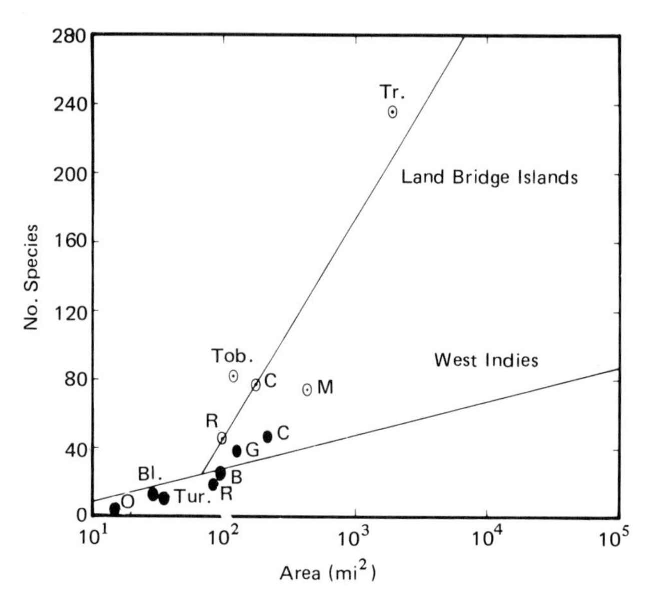


**Fig. 24–4.** Species-area regressions for Neotropical land bridge islands and the West Indies. The latter was taken directly from Terborgh (1973). Open circles pertain to land bridge islands: *Tr*, Trinidad; *M*. Margarita; *C*, Coiba; *Tob*, Tobago; *R*, Rey. Solid dots give data for oceanic islands near the Central or South American mainlands. With the exception of Grenada, these islands were not used in computing the regression line for the West Indies. *C*, Cozumel; *G*, Grenada; *B*, Bonaire; *R*, Roatan; *Tur*, Turneffe; *Bl*, Blanquilla; *O*, Orchilla.

view is supported by very similar findings in a comparison of oceanic and land bridge islands in the vicinity of New Guinea (Diamond, 1972).

We must take three considerations into account in reconstructing the kinetics of species loss from land bridge islands. First, it is necessary to have estimates of the number of species present at the time land connections were severed. These can be derived from interpolation between figures for the resident bird faunas of forested mainland localities of varying areas (Table 24–1). Because the number of species on an island at equilibrium (when immigration and extinction are balanced) is several to many times fewer than on a comparable portion of mainland, relatively crude estimates are satisfactory.

Second, we must consider the effect of immigration from the mainland in retarding the process of faunal dilution by extinction. At the moment of separation, an incipient island is in a state of equilibration with the adjacent mainland, which implies a very low immigration rate. After isolation has been imposed by the intervention of a stretch of water, the immigration kinetics become similar to those of any other island. Examination of the faunas of New World islands that have not had a recent land connection shows that their colonists have been drawn from an extremely limited group of mainland species. With few exceptions these oceanic islands lack any member of numerous families that are of considerable prominence in similar mainland habitats. These include the tinamous, guans, jacamars, puffbirds, barbets, toucans, woodcreepers, ovenbirds, antbirds, and manakins.

**Table 24–1.** Number of land bird species inhabiting various insular and continental landmasses of defined area.

| Island | Area ($mi^2$) | No. Species 10,000 yr B.P. | No. Species Now | Reference |
|---|---|---|---|---|
| Land bridge islands | | | | |
| Trinidad | 1,864 | 380 | 236 | Herklots (1961) |
| Margarita | 444 | 320 | 74 | Yepez and Benedetti (1940) |
| Coiba | 175 | 250 | 78 | Wetmore (1957) |
| Tobago | 116 | 300 | 82 | Herklots (1961) |
| Rey | 96 | 225 | 46 | MacArthur et al. (1972) |
| Escudo de Veraguas | 2.3 | 190 | 9 | Wetmore (1959) |
| Oceanic islands | | | | |
| Cozumel | 206 | | 47 | Paynter (1955); Bond (1961, 1963) |
| Grenada | 120 | | 38 | Bond (1971) |
| Bonaire | 95 | | 26 | Phelps and Phelps (1951) |
| Roatan | 78 | | 18 | Bond (1937) |
| Turneffe | 36 | | 11 | Bond (1954) |
| Blanquilla | 29 | | 13 | Phelps (1948) |
| Orchilla | 15 | | 5 | Phelps and Phelps (1959) |
| Mainland control areas | | | | |
| Surinam | 55,600 | | 494 | Haverschmidt (1968) |
| Panama Canal Zone | 362 | | 330 | Eisenmann and Loftin (1971) |
| Rancho Grande Natl. Pk., Ven. | 337 | | 401 | Schafer and Phelps (1954) |
| Barro Colorado Is., Panama | 6.5 | | 205 | E. Willis, personal communication |
| Hda. Luisiana, Peru | 2.0 | | 255 | Terborgh and Weske (1969) |

One must conclude that the species in these families are extremely loath to crossing water, however modest the distance. The presence of members of many of these families on land bridge islands strongly suggests the notion that they represent a residual element that has survived without secondary reinvasion since the time of separation from the mainland. Such direct evidence, as well as the behavior of reasonable mathematical models, indicates that immigration has an extremely minor effect in retarding the rate of species loss from islands that contain well in excess of the equilibrium number of species.

We now come to the third, and most important point, about species loss from isolated ecosystems—the question of a predictive model. Given the present number of species on a land bridge island and estimates of the time of separation and the number of species it then contained, we wish to devise a model that provides a biologically reasonable account of the time course of species loss during the intervening period. For the sake of completeness we will begin by considering three such models; then grounds will be presented for throwing out two of them as being biologically unrealistic (Figure 24–5).

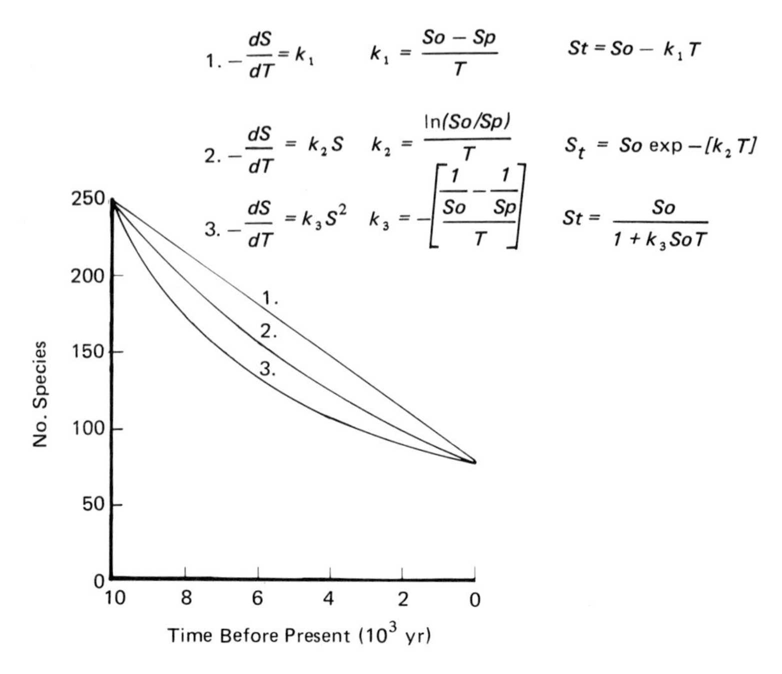


**Fig. 24–5.** Time course of post-Pleistocene species loss from Isla Coiba, Panama, in accordance with three hypothetical extinction models. An estimated 250 species are assumed to have been present on the island at the time of its separation from the mainland. For each model formulas are given that describe, respectively, the corresponding rate of species disappearance, extinction coefficient, and the number of species present $S_t$ after any period of elapsed time $T$. The remaining symbols are defined as follows: $S$, instantaneous species number; $S_0$, initial species number; $S_p$, present species number; $k_1$, $k_2$, $k_3$, extinction coefficients for the three models, respectively.

The first model states that a constant number of species are lost per unit time, a kinetics that implies a decreasing probability of survival for the remaining species as their competitors disappear from the community. This is wholly at odds with our knowledge of insular biogeography, which suggests, to the contrary that the members of species-poor communities have much greater survival potential than the members of comparatively rich communities (Mayr, 1965). In the second model the rate of species loss is proportional to the instantaneous number of species, under which conditions the survival potential of each species is independent of the size of the fauna. Again, this model contradicts our knowledge of communities since it denies the likelihood that interspecific competition plays a role in bringing about extinctions. The last model is the one we shall consider further since it includes an interactive term in $S^2$ that provides that species hasten each other's extinctions.

The probability of extinction now decreases with the number of species as would be anticipated. Integration of the equation over the presumed 10,000 years since the postglacial rise in sea level gives us an expression that can be solved to give an extinction coefficient $k_3$ for any particular island. Now, using these coefficients it is possible to solve for the presumed course of species loss. Table 24–2 gives figures for the computed number of extinctions over the first 100 and 1,000 years for the five Neotropical land bridge islands for which sufficient information is available. As might be anticipated, the extinction coefficient for Trinidad, a large island of 1,864 mi² is considerably lower than for the smaller islands of Coiba, Tobago, and Rey (Figure 24–6). Only Isla Margarita fails to fit the pattern in that it presently holds fewer species than expected. A likely reason is that Margarita is appreciably drier than the other islands and carries tall forest only on the elevated portions (Yepez and Benedetti, 1940).

The relation between the extinction coefficient $k_3$ and the area of land mass that has been cut off from immigration is the goal of our inquiry since it now allows us to make crude predictions of the effectiveness of forested wildlife

**Table 24–2.** Estimated species loss over first 100 and 1,000 years after isolation of five Neotropical land bridge islands.

| Island | No. Species Lost in $10^4$ Yr | $k_3$ | No. Species Lost 1st 100 Yr | No. Species Lost 1st 1,000 Yr | Percent of Initial Fauna Lost 1st Century |
|---|---|---|---|---|---|
| Trinidad | 144 | $1.606 \times 10^{-7}$ | 2.3 | 21.9 | 0.6 |
| Margarita | 246 | $1.039 \times 10^{-6}$ | 10.3 | 79.8 | 3.2 |
| Coiba | 172 | $8.821 \times 10^{-7}$ | 5.4 | 45.2 | 2.2 |
| Tobago | 218 | $8.862 \times 10^{-7}$ | 7.8 | 63.0 | 2.6 |
| Rey | 179 | $1.729 \times 10^{-6}$ | 8.4 | 63.0 | 3.7 |

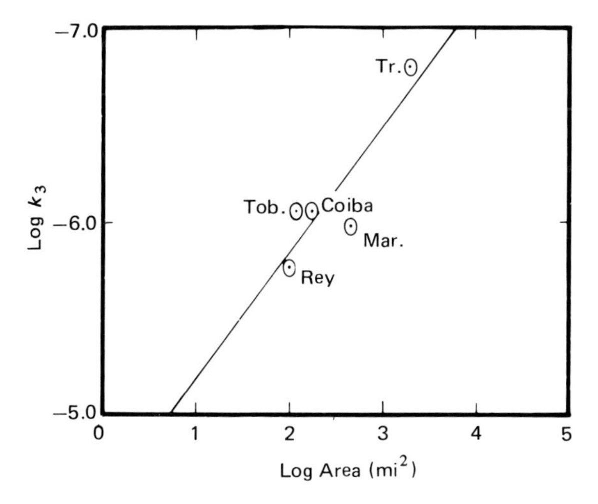


**Fig. 24–6.** Double log regression of extinction coefficient $k_3$ on island area. The relationship permits rough predictions of the rates of initial species loss from isolated "islands" of habitat of known area.

preserves. Let us now make such a prediction to test the worth of the method. This can be done for an already existing wildlife reserve, Barro Colorado Island in the Panama Canal Zone. The formation of Lake Gatun in conjunction with the construction of the canal left Barro Colorado as an emergent hilltop of some 6.5 $mi^2$ in extent. Since 1923 the island has been carefully protected from hunting and from destruction of the forest. Using its area of 6.5 $mi^2$, which gives from Figure 24–6 an estimated extinction coefficient of $8 \times 10^{-6}$, an initial species count of 205 resident birds (kindly provided by Dr. Edwin Willis), and an elapsed time of 50 years, we arrive at a predicted current fauna of 188.4 species. Thus, our simple model anticipates a loss of 16 to 17 species, while the records of Dr. Willis indicate that about 15 species have actually disappeared. In this lone available test the model has provided an acceptable account of reality.

## The Design of Faunal Preserves

As a final consideration we now wish to apply our knowledge to the problem of designing effective wildlife preserves. Both direct evidence from mainland colonization studies and inferred evidence from the faunas of oceanic islands point to an extremely low dispersal rate among tropical forest birds, such that distances of a few kilometers of unsuitable habitat frequently seem to pose insurmountable barriers to colonization. This leads us to the principle that, wherever possible, corridors should be left connecting forested areas so that the free interchange of populations is not interrupted. The low reproductive rates and exacting structural requirements of many species advise us that wildlife preserves should remain inviolate for the express purpose of preserving wildlife. Other forms of use such as hunting or lumbering would inevitably serve to accelerate the rate of extinction.

Reduction of extinction rates to acceptable levels (less than 1 percent of the initial species complement per century) requires reserves of substantial size, on the order of 1,000 $mi^2$. This means parks that are on the scale of the largest in the United States—Yellowstone, Grand Canyon, and Big Bend. Even within areas of this size it is questionable whether the largest predators such as the jaguar and harpy eagle could be saved for long. Nevertheless, our obligation to future generations demands that we make a concerted attempt.

## Acknowledgment

The fieldwork reported here received support at various times from the Chapman Fund of the American Museum of Natural History, the American Philosophical Society, the National Geographic Society, and the National Science

Foundation (GB-29332). The idea of using land bridge islands to estimate extinction rates comes from a recent paper by Jared Diamond (1972).

## References

Bond, J. 1937. Resident birds of the Bay Islands of Spanish Honduras. *Proc. Acad. Nat. Sci. Philadelphia* 88:353–364.

———. 1954. Birds of Turneffe and northern two cays, British Honduras. *Not. Nat. (Philadelphia)* 260:1–10.

———. 1961. Notes on the birds of Cozumel Island, Quintana Roo, Mexico. *Carib. J. Sci.* 1:41–47.

———. 1963. *Eighth Supplement to the Check-list of Birds of the West Indies*. Philadelphia: Acad. Nat. Sci.

———. 1971. *Birds of West Indies*. 2nd edit. Boston: Houghton Mifflin.

Broecker, W. S. 1966. Absolute dating and the astronomical theory of glaciation. *Science* 151:299–304.

Diamond, J. M. 1972. Biogeographic kinetics: Estimation of relaxation times for avifaunas of southwest Pacific islands. *Proc. Natl. Acad. Sci.* 69:3199–3203.

Eisenmann, E., and H. Loftin. 1971. *Field Checklist of Birds of the Panama Canal Zone Area*. 2nd edit. Florida Audubon Soc.

Flint, R. F. 1957. *Glacial and Pleistocene Geology*. New York: Wiley.

Greenway, J. C., Jr. 1967. *Extinct and Vanishing Birds of the World*. 2nd edit. New York: Dover.

Haverschmidt, F. 1968. *Birds of Surinam*. London: Oliver and Boyd.

Herklots, G. A. C. 1961. *The Birds of Trinidad and Tobago*. London: Collins.

MacArthur, R. H., J. M. Diamond, and J. R. Karr. 1972. Density compensation in island faunas. *Ecology* 53:330–342.

MacArthur, R. H., and E. O. Wilson. 1963. An equilibrium theory of island biogeography. *Evolution* 17:373–387.

Mayr. E. 1965. Avifauna: Turnover on islands. *Science* 150:1587–1588.

Mercer, J. H. 1972. Chilean glacial chronology 20,000 to 11,000 carbon-14 years ago: Some global comparisons. *Science* 176:1118–1120.

Paynter, R. A., Jr. 1955. The ornithogeography of the Yucatan Peninsula. *Bull. Peabody Mus. Nat. Hist.* 9:1–347.

Phelps, W. H., Jr. 1948. Las aves de la isla La Blanquilla y de los morros El Fondeadero y La Horquilla del Archipelago de los Hermanos. *Bol. Soc. Ven. Cien. Nat.* 71:85–118.

Phelps, W. H., and W. H. Phelps, Jr. 1951. Las aves de Bonaire. *Bol. Soc. Ven. Cien. Nat.* 77:161–187.

———. 1959. Las aves de la isla La Orchilla. *Bol. Soc. Ven. Cien. Nat.* 93:252–266.

Schafer, E., and W. H. Phelps. 1954. Las aves del Parque Nacional "Henri Pittier" (Rancho Grande) y sus funciones ecologicas. *Bol. Soc. Ven. Cien. Nat.* 83: 1–171.

Skutch, A. F. 1954, 1960. Life histories of Central American birds. *Cooper Ornithol. Soc. Pacific Coast Avifauna*. Nos. 31 and 34. Berkeley, Calif.

Terborgh, J. 1973. Chance, habitat and dispersal in the distribution of birds in the West Indies. *Evolution* 27:338–349.

———, and J. S. Weske. 1969. Colonization of secondary habitats by Peruvian birds. *Ecology* 50:765–782.

Wetmore, A. 1957. The birds of Isla Coiba, Panama. *Smithsonian Misc. Coll.* 134(9): 1–105.

———. 1959. The birds of Isla Escudo de Veraguas, Panama. *Smithsonian Misc. Coll.* 139(2):1–27.

Yepez, A. F., and F. L. Benedetti. 1940. Las aves de Margarita. *Bol. Soc. Ven. Cien. Nat.* 43:91–132.

# CHAPTER 25

# A Comparison of Soils Under Agriculture and Forests in San Carlos, Costa Rica

JULIA ELIZABETH KREBS

## Introduction

Increasing population in many tropical countries has encouraged settlement of forest lands. While there is little published data on the ecological impact of settlement, some ecologists have suggested that conversion of forest to agricultural lands may have lasting deleterious effects. For example, Holdridge (1959) and Tosi (1964) have postulated that the change from forest to permanent agriculture in Costa Rica might cause rapid deterioration of the soil. The logical basis for this conclusion is well established. High temperatures cause rapid decomposition of soil and litter organic matter, and high rainfall may cause rapid leaching of nutrients from the soil (Aubert and Tavernier, 1972). Thus removal of forest cover could conceivably result in a decline in soil nutrients, since in the intact forest the nutrients are taken up by vegetation immediately upon their release from decaying organic matter (Richards, 1952). The objective of this work was to compare soil characteristics of permanent agricultural fields and forest. The area chosen for the study was a farming community centered around the Centro Rural Metodista along a ridge in the Canton of San Carlos, 8 km north of Cuidad Quesada (Figure 25–1). Soils under three different crops—sugar cane, coffee, and pasture—were studied. Since forests in this region of Costa Rica were first cleared about 22 years ago by settlers moving from the south, fields ranging in age from 22 years to the present were available for study.

A preliminary survey in April 1971 showed that soils along a ridge (500 m elevation) bisecting a farming community were similar in character. The soil texture was sandy loam to loam and the color was dark yellow-brown (MCS 10 YR 3/4). The soil is volcanic in origin with alluvial material overlying lateritic deposits (Anke Neumann, personal communication). Weather data from Cuidad Quesada indicated that the average temperature was 23° and

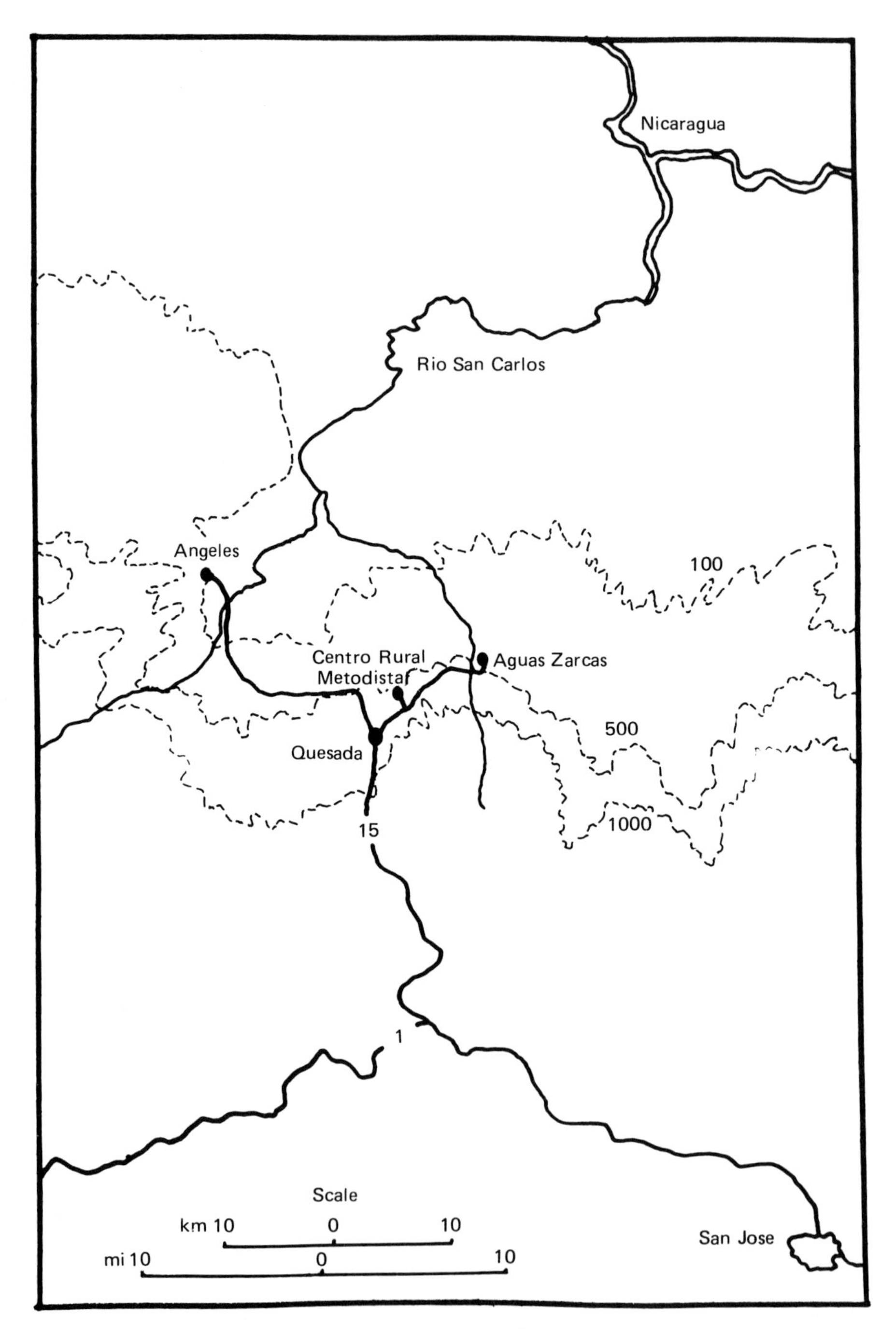


**Fig. 25–1.** Map of San Carlos, Costa Rica. capital of country; • towns; — roads; — rivers; 500 contour lines.

the average annual rainfall was 4,700 mm (20-yr average). On an average it rains 253 days/yr.

## Methods

Since the preliminary survey showed the soils of the area to be relatively homogeneous, sampling sites were chosen from forest and fields of different ages under different crops without considering site effects. Only fields under one crop from the time they were cleared from the forest to the present were chosen for sampling.

The fields were divided into four sections, and a composite of five samples was taken at each of three depths: 0 to 10 cm, 10 to 20 cm, and 20 to 30 cm. Samples were taken with a bucket auger, air-dried, and analyzed within three months.

The laboratory analyses were carried out in the field and in the Department of Agronomy laboratories at the University of Georgia and partially by the soil testing laboratory of the Agricultural Extension Service.

The pH was measured within 24 hr of sampling in a 1:2 suspension of soil in water. The percentage organic matter was determined by the Walkley-Black method (Jackson, 1956), and mechanical analyses by the Hydrometer method (Jackson, 1956). The percentage nitrogen was found by the Kjeldahl method (Jackson, 1956). Ammonium acetate was used to extract potassium, calcium, and magnesium, and concentrations were determined by a modified method of Flannery and Steckel (1967). Alumnium was extracted with the North Carolina extractant, 0.05N HCl and 0.025N $H_2SO_4$, and the concentration determined by atomic absorption spectrophotometry. A multivariate analysis of variance (Morrison, 1967) was used to find differences among factors. The effects of fields (or forests) and depth were determined using the Sheffe method of multiple comparisons. The only correlations reported are those significant at the 99.5 percent level.

## Results and Discussion

The basic approach in this study is to compare the soil characteristics for a crop on different aged fields with that of natural forest. In some cases there is a difference between fields and forest, but not among fields. In others there is a change with age of field; in still others the variation between field and forest and among fields shows no consistent trend. Furthermore, the trends vary among crops. I will consider the soil characteristics in the following order: carbon, nitrogen, potassium, pH, calcium, magnesium, and aluminum.

The percentage of organic carbon generally was different in forest and fields (Table 25–1), with fields having lower carbon levels. There was no

**Table 25–1.** Percent organic carbon of soils from forests and fields in San Carlos, Costa Rica (average of three depths).

| Land Use | | Age of Field (yr) | | | | | |
|---|---|---|---|---|---|---|---|
| Forest | | 1 | 4 | 9 | 15 | 16 | 22 |
| 4.45[a] | Sugar cane | 3.53[b] | 3.59[b] | 3.66[b] | 3.55[b] | — | 3.46[b] |
| 4.45[a] | Coffee | — | 4.42[a] | — | — | 3.79[b] | — |
| 4.45[a] | Pasture | — | 3.55[c] | — | 4.05[b] | — | — |

N.B.: Values in the same row with the same letter following are not different from each other.

difference in percentage of organic carbon among sugar cane fields. The older coffee plantation had a lower percentage of organic carbon than the forest, but there was no difference between the younger plantation and the forest. Both pastures were significantly lower in percentage of organic carbon than the forest, and the younger pasture was significantly lower than the older.

Popenoe (1960) in Guatemala and Brams (1971) and Nye and Greenland (1964) in West Africa also found a decrease in organic carbon from forest to fields. These results would be expected since less organic matter is returned to the soil under crops. For example, the tropical forest has an annual litter fall of about 11 metric tons/ha/yr (Golley, 1972), while only about 7.9 tons/ha of litter from sugar cane could be returned annually (Van Dillewijn, 1952; FAO, 1967). The removal of organic carbon in crops would also act to reduce the carbon level in field soils.

Less nitrogen also was usually found under fields than under the forest (Table 25–2); however fields also differed among themselves. The one-year-old sugar cane field had a significantly lower percentage of nitrogen than all but the four-year-old field. The 15-year-old sugar cane field had a significantly higher percentage of nitrogen than all but the nine-year-old field. These latter two fields had been fertilized yearly with nitrogen, more than likely accounting for the observed differences. However, the older coffee plantation had been fertilized and the younger had not, and the younger plantation had a higher

**Table 25–2.** Percent nitrogen of soils from forests and fields in San Carlos, Costa Rica (average of three depths).

| Land Use | | Age of Field (yr) | | | | | |
|---|---|---|---|---|---|---|---|
| Forest | | 1 | 4 | 9 | 15 | 16 | 22 |
| 0.52[a] | Sugar cane | 0.41[d] | 0.42[cd] | 0.45[bc] | 0.47[b] | — | 0.43[c] |
| 0.52[a] | Coffee | — | 0.54[a] | — | — | 0.46[b] | — |
| 0.52[a] | Pasture | — | 0.35[b] | — | 0.53[a] | — | — |

N.B.: Values in the same row with the same letter following are not different from each other.

percentage of nitrogen. The older pasture had a significantly higher percentage of nitrogen than the younger one, and was not different from forest.

Since the soil type is assumed to be the same under the fields and forest and since the sites are all on the same ridge with approximately the same topography, the data from the forest soil and the soil of the fields of different ages can be treated as if they came from one field that had been followed over a period of several years. Reasoning in this manner one could say that the percentages of nitrogen and organic matter decreased from forest to fields, but remained relatively stable over time. There is no continuous decrease in these parameters as the fields get older.

This pattern was not so clear for potassium (Table 25–3). The one-year-old field was significantly higher in potassium concentration than the forest and the other fields, and the four-year-old field was significantly lower. Finally, the older fields were not significantly different from the forest, nor were the coffee plantations significantly different from the forest. The four-year-old pasture had a significantly lower potassium concentration than the forest, but the older pasture was not significantly different from the forest.

**Table 25–3.** Concentration (in ppm) of potassium in soils from forests and fields in San Carlos, Costa Rica (average of three depths).

| | Land Use | Age of Field (yr) | | | | | |
|---|---|---|---|---|---|---|---|
| Forest | | 1 | 4 | 9 | 15 | 16 | 22 |
| 134[ab] | Sugar cane | 149[a] | 62[c] | 143[ab] | 135[ab] | — | 117[b] |
| 134[a] | Coffee | — | 91[a] | — | — | 117[a] | — |
| 134[a] | Pasture | — | 99[b] | — | 160[a] | — | — |

N.B.: Values in the same row with the same letter following are not different from each other.

It is difficult to interpret the potassium data. The youngest sugar cane field had been cleared from forest, burned, and planted within the year prior to the sampling. Burning of trees results in an increase in potassium concentration in the soil (Popenoe, 1960). Furthermore, the nine- and fifteen-year-old sugar cane fields and the sixteen-year-old coffee plantation had been fertilized yearly with potassium from the time they were planted. But it is unclear why the four-year-old fields had low potassium concentrations compared to older fields or forest.

Some of the soil parameters showed a continuous decrease with length of time the fields had been under cultivation. For example, pH in the sugar cane fields (Table 25–4) decreased with length of time the field had been under cultivation. A linear regression of pH of the top 10 cm samples and age of field, including forest soil, (Figure 25–2) was highly significant. In contrast, there was no difference between soil pH in the coffee plantations (Table 25–4) or between the plantation soil and forest. The four-year-old pasture has a

**Table 25–4.** pH of soils from forests and fields in San Carlos, Costa Rica (average of three depths).

| Forest | Land Use | Age of Field (yr) 1 | 4 | 9 | 15 | 16 | 22 |
|---|---|---|---|---|---|---|---|
| 5.41[a] | Sugar cane | 4.91[d] | 5.54[a] | 5.22[b] | 5.08[bcd] | — | 5.17[bc] |
| 5.41[a] | Coffee | — | 5.20[a] | — | — | 5.16[a] | — |
| 5.41[a] | Pasture | — | 5.09[b] | — | 5.40[a] | — | — |

N.B.: Values in the same row with the same letter following are not different from each other.

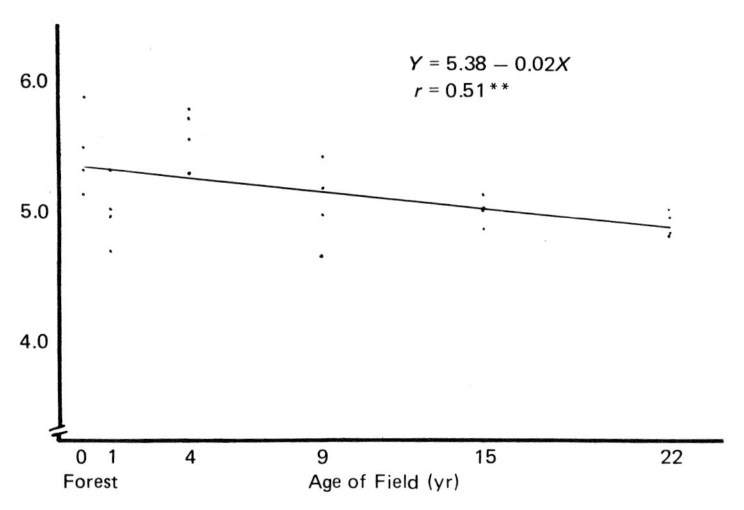


**Fig. 25–2.** Linear regression of pH on age of sugar cane fields near Ciudad Quesada, San Carlos, Costa Rica.

significantly lower pH than the older pasture, which was not different from the forest.

Calcium and magnesium showed the same trend of a decrease in concentration with length of time under cultivation in sugar cane (Figures 25–3 and 25–4). Calcium concentration (Figure 25–3) decreased from an average of 591 ppm in the forest to 165 ppm in the oldest field. It is interesting that lime that had been applied to the nine-year-old and fifteen-year-old sugar cane fields did not seem to reverse the observed trend. Neither coffee plantation was significantly different in calcium concentration from the forest (Table 25–5). However, the older coffee plantation had had lime applied and it was significantly different from the younger plantation. The younger pasture had a significantly lower concentration of calcium than the forest or the older pasture. The latter two were not different from each other.

Magnesium concentration in the top soil of sugar cane fields also showed a significant decrease with length of time the field had been under cultivation (Figure 25–4). The decrease was not so evident as was the case with calcium.

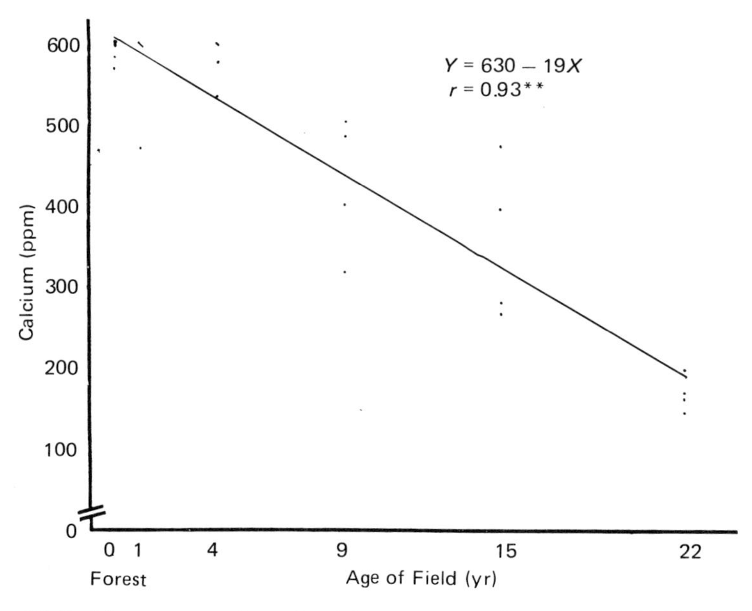


**Fig. 25–3.** Linear regression of calcium on age of sugar cane fields near Ciudad Quesada, San Carlos, Costa Rica.

**Fig. 25–4.** Linear regression of magnesium on age of sugar cane fields near Ciudad Quesada, San Carlos, Costa Rica.

There was also a decrease from forest to fields under coffee and pasture (Table 25–6).

The decrease in cations is not unexpected because they are removed from the system with the harvest of the vegetation. If they are not replaced by weathering or fertilization, their concentration will decrease in the soil. Furthermore, the decrease in percentage organic matter adversely affects the capacity of the soil to hold cations.

Aluminum concentration shows an opposite pattern of an increase in concentration with age of field (Figure 25–5). Possibly aluminum is replacing

**Table 25–5.** Concentration (in ppm) of calcium in soils from forests and fields in San Carlos, Costa Rica (average of three depths).

| Forest | Land Use | Age of Field (yr) 1 | 4 | 9 | 15 | 16 | 22 |
|---|---|---|---|---|---|---|---|
| 384[abc] | Sugar cane | 421[ab] | 462[a] | 306[cd] | 234[d] | — | 134[e] |
| 384[ab] | Coffee | — | 304[b] | — | — | 451[a] | — |
| 384[a] | Pasture | — | 257[b] | — | 362[a] | — | — |

N.B.: Values in the same row with the same letter following are not different from each other.

**Table 25–6.** Concentration (in ppm) of magnesium in soils from forests and fields in San Carlos, Costa Rica (average of three depths).

| Forest | Land Use | Age of Field (yr) 1 | 4 | 9 | 15 | 16 | 22 |
|---|---|---|---|---|---|---|---|
| 137[ab] | Sugar cane | 145[a] | 134[abc] | 121[bcd] | 105[d] | — | 65[e] |
| 137[a] | Coffee | — | 89[b] | — | — | 87[b] | — |
| 137[a] | Pasture | — | 104[b] | — | 92[c] | — | — |

N.B.: Values in the same row with the same letter following are not different from each other.

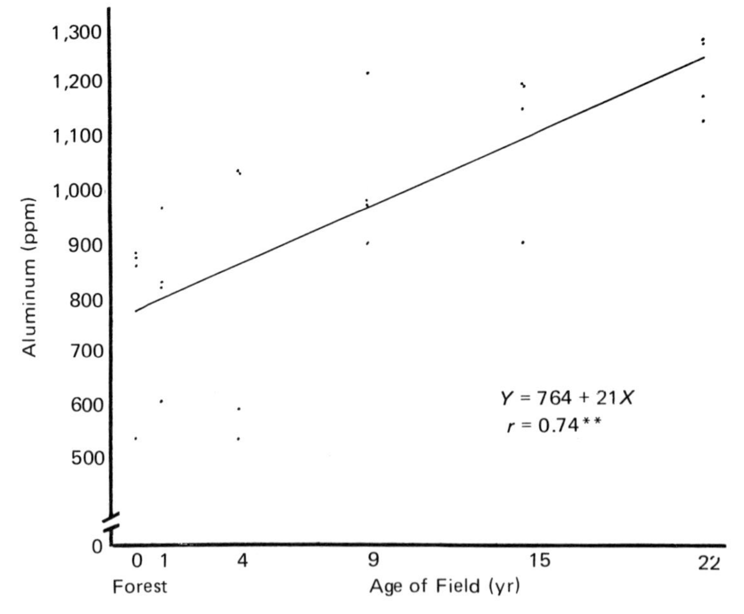


**Fig. 25–5.** Linear regression of aluminum on age of sugar cane fields near Ciudad Quesada, San Carlos, Costa Rica.

calcium and magnesium on the exchange sites. A highly significant negative correlation between aluminum and calcium (— 78 percent) was observed. In these moderately acid soils aluminum forms alumino-hydroxy ions and frees hydrogen ions making the soil more acid, which explains the slight increase in acidity with increasing age of field.

## Conclusions

The trends of nutrient concentration in soils of permanent agricultural fields can be summarized as follows. Conversion of forest to field results in a decline in soil organic matter, nitrogen, pH, calcium, and magnesium and an increase in aluminum. Continued agriculture on the same field results in a continued decline in pH, calcium, and magnesium and an increase in aluminum. The percentages of organic matter and nitrogen do not usually decline with the age of field. No trend with age was observed for potassium concentration.

These trends may be partly explained by changes in litter production, rate of leaching, soil weathering, and removal of nutrients in harvested crops. The reduction in biomass from forest to fields should result in a decrease in the amount of organic matter returned to the soil. Nutrients might be replaced by fertilization, as observed for nitrogen and potassium, but fertilization was not always successful in raising nutrient concentration. Aluminum may replace some of the cations removed by the crop and also may cause a slight increase in acidity.

In this area in San Carlos, Costa Rica, where sugar cane, coffee, and pasture are grown on a volcanic soil, the soil does not seem to be seriously deteriorating with permanent agriculture. However, the observed trends suggest that in the future farmers may have to pay more attention to fertilization and soil management.

## References

Aubert, G., and R. Tavernier. 1972. Soil Survey. *In Soils of the Humid Tropics*. Washington, D.C.: Committee Trop. Soils, Agric. Board, Nat. Res. Council, Natl. Acad. Sci.

Brams, E. A. 1971. Continuous cultivation of West African soils: Organic matter diminution and effects of applied lime and phosphorus. *Plant and Soil* 35:401–414.

Food and Agricultural Organization of the United Nations. 1967. *Production Yearbook*. Vol. 20. Rome.

Flannery, R. L., and J. E. Steckel. 1967. Simultaneous determinations of calcium, potassium, magnesium, and phosphorus in soil electrodialyzates by auto analysis. In *Soil Testing and Plant Analysis*. Part 1, pp. 137–150. Madison, Wis.: Soil Sci. Soc. Am. Spec. Am. Special Publ. No. 2, Soil Sci. Soc. Am.

Golley, F. B. 1972. Energy flux in ecosystems. In *Ecosystem Structure and Function*, John A. Wiens, ed. pp. 69–90. Corvallis: Oregon State Univ. Press.

Holdridge, L. R. 1959. Ecological indicators of the need for a new approach to tropical land use. In *Symposia Interamericana No. 1*. pp. 1–12. Turrialba, Costa Rica: Inter-American Institute of Agricultural Science.

Jackson, M. L. 1956. *Soil Chemical Analysis—Advanced Course*. Madison: Dept. Soil Sci., Univ. Wisconsin.

Morrison, Donald F. 1967. *Multivariate Statistical Methods*. New York: McGraw-Hill.

Nye, P. H., and Greenland, D. J. 1964. Changes in the soil after clearing tropical forest. *Plant and Soil* 21(1):101–112.

Popenoe, Hugh. 1960. Effects of shifting cultivation on natural soil constituents in Central America. Ph.D. thesis, Univ. Florida, Gainesville.

Richards, P. W. 1952. *The Tropical Rain Forest*. Cambridge: Cambridge Univ. Press.

Tosi, J. A., Jr. 1964. Climatic control of terrestrial ecosystems: A report on the Holdridge model. *Econ. Geog.* 40:178–181.

Van Dillewijn, C. 1952. *Botany of Sugar Cane*. Waltham, Mass.: Chronica Botanica.

# INDEX